Functions of Complex Variables

FUNCTIONS OF COMPLEX VARIABLES

An Introduction

ZANE C. MOTTELER
Michigan Technological University

Intext Educational Publishers   New York

Copyright © 1975 by Thomas Y. Crowell Company, Inc.

All Rights Reserved

Except for use in a review, the reproduction or utilization of this work in any form or by any electronic, mechanical, or other means, now known or hereafter invented, including photocopying and recording, and in any information storage and retrieval system is forbidden without the written permission of the publisher.

Published simultaneously in Canada by Fitzhenry & Whiteside, Ltd., Toronto.

Library of Congress Cataloging in Publication Data

Motteler, Zane C.
    Functions of complex variables.

    Includes index.
    1. Functions of complex variables.   I. Title.
QA331.M85   1975   515'.9   74-34329
ISBN 0-7002-2471-8

Intext Educational Publishers
666 Fifth Avenue
New York, New York 10019

TO MARILYNN
my wife
my love

Contents

Preface   *ix*

Chapter 1
Introduction to Complex Numbers and Functions   *1*

1.1. An extension of the real number system   *2*
1.2. The algebra and geometry of complex numbers   *10*
1.3. Powers and roots of complex numbers   *19*
1.4. Algebraic functions of a complex variable   *24*
1.5. Transcendental functions of a complex variable   *34*

Chapter 2
Continuity, Differentiation, and Integration of Complex Functions   *43*

2.1. Continuity and limits of functions of a complex variable   *44*
2.2. Differentiable and analytic functions   *57*
2.3. Integration of functions of a complex variable   *68*
2.4. Integral theorems and higher derivatives   *87*
2.5. The maximum modulus theorem, harmonic functions, and Goursat's theorem   *99*

Chapter 3
Taylor and Laurent Series and the Calculus of Residues   *111*

3.1. Introduction to sequences and series   *111*
3.2. Power series and Taylor's theorem   *126*
3.3. Laurent series expansions for analytic functions with isolated singularities   *142*
3.4. The calculus of residues   *154*
3.5. Applications of the calculus of residues   *165*

Chapter 4
The Elements of Conformal Mapping   *183*

4.1. Analytic and univalent mappings   *183*
4.2. Some examples of conformal mappings   *195*
4.3. The Schwarz-Christoffel transformation   *204*

Chapter 5
Two-dimensional Incompressible Fluid Flow *219*

- **5.1.** Derivation of the equations and their implications *220*
- **5.2.** Certain basic flows *231*
- **5.3.** Principal values; hydrodynamics as an aid to conformal mapping *244*
- **5.4.** Flow past an obstacle; the Kutta-Joukowski airfoil theory *253*
- **5.5.** Free streamlines; contraction coefficients *265*

Appendix A. Some Results from Real Calculus *287*

Appendix B. Odd Numbered Answers *293*

Index *305*

Preface

An alternative title for this book could have been "The Complex Foundations of Two-Dimensional Hydrodynamics." However, the impression might be given of a specialized, highly rigorous treatise, which the book is not. The actual title, on the other hand, suggests a standard introductory complex variables course. But the book is not that either.

It offers an alternative to the usual one-semester complex variables course on the junior or senior level. It differs from standard texts in that its whole burden and purpose is to reach, in a single semester, the point where one can study more than superficially a classical application of some significance and interest, namely the hydrodynamics in the last chapter. And almost every decision on what to include or omit was dictated by this goal. Many standard results are absent, while, on the other hand, there are topics included here that one does not find in most elementary texts.

One would like to assume that prior to his first encounter with complex variables, the student has had substantial exposure to calculus, advanced calculus, and differential equations. It is a sad but true fact, though, that many students with only a calculus background take complex variables. Sections 2.1 and 3.1, which are really more concerned with advanced calculus than complex analysis, and the Appendix, have been included for these students. For an unusually well-prepared class these details need only be briefly reviewed.

The book was written with the firm belief that there is entirely too much of the "you epsilon, me delta" approach to mathematics on the undergraduate level. Rigor occurs naturally and necessarily in many undergraduate courses (modern algebra, intermediate analysis, advanced calculus). Too much of it obscures the material in the more applications-oriented courses such as complex analysis, and makes it difficult to reach the applications, which often supply strong motivation. While I have attempted not to downgrade rigor to the point of sloppiness, nevertheless much in the way of completeness and detail has been sacrificed in order to maintain a certain pace. When a proof has been omitted or abbreviated, I have pointed out this fact. With the help of the appendix, which contains in outline form some of the material from the calculus of the reals to which I make the most frequent appeal, and with a little work, the student can supply the details needed to make many of the proofs completely rigorous.

If a student is going on to graduate studies he will see complex analysis treated in all its rigor. If, on the other hand, he is a science or engineering major or a terminal mathematics major, then an understanding of at least one nontrivial but basic application is the most important thing that a complex analysis course can impart to him.

Except for the last chapter, the contents of this book are relatively standard topics, although treated concisely and on occasion with a new twist. Chapter 1, for example, embodies a standard treatment of the arithmetic of complex numbers and functions. I place considerable emphasis on polynomials and rational functions as representatives of the classes of analytic and meromorphic functions which are discussed later. I spend considerable effort in an attempt to motivate and justify the definition of the complex exponential, and eventually prove that the usual definition is the only possible one. It is important that students realize that things work out as well as they do not totally by accident.

Chapter 2 is a treatment of the calculus of complex functions, leading to all the usual results—Cauchy's and Morera's theorems, the Cauchy-Riemann equations, etc. The limit and continuity results in Section 2.1 can be skipped over rather rapidly if students are familiar with these results in the real case. Even here I do try to prove the theorems with a minimum of epsilonics, by frequent use of the "sandwich principle" and "big-O, little-o" notation, as well as by appealing to (hopefully) known results from the real calculus. By using the concept of the antiderivative I hope to impart an early facility with complex integration and to reinforce the notion that integration and differentiation are inverse operations. By frequent use of polynomials as examples and prototypes of analytic functions, I anticipate the development of series ("generalized polynomials") in the following chapter. Chapter 2 closes with some interesting results (the maximum modulus theorem and Goursat's proof of Cauchy's theorem) and a brief discussion of harmonic functions. In a course at this level I do not recommend teaching Goursat's proof. However, the proof of Cauchy's theorem using Green's theorem earlier in the chapter is more transparent, and is sufficiently general for the purposes of this book.

In Chapter 3, I discuss Taylor and Laurent series and the calculus of residues. Here the treatment is fairly standard, although I try to probe the limits of the theory by lavish use of examples. Chapter 3 leads right into Chapter 4, which discusses conformal mapping up to and including the Schwartz-Christoffel transformation. Many complex variables courses (regardless of the text used) end at or before this point, giving the students little idea why such an edifice was built at such expense.

Hence Chapter 5. This is an elementary treatment of some of the significant topics in two dimensional fluid flow. Because the problems in this chapter (and in the last portion of Chapter 4) are more difficult and involved than in the earlier chapters, I drop the division into graded sections of "exercises" (just that), "problems" (less routine), and "proofs" (which range from trivial to tough).

I owe this book in a large sense to the inspiration of my mentor, Gordon Latta, from whom I had my first complex variables course. He taught this type of course and did it far better than I. In reviewing this manuscript he made helpful and thought-provoking suggestions. I also thank my wife, Marilynn, who typed the early drafts and whose support and help during the writing over many years and through the several drafts has kept me from giving it up entirely. My longtime student Colleen Roe worked the problems and sought out and detected many errors. The flawless typing of Marie Dennis turned the final draft into very nearly a work of art. My colleague Michael Gilpin taught out of one of the early versions with considerable difficulty and came up with some valuable ideas. Helpful suggestions from many reviewers have made the book much better than it started out to be. To all these and to others who have helped with this project I acknowledge my indebtedness.

Chapter **1**

Introduction to Complex Numbers and Functions

The human mind is unusually resistant to new ideas about numbers. Perhaps this is the reason that the solutions of the equation

$$x^2 + 1 = 0$$

are still called "imaginary." It is interesting to note that only in very recent historical times was the existence even of negative numbers accepted, such as the solution of the much simpler equation

$$x + 1 = 0.$$

The study of complex numbers began and flourished almost simultaneously with the recognition that negative numbers are *bona fide* objects for study. In the ninth century, at a time when there was considerable controversy about the "reality" of negative numbers, the great Arabian mathematician Mahavira observed that negative numbers do not have square roots, at least not among the numbers as we know them. However, by the sixteenth century Cardan had developed a formula for the roots of a cubic in which he used a symbol for the "nonexistent" square root of a negative number. His attitude was that if something is mathematically useful and leads to a correct result, then why bother with philosophical speculations as to its meaning? This attitude persists even today, primarily among those

2 INTRODUCTION TO COMPLEX NUMBERS AND FUNCTIONS

people who see mathematics as a tool and not as an object for study for its own sake.

Eventually, Gauss in 1831, and Hamilton a few years later, showed that the system of complex numbers could be developed quite mathematically as an extension of the system of real numbers, so that square roots of negative numbers need no longer be considered any more "imaginary" than others. Even so, some years later, Cauchy, one of the pioneers in the study of complex analysis, still considered the square roots of negative numbers to be figments of the imagination. This idea persists today among the many people who manipulate formally with complex numbers without really stopping to think about what they are doing.

It is in the hope that the reader has a bit more than the normal curiosity that I write this book. We shall begin by attempting to gain an insight into the construction of the complex number system as an extension of the reals, not just by pulling a formal axiom system out of the air and showing that it works, but by using intuition tempered by hindsight (we know what we are looking for!) to see why the particular formulation normally used is eminently reasonable.

1.1. AN EXTENSION OF THE REAL NUMBER SYSTEM

The point of departure for our study of complex analysis is the equation

$$x^2 + 1 = 0.$$

This equation has no solution in the real numbers, since if x is real we must have $x^2 \geq 0$, and hence $x^2 + 1 > 0$. When confronted with a situation like this, a mathematician's normal tendency is to seek some extension of our real number system in which the equation $x^2 + 1 = 0$ *does* have a solution—just as the study of negative numbers can be said to arise from the study of the equation $x + 1 = 0$, of the nonintegral rationals from $2x - 1 = 0$, and of the irrationals from $x^2 - 2 = 0$.

The incurious will simply say that one should postulate a solution of $x^2 + 1 = 0$ and give it a name, say, "i," and proceed from there as if "i" acted like any other algebraic quantity. This approach has the defect that the mysterious "i" is something unique and set apart, that it is somehow less "real" than other numbers. After all, it is represented by a letter, the initial letter of the word "imaginary," and it appears not to have digits or a decimal expansion, unlike those other famous letters "e" and "π." It is a totally mysterious and elusive entity.

The purpose of this section is to convince the student that "i" is no less real than, say, $\sqrt{2}$. Very few students are willing to doubt the existence of $\sqrt{2}$; after all, one can construct a segment of length $\sqrt{2}$ with ruler and compasses, and by well-known methods, calculate $\sqrt{2}$ to any desired degree of accuracy. Yet $\sqrt{2}$ is not rational—that is, it is not the quotient of two integers—and the discovery of this fact caused considerable trauma to ancient Greek mathematics. It forced the Greeks to enlarge their numerical horizons; in modern terminology, they had to adjoin $\sqrt{2}$ to the field of the rationals. We shall see what this means as soon as we codify what we know about the set $\mathbb{R}$ of real numbers. Two *binary operations* "+" and "×" are defined on $\mathbb{R}$ and satisfy the following axioms:

1. *Closure.* If a and b are in $\mathbb{R}$, so are $a + b$ and $a \times b$.
2. *The Associative Law.* $(a + b) + c = a + (b + c)$ and $(a \times b) \times c = a \times (b \times c)$, for all a, b, c in $\mathbb{R}$.
3. *The Units.* There exist two numbers 0 and 1 ($0 \neq 1$) in $\mathbb{R}$ such that $a + 0 = a$ and $a \times 1 = a$, for all a in $\mathbb{R}$.
4. *The Inverses.* Given a in $\mathbb{R}$, there exists an element $-a$ in $\mathbb{R}$ such that $a + (-a) = 0$. If $a \neq 0$, there exists a^{-1} in $\mathbb{R}$ such that $a \times a^{-1} = 1$.
5. *The Commutative Law.* $a + b = b + a$ and $a \times b = b \times a$ for all a and b in $\mathbb{R}$.
6. *The Distributive Law.* $a \times (b + c) = (a \times b) + (a \times c)$ for all a, b, and c in $\mathbb{R}$.

Any set of objects on which two "binary operations" are defined, such that the above axioms are satisfied, is called a *field*. $\{\mathbb{R}, +, \times\}$, that is, the set $\mathbb{R}$ together with the two operations which endow $\mathbb{R}$ with the field properties, is not the only field around, of course. The set of rational numbers $\mathbb{Q}$, that is, all numbers of the form m/n ($n \neq 0$, m and n integers), also forms a field under the operations "+" and "×." The Greeks, as noted above, were nonplused to discover that $\mathbb{Q}$ did not contain all numbers, since it lacked $\sqrt{2}$, for instance. Note that $\mathbb{Q}$ is a subset of $\mathbb{R}$, is a field under the same operations as $\mathbb{R}$, and has the same identity elements 0 and 1; hence it is called a *subfield* of $\mathbb{R}$.

If one begins with the field $\mathbb{Q}$ and attempts to solve the equation $x^2 - 2 = 0$, then the equation has no solution. In order to create a field containing $\mathbb{Q}$ as a subfield in which the equation $x^2 - 2 = 0$ has a solution, one undergoes a process which is called "adjoining the square root of two to $\mathbb{Q}$," the end result of which is a field called $\mathbb{Q}(\sqrt{2})$. This process can be carried out using only the familiar arithmetic properties of $\mathbb{Q}$, without introducing enigmatic symbols such as "$\sqrt{\ }$," in the following manner.

We shall consider *ordered pairs* $\langle a, b \rangle$ of elements in $\mathbb{Q}$ and define an arithmetic on them which is consistent with that on $\mathbb{Q}$. This is quite analogous to the way in which we define rational numbers as pairs of integers n/m ($m \neq 0$) and develop arithmetic on them in such a way that arithmetic on rationals of the form $n/1$ looks just like integer arithmetic. Namely, we shall identify an ordered pair of the form $\langle a, 0 \rangle$ with the rational number a,

4 INTRODUCTION TO COMPLEX NUMBERS AND FUNCTIONS

and, of course, define arithmetic so that

$$\langle a, 0 \rangle + \langle b, 0 \rangle = \langle a + b, 0 \rangle$$
$$\langle a, 0 \rangle \times \langle b, 0 \rangle = \langle a \times b, 0 \rangle.^* \qquad (1.1)$$

Then ordered pairs with zero second element, together with their arithmetic, are really indistinguishable from the rationals and their arithmetic. (The student should verify this statement.) The fact that the equation $x^2 - 2 = 0$ has no rational solution is now equivalent to saying that $\langle x, 0 \rangle \times \langle x, 0 \rangle$ (or $\langle x, 0 \rangle^2$) is never equal to $\langle 2, 0 \rangle$. Thus, we shall have to look among other ordered pairs for the quantity whose square is $\langle 2, 0 \rangle$. The simplest of these is the pair $\langle 0, 1 \rangle$, so let us simply define

$$\langle 0, 1 \rangle \times \langle 0, 1 \rangle = \langle 0, 1 \rangle^2 = \langle 2, 0 \rangle.^\dagger \qquad (1.2)$$

Addition of two ordered pairs is defined as

$$\langle a, b \rangle + \langle c, d \rangle = \langle a + c, b + d \rangle. \qquad (1.3)$$

From Equations (1.1), (1.2), and (1.3) and the axioms of a field, it is not hard to show that the law of multiplication must be

$$\langle a, b \rangle \times \langle c, d \rangle = \langle ac + 2bd, ad + bc \rangle.$$

In the above system of ordered pairs of rationals, $\langle 0, 1 \rangle$ is the square root of $\langle 2, 0 \rangle$; thus, the number which is *identified with the rational number* 2 possesses a square root. In fact, it is not hard to see that the set of all expressions of the form $a + b\sqrt{2}$, where a and b are rational, has the same arithmetic as the set of all ordered pairs of rationals $\langle a, b \rangle$ described above. Either set, under the operations $+$ and $\times$, forms a field. This field is denoted by $\mathbb{Q}(\sqrt{2})$ and is said to be the field obtained by adjoining $\sqrt{2}$ to $\mathbb{Q}$. Note that $\mathbb{Q}$ is a subfield of $\mathbb{Q}(\sqrt{2})$ and that $\mathbb{Q}(\sqrt{2})$ in turn is a subfield of $\mathbb{R}$.

The above process no doubt seems an unusual or abstruse method for arriving at an idea of $\sqrt{2}$ without use of the square root symbol. But the point of the method is that we can define something which behaves like $\sqrt{2}$ purely in terms of the rationals and their arithmetic. In a like manner, we can define something in terms of the real numbers alone which behaves like $\sqrt{-1}$. The analogy with the above construction is quite close. We first consider the set $\mathbb{C}$ of all ordered pairs of *reals* (x, y) (using round brackets to distinguish this set of pairs from the other). We identify the reals as a subset of $\mathbb{C}$ by identifying the real number x with the pair $(x, 0)$ and demanding as before that the arithmetic be such as to satisfy

$$(x_1, 0) + (x_2, 0) = (x_1 + x_2, 0); \qquad (x_1, 0) \times (x_2, 0) = (x_1 x_2, 0).$$

---

* Equality of two ordered pairs, $\langle a, b \rangle = \langle c, d \rangle$, of course means that $a = c$ and $b = d$.
† One could define *any* ordered pair with nonzero second element to have square $\langle 2, 0 \rangle$. Each possible way leads to a different rule for multiplication. Of all these, $\langle 0, 1 \rangle$ is the simplest choice.

1.1 AN EXTENSION OF THE REAL NUMBER SYSTEM 5

The real number -1 has no square root, which is to say that $(-1, 0)$ is not the square of anything of the form $(x, 0)$. Thus, as previously, the square root of $(-1, 0)$ must be a pair with nonzero second element. We arbitrarily choose $(0, 1)$, the simplest such pair, and define

$$(0, 1) \times (0, 1) = (-1, 0).$$

The set $\mathbb{C}$ of ordered pairs of reals, with the arithmetic which arises out of the above definitions, together with the definition of addition

$$(x_1, y_1) + (x_2, y_2) = (x_1 + x_2, y_1 + y_2), \tag{1.4}$$

will form a field, as we shall see, which we shall call the *field of complex numbers*. A complex number of the form $(x, 0)$ is called a *(pure) real complex number*, and one of the form $(0, y)$ is called a *(pure) imaginary complex number*. Likewise, the real number x is said to be the *real part* of (x, y), and the real number y is said to be the *imaginary part* of (x, y).

To develop the arithmetic of $\mathbb{C}$ completely, we need only know how multiplication goes. To determine this, we first note that if α is a real number, then it seems reasonable that

$$\alpha(x, y) = (\alpha x, \alpha y).$$

This is certainly true if $\alpha = n$, an integer, since

$$\underbrace{(x, y) + (x, y) + \cdots + (x, y)}_{n \text{ times}} = n(x, y) = (nx, ny), \tag{1.5}$$

as follows from (1.4) by a simple induction proof. Equation (1.5) immediately leads to

$$(x, y) = \frac{1}{n}(nx, ny)$$

which implies that

$$\frac{1}{n}(x, y) = \left(\frac{x}{n}, \frac{y}{n}\right). \tag{1.6}$$

By combining (1.5) and (1.6), we have

$$\frac{m}{n}(x, y) = \left(\frac{m}{n}x, \frac{m}{n}y\right)$$

for all rationals m/n. Hence, it is quite reasonable to insist that

$$\alpha(x, y) = (\alpha x, \alpha y) \tag{1.7}$$

for all real α. By using this fact plus the low of addition (1.4), it is possible to write any complex number (x, y) as

$$(x, y) = x(1, 0) + y(0, 1).$$

6 INTRODUCTION TO COMPLEX NUMBERS AND FUNCTIONS

We can now derive a formula for complex multiplication as follows:

$$\begin{aligned}(x_1, y_1) \times (x_2, y_2) &= [x_1(1,0) + y_1(0,1)] \times [x_2(1,0) + y_2(0,1)] \\ &= x_1 x_2 (1,0) \times (1,0) + y_1 x_2 (0,1) \times (1,0) + x_1 y_2 (1,0) \\ &\quad \times (0,1) + y_1 y_2 (0,1) \times (0,1) \\ &= x_1 x_2 (1,0) + (x_1 y_2 + x_2 y_1)(0,1) + y_1 y_2 (-1, 0) \\ &= (x_1 x_2 - y_1 y_2)(1,0) + (x_1 y_2 + x_2 y_1)(0,1) \\ &= (x_1 x_2 - y_1 y_2, x_1 y_2 + x_2 y_1).\end{aligned}$$

We want $\mathbb{C}$ to be a field, so many of the above steps are forced by our insistence that the field axioms hold. The second step is due to the distributive law; the third follows since

$$(1,0) \times (1,0) = (1,0);$$
$$(1,0) \times (0,1) = (0,1) \times (1,0) = (0,1);$$
$$(0,1) \times (0,1) = (-1, 0).$$

The middle term in the third line comes from applying the distributive law to the $(0, 1)$ terms.

It is now a relatively simple matter to verify that $\mathbb{C}$ is a field under "$+$" and "$\times$" as defined above. The associative and commutative laws for addition are obvious, and are relatively easy to prove for multiplication (see Proofs 1 and 2 at the end of the section). The distributive law is proved as follows: first,

$$\begin{aligned}(x_1, y_1)[(x_2, y_2) + (x_3, y_3)] &= (x_1, y_1)(x_2 + x_3, y_2 + y_3) \\ &= [x_1(x_2 + x_3) - y_1(y_2 + y_3), x_1(y_2 + y_3) + y_1(x_2 + x_3)].\end{aligned}$$

Next,

$$\begin{aligned}(x_1, y_1)&(x_2, y_2) + (x_1, y_1)(x_3, y_3) \\ &= (x_1 x_2 - y_1 y_2, x_1 y_2 + y_1 x_2) + (x_1 x_3 - y_1 y_3, x_1 x_3 + y_1 x_3) \\ &= (x_1 x_2 - y_1 y_2 + x_1 x_3 - y_1 y_3, x_1 y_2 + y_1 x_2 + x_1 y_3 + y_1 x_3).\end{aligned}$$

The last expressions in the two calculations are equal because of the commutative law of addition in $\mathbb{R}$ and the distributive law in $\mathbb{R}$. Once again $\mathbb{C}$ enjoys certain properties by dint of their truth for $\mathbb{R}$, a circumstance which we shall notice frequently as we continue our studies.

Clearly, $(0, 0)$ and $(1, 0)$ are the additive and multiplicative identities in $\mathbb{C}$, corresponding to 0 and 1 in $\mathbb{R}$. Further, $(-x, -y)$ will be the additive inverse for (x, y).

Let us now verify that every element (x, y) of $\mathbb{C}$ which is not equal to $(0, 0)$ has a unique multiplication inverse, i.e., a complex number (u, v) for

1.1. An Extension of the Real Number System

which
$$(x, y) \times (u, v) = (1, 0).$$
By the multiplication law,
$$(x, y) \times (u, v) = (xu - yv, yu + xv),$$
and for this to equal $(1, 0)$ we must have
$$xu - yv = 1,$$
$$yu + xv = 0.$$
If we multiply the first of these equations by x and the second by y and add, we obtain
$$(x^2 + y^2) u = x$$
or
$$u = \frac{x}{x^2 + y^2}.$$
The division can be performed since x and y are not both zero. By a similar calculation we obtain
$$v = -\frac{y}{x^2 + y^2},$$
so that
$$(x, y)^{-1} = \left(\frac{x}{x^2 + y^2}, -\frac{y}{x^2 + y^2} \right), \tag{1.8}$$
where the exponent "-1" signifies the multiplicative inverse.

It would be surprising indeed if we could not now solve our crucial equation $x^2 + 1 = 0$ in the field $\mathbb{C}$ of complex numbers, since after all we constructed $\mathbb{C}$ for this very purpose. In order to solve this equation, let us rewrite it as
$$(x, y)^2 + (1, 0) = (0, 0) \tag{1.9}$$
where (x, y) is the unknown complex solution and $(x, y)^2 = (x, y) \times (x, y) = (x^2 - y^2, 2xy)$. Then we have
$$(x^2 - y^2 + 1, 2xy) = (0, 0),$$
which is true if and only if *real* solutions of the system
$$x^2 - y^2 + 1 = 0,$$
$$2xy = 0$$
can be found. The second equation implies that either x or y is zero. If y were zero, the first equation would be $x^2 + 1 = 0$, which has no solution in $\mathbb{R}$; hence we must have $x = 0$ and $-y^2 + 1 = 0$, or $y = \pm 1$. Thus $(0, 1)$

8 INTRODUCTION TO COMPLEX NUMBERS AND FUNCTIONS

and $(0, -1)$ are the only two solutions of the complex equation (1.9). We conclude from this that our extension of the field $\mathbb{R}$ to the field $\mathbb{C}$ is a success, in that (1) $\mathbb{R}$ is a subfield of $\mathbb{C}$ and (2) the equation $x^2 + 1 = 0$, which cannot be solved in $\mathbb{R}$, can be solved in $\mathbb{C}$ when each number in it is interpreted as a number in $\mathbb{C}$.

The notation (x, y) for a complex number as an ordered pair of reals is not the one normally used by mathematicians, and we shall soon abandon it. As remarked previously, it is customary to use the letter "i" to represent the pure imaginary complex number $(0, 1)$ (the "square root of minus one"), so that a complex number $z = (x, y)$ can be written

$$z = (x, y) = (x, 0) + (0, y) = (x, 0) + (y, 0)(0, 1)$$

and thus

$$z = (x, y) = (x, 0) + i(y, 0).$$

Since we identify $(x, 0)$ with the real number x and $(y, 0)$ with the real number y, it is shorter and more convenient to write

$$z = x + iy.$$

With the convention that $i^2 = -1$, complex numbers obey the usual rules of algebra, and we can easily verify the reasonableness of our arithmetic for complex numbers, which in the new notation is summarized by the following:

$$(x_1 + iy_1) + (x_2 + iy_2) = (x_1 + x_2) + i(y_1 + y_2),$$

$$-(x + iy) = (-x) + i(-y) = -x - iy,$$

$$(x_1 + iy_1)(x_2 + iy_2) = (x_1 x_2 - y_1 y_2) + i(x_1 y_2 + x_2 y_1),$$

$$(x + iy)^{-1} = \frac{x - iy}{x^2 + y^2}.$$

It is always possible to begin a course in complex variables by simply defining the quantity "i" as the (imaginary) square root of minus one and proceeding from there using the rules of algebra. The reader might well ask why we have gone to so much trouble to work up complex arithmetic in another way. The answer is that we are attempting to show here that "i" is no more imaginary than "$\sqrt{2}$"; it is an element of a field of numbers which can be constructed by building upon a system with which we are already familiar in a mathematically reasonable way. But old ideas do die hard, and thus the term "imaginary" is with us right up to the present day.

A. Exercises

1 Add and multiply the following pairs of complex numbers:

a $(2, 1), (2, -1)$ **b** $(4, 7), (-3, -2)$
c $(-6, 2), (1, 12)$ **d** $(-4, 3), (-4, -3)$
e $(3, 1), (1, 3)$ **f** $(\frac{1}{3}, \frac{3}{4}), (-\frac{1}{2}, \frac{2}{3})$

1.1. An Extension of the Real Number System

g $(-5, \frac{7}{4}), (-3, 1)$
h $(\frac{1}{2}, \frac{3}{2}), (\frac{1}{2}, -\frac{3}{2})$
i $(2, 2), (3, 3)$
j $(-2, 3), (-2, 3)$
k $(1/\sqrt{2}, -3/\sqrt{2}), (1/\sqrt{2}, 3/\sqrt{2})$
l $(\frac{4}{3}, -\frac{12}{7}), (\frac{4}{3}, -\frac{12}{7})$

2 Add and multiply the following pairs of complex numbers:

a $5 + 3i, 4 - 7i$
b $25i, 8 + 4i$
c $2 + i, i - 2$
d $7 + 3i, 7 + 3i$
e $7 + 3i, 7 - 3i$
f $7 + 3i, 3 + 7i$
g $3 + 4i, 3/25 - 4/25i$
h $1 + i, 1 + 3i$
i $7 + 31i, 17$
j $2 - 4i, 1/10 + 1/5i$
k $1/2 + 2i, 1 + 3/2i$
l $-2/3 + 3/4i, -4/5 - 1/3i$

3 Find the multiplicative inverses of the following complex numbers:

a i
b $1 + i$
c $2 + 3i$
d 14
e $14 - i$
f $-6 + 2i$
g $1/\sqrt{2} + 1/\sqrt{2}i$
h $1/2 - \sqrt{3}/2i$
i $-4/5 + 3/5i$
j $5/13 + 12/13i$
k $-1/2 - 1/2i$
l $7 - 2/3i$

4 Divide the first complex number of each of the following pairs by the second:

a $1 + i, 1 - i$
b $i, 2 + 3i$
c $2 - i, 3 + 2i$
d $2 + 2i, 3 - i$
e $-2 - i, -5 - 3i$
f $1 - i, -2 + i$

5 Carry out the indicated arithmetic operations.

a $\dfrac{2 + 3i}{2 - 3i} + (3 + 4i) 2i$

b $\dfrac{1}{\cos \theta + i \sin \theta}$

c $\dfrac{1}{7 + i} + \dfrac{1}{7 - i}$

d $\dfrac{1}{-3 - 2i} + \dfrac{1}{3 - 2i}$

e $(\cos \theta + i \sin \theta)^2$
f $(4 - 3i)(7 + i) + (4 + 3i)(7 - i)$
g $(11 - 12i)(11 + 12i) + (-7 + 4i)(7 + 4i)/i$
h $5(7 - 2i) + (3 - 4i)(11 + 5i) - (7 - 2i)(3i + 7)$

6 Show that the indicated numbers are solutions of the given equations:

a $1 + i; z^2 - 3z + 3 + i = 0$

b $5 - 3i; \dfrac{4}{z-4} - \dfrac{3}{z+1} = i$

c $2 - 2i; z^3 + 2iz^2 - (3 - 4i)z - 2 + 2i = 0$

d $3 + i; z^3 + (1 - i)z^2 - (7 + 4i)z - 5(3 + i) = 0$

7 Find the other solutions of each of the equations in Problem 6.

B. Problems

Develop formulas for addition and multiplication for ordered quadruples of real numbers (a, b, c, d), where real numbers are to be identified with quadruples of the form $(a, 0, 0, 0)$ and where

$$(0, 1, 0, 0)(0, 1, 0, 0) = (-1, 0, 0, 0),$$
$$(0, 0, 1, 0)(0, 0, 1, 0) = (-1, 0, 0, 0),$$
$$(0, 0, 0, 1)(0, 0, 0, 1) = (-1, 0, 0, 0)$$

and further

$$(0, 1, 0, 0)(0, 0, 1, 0) = (0, 0, 0, 1),$$
$$(0, 0, 1, 0)(0, 0, 0, 1) = (0, 1, 0, 0),$$
$$(0, 0, 0, 1)(0, 1, 0, 0) = (0, 0, 1, 0).$$

Is multiplication commutative in this set? (This set of quadruples with multiplication defined by the above comprises the *quaternions* of Hamilton.)

C. Proofs

1 Prove that complex multiplication is commutative.
2 Prove that complex multiplication is associative.
3 Prove that $\mathbb{Q}$ is a field.
4 Prove that $\mathbb{Q}(\sqrt{2})$ is a field.

1.2. THE ALGEBRA AND GEOMETRY OF COMPLEX NUMBERS

We have seen that the complex numbers can readily be defined as ordered pairs of real numbers. As we know from analytic geometry, ordered pairs of real numbers (x, y) can also be used to represent points in the plane, "x" representing the distance along a horizontal axis (the "abscissa") and "y" being the distance along a vertical axis (the "ordinate"), each measured from the point of intersection of the axes (the "origin") to the point where a perpendicular dropped from (x, y) meets the respective axis (Figure 1.1). Thus according to the notation we have adopted, the point with coordinates

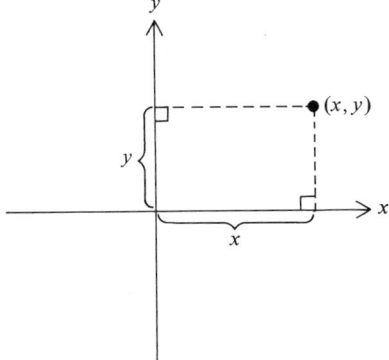

Figure 1.1

(x, y) can equally well be represented by the complex number $z = x + iy$. When we use this convention we shall call the x axis the *real axis*, since the abscissa of (x, y) is the real part of $x + iy$; similarly, we shall call the y axis the *imaginary axis*. We shall often call the (complex) plane simply the z plane, after the complex variable $z = x + iy$. (This method of depicting complex numbers geometrically is also called an Argand diagram.)

Of course, there exist alternative coordinate systems for representing points in the plane. The most commonly used of these is the polar coordinate system, in which the distance r of a point from the origin and the angle θ which the line to the point from the origin makes with a reference line (usually the positive x axis) are the two coordinates (Figure 1.2). The relationship between polar and Cartesian coordinates is the following:

$$x = r \cos \theta, \quad y = r \sin \theta$$

and

$$r = \sqrt{x^2 + y^2}, \quad \tan \theta = \frac{y}{x}.$$

In polar coordinates, the complex number $z = x + iy$ can be written $z = r(\cos \theta + i \sin \theta)$.* The quantity $r = |z| = \sqrt{x^2 + y^2}$, the length of the directed segment or vector from the origin to z, is called the *modulus* or *absolute value* of z. (Note that if z is real, $|z|$ is the absolute value in the usual sense.) The angle θ, which is the angle of inclination of the vector from the origin to the point z with the real axis, measured positive in the counterclockwise direction, is called the *argument of z*, usually abbreviated "arg z."

---

* We shall assume that the reader is familiar with the trigonometric functions sine, cosine, and tangent and with their basic properties and identities.

12 INTRODUCTION TO COMPLEX NUMBERS AND FUNCTIONS

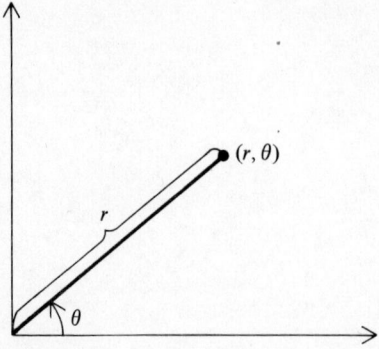

Figure 1.2

The argument of z is not defined if $z = 0$; otherwise, there are an infinite number of possible values which differ from one another by integer multiples of 2π. To avoid ambiguity, we normally choose the value of $\arg z$ which lies in the range $-\pi < \arg z \leqq \pi$.

Let us now proceed to a geometric interpretation of complex addition and multiplication. If $z_1 = x_1 + iy_1$ and $z_2 = x_2 + iy_2$, we know that $z_1 + z_2 = (x_1 + x_2) + i(y_1 + y_2)$, that is, to find the real part of the sum we add the real parts of the addends, and similarly for the imaginary part. If we consider a complex number as a *vector* with two components, then we see that complex addition is like the addition of vectors in two dimensions: $z_1 + z_2$ is the point at the end of the diagonal of the parallelogram determined by the vectors from the origin to z_1 and z_2 (Figure 1.3). Because of this, we shall often consider complex numbers not just as points in the plane, but as two dimensional vectors, a concept which has many useful applications.

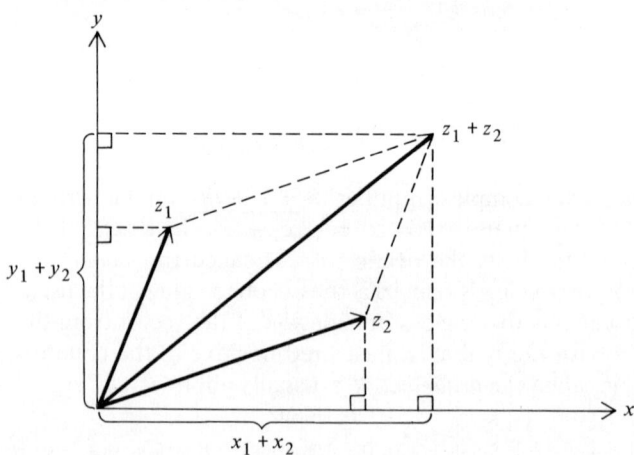

Figure 1.3

1.2. THE ALGEBRA AND GEOMETRY OF COMPLEX NUMBERS

For a geometric interpretation of complex multiplication, it is more transparent to consider the polar form of the two complex numbers $z_1 = r_1(\cos\theta_1 + i\sin\theta_1)$ and $z_2 = r_2(\cos\theta_2 + i\sin\theta_2)$. Thus,

$$z_1 z_2 = r_1 r_2 (\cos\theta_1 \cos\theta_2 - \sin\theta_1 \sin\theta_2)$$
$$+ i(\cos\theta_1 \sin\theta_2 + \cos\theta_2 \sin\theta_1)$$

By utilizing the well-known addition formulas for the trigonometric functions, we find

$$z_1 z_2 = r_1 r_2 [\cos(\theta_1 + \theta_2) + i\sin(\theta_1 + \theta_2)].$$

In other words, to multiply two complex numbers, we *multiply* their *moduli* and *add* their *arguments* (Figure 1.4). This addition of arguments when we multiply the two numbers leads us to an interesting speculation, for if we multiply two real powers of a positive real number a together, then the exponents add to

$$a^b \cdot a^c = a^{b+c}.$$

When we multiply $\cos b + i\sin b$ by $\cos c + i\sin c$, the angles add to

$$(\cos b + i\sin b)(\cos c + i\sin c) = \cos(b+c) + i\sin(b+c).$$

Any such analogous behavior of two different expressions in mathematics automatically prompts curiosity: Does this similarity of behavior imply that the expression $\cos\theta + i\sin\theta$ has some relationship to an exponential function? We shall consider this intriguing question in Section 1.5.

Let us now turn to the operations of subtraction and division.

First we shall consider the graphical interpretation of the expression $z_1 - z_2$. In Section 1.1 we saw that if $z_2 = x_2 + iy_2$, then $-z_2 = -x_2 - iy_2$.

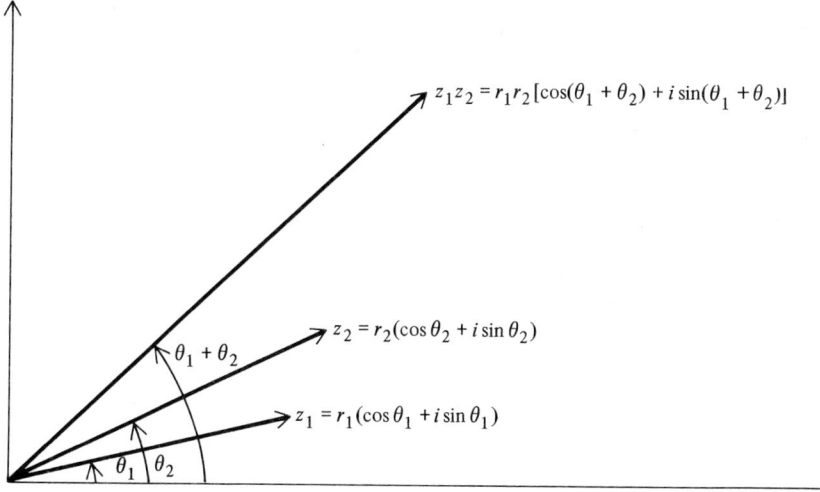

Figure 1.4

14 INTRODUCTION TO COMPLEX NUMBERS AND FUNCTIONS

Thus $-z_2$ is the reflection of z_2 through the origin. Since $z_1 - z_2 = z_1 + (-z_2)$, the parallelogram law shows us that the real and imaginary parts of the complex number $z_1 - z_2$ correspond to the components of the vector pointing from z_2 to z_1 (Figure 1.5).

Next let us consider the graphical interpretation of the inverse of a nonzero complex number z. From Equation (1.8) we have

$$z^{-1} = \frac{1}{z} = \frac{x}{x^2 + y^2} - \frac{iy}{x^2 + y^2},$$

which in polar coordinates is

$$z^{-1} = \frac{r(\cos \theta - i \sin \theta)}{r^2} = \frac{1}{r}[\cos(-\theta) + i \sin(-\theta)].$$

Thus in inverting a complex number we *invert its modulus and negate its argument* (Figure 1.6). z^{-1} lies on the reflection of the radius vector from the origin to z through the real axis; if z is outside the circle of unit radius about the origin, then z^{-1} is inside it, and conversely. In the case $r = 1$, we have

$$(\cos \theta + i \sin \theta)^{-1} = \cos(-\theta) + i \sin(-\theta).$$

Note the analogy between the behavior of this expression and that of a real exponential:

$$(a^b)^{-1} = a^{(-b)}.$$

This is the second time that we have had occasion to remark on this similarity behavior, and it will not be the last.

We shall now consider one further bit of terminology associated with

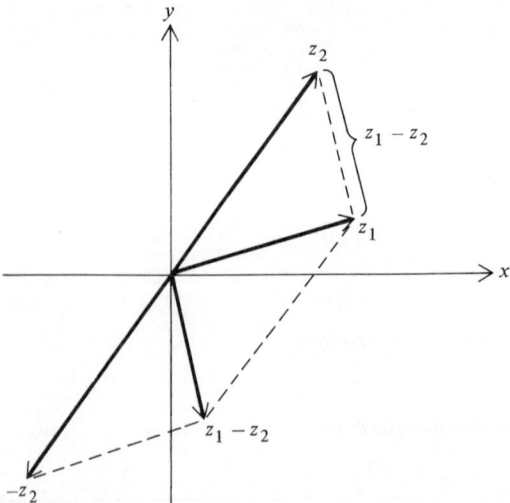

Figure 1.5

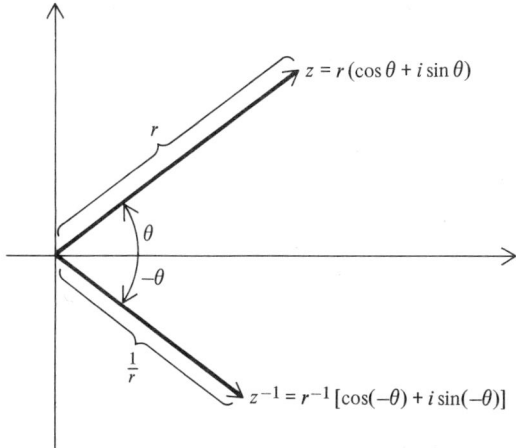

Figure 1.6

complex numbers:

1.2.1. Definition If $z = x + iy$ is a complex number, then the complex number $\bar{z} = x - iy$ is called the *conjugate of z*.

The conjugate of z is simply its reflection in the real axis (Figure 1.7). Several interesting and useful formulas arise from the use of the conjugate notation. First, $z\bar{z} = (x + iy)(x - iy) = x^2 + y^2 = |z|^2$. Thus the modulus of z is given by

$$|z| = (z\bar{z})^{1/2}.$$

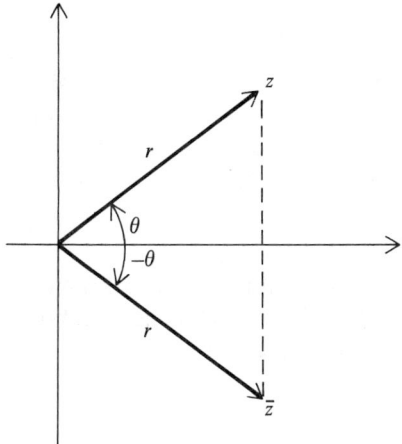

Figure 1.7

16 INTRODUCTION TO COMPLEX NUMBERS AND FUNCTIONS

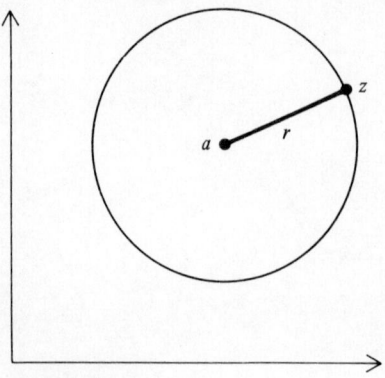

Figure 1.8

By using this fact, we can check our formula for z^{-1} which was so laboriously derived previously:

$$\frac{1}{z} = \frac{\bar{z}}{z\bar{z}} = \frac{\bar{z}}{|z|^2} = \frac{x - iy}{x^2 + y^2}.$$

Also, note that $z + \bar{z} = (x + iy) + (x - iy) = 2x$, and $z - \bar{z} = (x + iy) - (x - iy) = 2iy$, and hence

$$\text{Re}(z) = x = \frac{1}{2}(z + \bar{z}); \text{Im}(z) = y = \frac{1}{2i}(z - \bar{z}).$$

These formulas are useful for finding the real and imaginary parts and the moduli of especially complicated complex expressions.

As applications of this complex notation, let us consider describing various geometric loci in complex terms.

EXAMPLE 1 A circle with center a and radius $r > 0$ is the locus of all z which are a distance r away from the point a. As we saw in Figure 1.5, the distance from z to a is simply the modulus of the vector $z - a$, which points from a to z (Figure 1.8). Therefore the equation of this circle is given by

$$|z - a| = r.$$

EXAMPLE 2 An ellipse is the locus of all points the sum of whose distances from two fixed points, the foci, is constant. If the foci are at a and b and the constant sum is $l \geq |a - b|$, then the equation of the ellipse is

$$|z - a| + |z - b| = l.$$

An interesting application of the use of complex conjugate notation is the proof of the so-called triangle inequality:

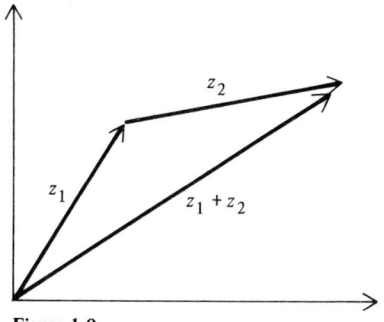

Figure 1.9

1.2.2. Theorem For any two complex numbers z_1 and z_2, we always have

$$|z_1 + z_2| \leq |z_1| + |z_2|.$$

Proof (This theorem is called the *triangle inequality* because it states the familiar result that one side of a triangle is not longer than the sum of the lengths of the other two; see Figure 1.9.) We prove this as follows:

$$\begin{aligned}|z_1 + z_2|^2 &= (z_1 + z_2)(\bar{z}_1 + \bar{z}_2) = z_1\bar{z}_1 + z_1\bar{z}_2 + z_2\bar{z}_1 + z_2\bar{z}_2 \\ &= |z_1|^2 + 2\,\mathrm{Re}(z_1\bar{z}_2) + |z_2|^2 \\ &\leq |z_1|^2 + 2|z_1\bar{z}_2| + |z_2|^2 \\ &= |z_1|^2 + 2|z_1||z_2| + |z_2|^2 \\ &= (|z_1| + |z_2|)^2.\end{aligned}$$

The result follows simply by taking the square roots. (We have used some elementary facts above which we leave to the student to prove in Problem 6.)

A. Exercises

1 Find the arguments and moduli of the following complex numbers, without using tables:

  a $1 + i, 1 - i, -1 + i, -1 - i$
  b $\sqrt{3} + i, 2 - 2\sqrt{3}\,i, -5 + 5\sqrt{3}\,i, -2\sqrt{3} + 2i$
  c $17, -23, 8i, -12i$

2 Find the arguments and moduli of the following complex numbers, using tables if necessary to find the arguments:

  a $4 - 3i$ **b** $-5 + 12i$ **c** $17 - 3i$
  d $11 + 4i$ **e** $-10 - 5i$ **f** $4 + 5i$

18 INTRODUCTION TO COMPLEX NUMBERS AND FUNCTIONS

g $256 + 36i$ h $-12 + 14i$ i $19 - i$
j $-7 - 8i$ k $2 + 6i$ l $4 + 6i$

3 Graph each pair of complex numbers and their sum:

a $2 + 3i, 3 + i$ b $7 - 5i, -5 + 3i$
c $-6 - 4i, 3 + 3i$ d $7 + i, -4 + i$
e $2 + 5i, 12 - 7i$ f $-5 - i, -7 - 12i$
g $13 - 2i, -13 + i$ h $3 - 17i, -2 + 4i$

4 Graph each pair of complex numbers and their product:

a $1 + i, 1 + i$ b $1 + \sqrt{3}i, \sqrt{3} + i$
c $2 + i, -3 + i$ d $3 + 4i, -4 - 2i$
e $-3 + 7i, -7 - 3i$ f $2 + 4i, 6 - 13i$
g $3 - 5i, 2 + 6i$ h $10 + i, -7 - 11i$

5 Show graphically that $-2 + 2i$ is a solution of $z^2 + 4z + 5 = 0$.

B. Problems

1 Find the complex equation of a straight line in the plane.

2 A parabola is the locus of all points whose distances from a fixed point and from a fixed line are equal. Find the complex equation of a parabola, if the line is parallel to the real axis.

3 A hyperbola is the locus of all points the difference of whose distances to two fixed points remains constant. Find the complex equation of a hyperbola.

4 Describe the following point sets in the complex plane:

a $0 < \arg z < \pi/3$
b $|z - a| < r; |z - a| > r; r$ real and positive
c $3 < z + \bar{z} \leq 5$
d $2 \leq \text{Im}(z) < 5$
e $\text{Re}(z) > 4$
f $a_1 \leq \arg z \leq a_2; |z| < 5$

5 Find complex relationships defining the following plane sets:

a All points outside the ellipse with foci at $z = 2$ and $z = 3i$, with semi-major axis 9.
b All points inside a semi-infinite strip running vertically to the right from the y axis, between $y = 5$ and $y = 27$.
c The line $y = -x$ and all points above it.
d All points to the right of the right half of the hyperbola $\frac{1}{4}x^2 - \frac{1}{9}y^2 = 1$.

6 Prove the following facts which were used in Theorem 1.2.2:

a For any complex z, $\text{Re}(z) \leq |z|$.
b $|z| = |\bar{z}|$.
c $|z_1 z_2| = |z_1||z_2|$.

C. Proof

Prove that $z_1 z_2 z_3 \cdots z_n = r_1 r_2 r_3 \cdots r_n [\cos(\theta_1 + \theta_2 + \cdots \theta_n) + i \sin(\theta_1 + \theta_2 + \cdots + \theta_n)]$ for all integer values of $n \geq 2$. Hence show that $z^n = r^n(\cos n\theta + i \sin n\theta)$.

1.3. POWERS AND ROOTS OF COMPLEX NUMBERS

Raising a complex number to an integral power is a simple enough matter. One need only multiply it by itself a sufficient number of times, and invert if the power is negative. Thus, if $z = x + iy$, then

$$z^2 = (x^2 - y^2) + 2xyi,$$
$$z^3 = (x^3 - 3xy^2) + (3x^2y - y^3)i,$$
$$z^4 = (x^4 - 6x^2y^2 + y^4) + (4x^3y - 4xy^3)i,$$

and so forth. However, these formulas have already begun to become unwieldy. A more compact form (and a more useful one) can be obtained if we write z in its polar representation $r(\cos \theta + i \sin \theta)$. Then the assigned proof at the end of the last section claims that

$$z^n = r^n(\cos n\theta + i \sin n\theta)$$

for all nonnegative integers n. This result can easily be verified by induction. The above formula also holds for negative integers, since we have already shown that

$$z^{-1} = r^{-1}[\cos(-\theta) + i \sin(-\theta)];$$

and another straightforward induction proof leads to the conclusion that

$$z^{-n} = r^{-n}[\cos(-n\theta) + i \sin(-n\theta)].$$

We shall need these formulas not because it is a computationally difficult process to calculate z^n, but rather because they will aid us greatly in the reverse process, namely in extracting nth roots of z. For example, if we square the number $(1 - i)$, we obtain the number $-2i$. Hence one value of $(-2i)^{\frac{1}{2}}$ is $1 - i$. As in the real case, we see that $-(1 - i) = -1 + i$ is also a value of $(-2i)^{\frac{1}{2}}$. There is a real ambiguity here in the meaning of the notation $z^{\frac{1}{2}}$ when z is complex, because we can no longer meaningfully speak of the positive and negative square root: For real positive x, the notation $x^{\frac{1}{2}}$ always means the positive square root of x. But there is no such thing as a positive or negative nonreal complex number. The situation becomes more complicated with higher roots; for example, $(1)^{\frac{1}{4}}$ obviously has $1, -1, i$, and $-i$ as possible values.

We shall now develop a systematic process for determining the nth roots of a complex number, which we shall do simply by reversing the process of obtaining the nth power. Setting $z = r(\cos\theta + i\sin\theta)$, and assuming of course that $r \geq 0$, we wish then to find the modulus and argument of $w = z^{1/n} = \rho(\cos\phi + i\sin\phi)$. Since w is an nth root of z, then z is the nth power of w, and we have $w^n = z$, so that, by using our formula for the nth power of a complex number,

$$z = w^n = \rho^n(\cos n\phi + i\sin n\phi) = r(\cos\theta + i\sin\theta).$$

We must express the unknowns ρ and ϕ in terms of the known modulus r and argument θ of z. For two complex numbers to be equal, their moduli must be equal, and their arguments can differ only by an integer multiple of 2π. We have the following two equations as a result:

$$\rho^n = r, \qquad n\phi = \theta + 2k\pi \qquad (k \text{ an integer}).$$

By solving for the two unknowns, we find that

$$\rho = \sqrt[n]{r}, \qquad \phi = \frac{\theta}{n} + \frac{2k\pi}{n},$$

where $\sqrt[n]{r}$ represents the unique nonnegative nth root of the nonnegative real number r.

From this formula we see that the modulus ρ of w, any nth root of z, is simply the real, nonnegative nth root of the modulus r of z. The argument ϕ of w is not unique, but is obtained by dividing any argument θ of z by n, then adding one of many possible fractional multiples of 2π. (This is the mathematical counterpart of dividing a pie into n equal slices. See Figure 1.10) This gives us

$$z^{1/n} = \sqrt[n]{r}\left[\cos\left(\frac{\theta}{n} + \frac{2k\pi}{n}\right) + i\sin\left(\frac{\theta}{n} + \frac{2k\pi}{n}\right)\right]$$

$$k = 0, 1, 2, \ldots, n-1.$$

Since the sine and cosine are functions which are periodic with period 2π, any further values of k will give roots which are repetitions of some of the ones listed above. For the values $k = 0, 1, 2, \ldots, n-1$, we get precisely n different nth roots. (This seems reasonable if we recall that every real positive number has two square roots, a positive and a negative one.) Hence we have shown that *every complex number (except zero) has exactly n distinct nth roots.*

EXAMPLE 3 Find the fourth roots of $1 + i$ (Figure 1.10). Since $|z| = r = \sqrt{2}$, and $\arg z = \tfrac{1}{4}\pi$ (by choosing the smallest nonnegative value), we have

$$z^{\frac{1}{4}} = \sqrt[8]{2}\left[\cos\left(\frac{\pi}{16} + \frac{k\pi}{2}\right) + i\sin\left(\frac{\pi}{16} + \frac{k\pi}{2}\right)\right],$$

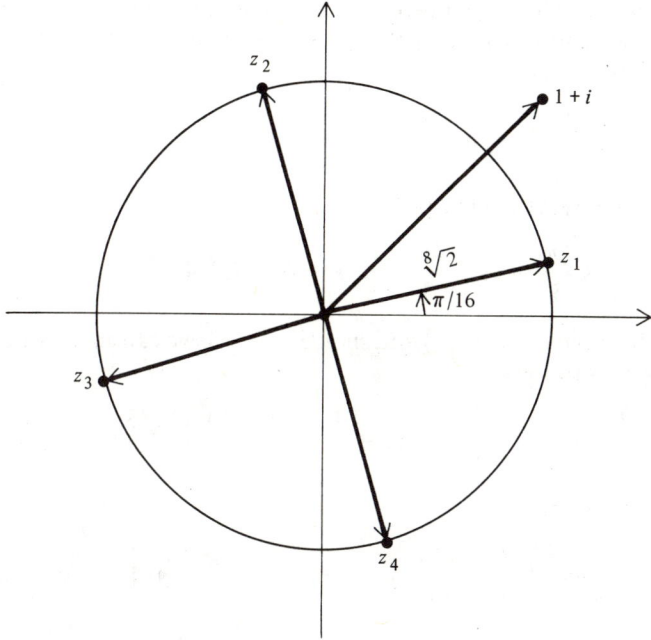

Figure 1.10

for $k = 0, 1, 2, 3$; and by writing out these four roots, we obtain

$$z_1 = \sqrt[8]{2}\left(\cos\frac{\pi}{16} + i\sin\frac{\pi}{16}\right),$$

$$z_2 = \sqrt[8]{2}\left(\cos\frac{9\pi}{16} + i\sin\frac{9\pi}{16}\right),$$

$$z_3 = \sqrt[8]{2}\left(\cos\frac{17\pi}{16} + i\sin\frac{17\pi}{16}\right),$$

$$z_4 = \sqrt[8]{2}\left(\cos\frac{25\pi}{16} + i\sin\frac{25\pi}{16}\right).$$

By examining the graph we see that the nth roots are equally spaced around the circumference of the circle around the origin of radius $r^{1/n}$, and they divide the circle into n equal parts. The first root occurs on this circle at angle θ/n, one nth of the argument of z. Each subsequent root has an argument $2\pi/n$ larger than the preceding one.

EXAMPLE 4 The nth roots of 1. This number has modulus 1 and argument 0; hence all the nth roots of 1 have modulus 1, i.e., they lie on the unit circle

22 INTRODUCTION TO COMPLEX NUMBERS AND FUNCTIONS

centered at the origin. Their arguments lie $2\pi/n$ radians apart along the circumference of the unit circle, with the argument of the first being 0 (1 is always an nth root of 1). Thus

$$1^{1/n} = \cos\frac{2k\pi}{n} + i\sin\frac{2k\pi}{n}, \quad k = 0, 1, \ldots, n-1.$$

We find, for example, this formula that

$$1^{1/6} = \cos\frac{k\pi}{3} + i\sin\frac{k\pi}{3}, \quad k = 0, 1, 2, 3, 4, 5,$$

and, by using the fact that $\cos \pi/3 = \frac{1}{2}$ and $\sin \pi/3 = \frac{1}{2}\sqrt{3}$, we can enumerate the sixth roots of 1 as follows:

$$1, \frac{1}{2} + \frac{\sqrt{3}}{2}i, -\frac{1}{2} + \frac{\sqrt{3}}{2}i, -1, -\frac{1}{2} - \frac{\sqrt{3}}{2}i, \frac{1}{2} - \frac{\sqrt{3}}{2}i.$$

Now, for any complex z,

$$z^{1/n} = r^{1/n}\left[\cos\left(\frac{\theta}{n} + \frac{2k\pi}{n}\right) + i\sin\left(\frac{\theta}{n} + \frac{2k\pi}{n}\right)\right]$$

$$= r^{1/n}\left(\cos\frac{\theta}{n} + i\sin\frac{\theta}{n}\right)\left(\cos\frac{2k\pi}{n} + i\sin\frac{2k\pi}{n}\right);$$

we immediately perceive that

$$z^{1/n} = w\omega_k, \quad k = 1, 2, \ldots, n,$$

where w is the nth root of z with smallest possible nonnegative argument, and $\omega_1, \omega_2, \ldots, \omega_n$ are the nth roots of 1. This alternate method of listing the nth roots shows that we have, in fact, reduced the problem of finding all the nth roots of a general complex number z to the problem of finding one root of z and the nth roots of 1.

Let us note for future reference that we have in this section actually proved the following theorem:

1.3.1. Theorem *If a is any nonzero complex number and n is any positive integer, then the polynomial equation $z^n + a = 0$ has exactly n complex solutions.*

We shall later prove that *every* polynomial equation of degree n with complex coefficients has exactly n complex solutions. Theorem 1.3.1 at least lends plausibility to the contention that this might be the case.

For future reference we should also note that in Example 4 we have uncovered another instance of the exponential type behavior of $\cos \theta +$

$i \sin \theta$. For, in raising this number to the $1/n$ power, one of the steps which we performed was to multiply its argument by $1/n$.

A. Exercises

1 Find the square roots of the following numbers:

    a 25
    b $9i$
    c $2 + 2i$
    d $2 - 2i$
    e $-4 + 3i$
    f $-5 - 12i$
    g $\frac{1}{2} - \frac{1}{2}\sqrt{3}i$
    h $4 + 7i$
    i $-2 - 6i$
    j $12 - 5i$
    k $39i$
    l $-2 + 4i$

2 Find the solution(s) of the equation $z^a = b$, where a and b are as follows:

    a $5, 32 - 32i$
    b $7, -8 + 8\sqrt{3}i$
    c $3, 8i$
    d $\frac{1}{4}, 2 + 3i$
    e $\frac{1}{6}, -1 + i$
    f $\frac{1}{5}, -32 + 32i$

B. Problems

Assuming that the formula for the solution of a quadratic equation is valid in the complex domain (Cf. Proofs # 1), solve the following quadratics:

1 $z^2 + (1 - i)z - 3i = 0$
2 $3iz^2 + 4z - 2i = 0$
3 $z^2 + iz + i = 0$
4 $z^2 - (3 + 2i)z + 1 + 3i = 0$
5 $z^2 - (2 + 8i)z + 16i = 0$
6 $z^2 - (4 + 3i)z + 7 + i = 0$
7 $2z^2 - (5 + 6i)z - 2 + 6i = 0$
8 $9iz^2 + 2iz + 10i = 0$

C. Proofs

1 Prove that the quadratic equation $az^2 + bz + c = 0$, where $a \neq 0$, b, and c are complex, has two (possibly equal) solutions; and that they are given by the quadratic formula
$$z_1, z_2 = \frac{-b \pm \sqrt{b^2 - 4ac}}{2a}$$
with either value for the square root.

2 Given a positive integer n, and nth root ω of 1 is said to be a *primitive* nth root of of 1 if all the nth roots of 1 can be obtained from it by raising it to the powers $1, 2, 3, \ldots, n$. Prove that the nth root of 1 with smallest positive argument is primitive.

3 Prove that ω_k is a primitive nth root of 1 if and only if n and k are relatively prime, that is, n and k have no common divisors other than 1.

$$\left(\text{Recall from the text that } \omega_k = \cos\frac{2\pi k}{n} + i\sin\frac{2\pi k}{n}\right).$$

24 INTRODUCTION TO COMPLEX NUMBERS AND FUNCTIONS

4 Prove that the problem of finding the nth roots of a complex number z can be reduced to the problem of finding a single nth root of z and a primitive nth root of 1.

5 a In trying to define the expression $z^{m/n}$, one might write it as either $(z^m)^{1/n}$ or $(z^{1/n})^m$. Assuming that m and n have no common factors, show that both of these expressions assume the same values, and hence that $z^{m/n}$ is well defined and assumes n distinct values.

HINT: You will need to use the fact that if m and n have no common factors other than ± 1, then there exist integers k and l such that $mk + nl = 1$.

b Use the result of this proof to solve the equation $z^{5/3} = -16\sqrt{2}(1 + i)$.

6 Prove the binomial theorem—that is, that for any two complex numbers z and a and any positive integer n, we have

$$(z + a)^n = \sum_{k=0}^{n} \binom{n}{k} z^{n-k} a^k = z^n + n z^{n-1} a + \cdots,$$

where

$$\binom{n}{k} = \frac{n!}{k!(n-k)!}.$$

1.4. ALGEBRAIC FUNCTIONS OF A COMPLEX VARIABLE

By an *algebraic function* we shall mean an expression (involving complex constants and a complex variable z) which contains only algebraic operations: addition, subtraction, multiplication, division, and the extraction of roots.* Let us make this concept more precise by considering the most basic example.

Let $a_0, a_1, \ldots, a_n (a_n \neq 0)$ be given complex constants. Then an expression of the form

$$a_n z^n + a_{n-1} z^{n-1} + \cdots + a_1 z + a_0$$

is called a polynomial of degree n in z. Clearly if we replace the symbol "z" by any complex number z_0 and perform the indicated calculations, then a unique complex number will result, which we call "the value of the polynomial at z_0." We often use a notation like $p(z)$ to denote the polynomial and its dummy variable and $p(z_0)$ to denote its value at z_0.

A polynomial is one example of what we are going to call a function. By the time a student has reached this point in his studies, he usually has a very good working knowledge of what a function is: some sort of a black box (usually a formula) into which we can insert one number and from which we then extract a number (or numbers, or nothing) in return. Quite possibly

---

* More generally, w is said to be an algebraic function of z if there is a nontrivial polynomial P in two variables such that $P(z, w) \equiv 0$. We shall not need all this generality, and shall limit our consideration to the functions mentioned above.

we may extract no number in return, for example, if we feed $x = 0$ into $1/x$; or we may extract more than one number, for example by feeding $x = \frac{1}{2}$ into a function defined implicitly by an equation such as $x^2 + y^2 = 1$.

Clearly one can be very rigorous or very sloppy in one's definition of a function, and there is lots of territory between the two extremes. We shall content ourselves with a very utilitarian definition in the confident hope that the student will learn to understand the concept of function more by working with functions than by studying definitions.*

1.4.1. Definition Let A be a collection of points in the complex plane. A rule which assigns to each point z in A at least one complex number w is called a *function of a complex variable*. If we denote the rule by a letter "f", then whenever the value w which is assumed at z is unique, it is denoted by $w = f(z)$. The set A is called the *domain of f*; the set of all numbers assigned by f to elements in A is called the *range of f* and is often denoted by $f(A)$.

EXAMPLE 4 $f(z) = 3z + 2$ is a function which is defined for all values of z; hence its domain is the entire complex plane. The expression $3z + 2$ takes on every complex value w precisely once [namely, at the point $z = \frac{1}{3}(w - 2)$]; hence the range of f is also the entire complex plane.

Normally, we shall consider only those functions like the above one which assign *exactly one* complex number w to each z in A. A function with this property is said to be *single-valued*. Any function which is not single-valued is said to be *multiple-valued*. While many multiple-valued functions will be useful in our studies, we shall virtually always specify which one of the several values we wish to use, thuse artificially creating a single-valued function.

EXAMPLE 5 The function $f(z) = z^{\frac{1}{2}}$ is obviously a double-valued function, except at $z = 0$, since, as we saw in the previous section, every nonzero complex number has two square roots, and it is by no means clear whether $z^{1/2}$ denotes one, or the other, or both. In fact, let $z_0 = r_0(\cos \theta_0 + i \sin \theta_0)$ (where $0 \leqq \theta_0 < 2\pi$) be some given point and suppose we choose to let

$$z_0^{\frac{1}{2}} = r_0^{\frac{1}{2}}(\cos \tfrac{1}{2} \theta_0 + i \sin \tfrac{1}{2} \theta_0);$$

in other words, we have chosen to define $z_0^{\frac{1}{2}}$ to be that square root of z_0 which has the smallest nonnegative argument. Now, if we let z_0 move just *very slightly* and return to its former position, then $z_0^{\frac{1}{2}}$ returns to its former value. However, if we allow z_0 to traverse a counterclockwise path around the origin (Figure 1.11) before returning to the same place, then $\arg z_0$ has

---

* In support of this view, I quote Thomas à Kempis, who said, "I had rather feel contrition than know the definition thereof."

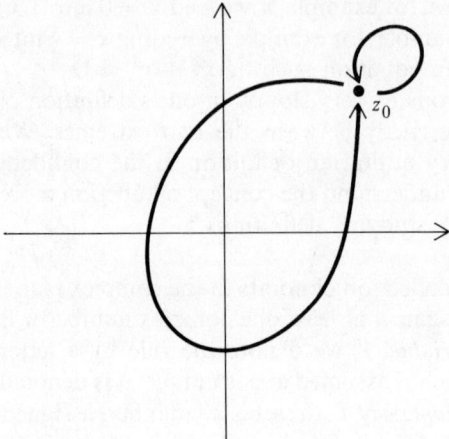

Figure 1.11

increased by 2π. Hence we now have
$$z_0^{\frac{1}{2}} = r_0^{\frac{1}{2}}[\cos \tfrac{1}{2}(\theta_0 + 2\pi) + i \sin \tfrac{1}{2}(\theta_0 + 2\pi)],$$
which is obviously a different value.

    The crux of the whole matter, of course, is a concept which we have not defined yet: *continuity*. We shall discuss and define precisely what we mean by continuity in a future section. For the moment we shall merely think of a continuous function as one which has no sudden jumps or breaks. Thus, although we can uniquely define a single-valued function $f(z) = z^{\frac{1}{2}}$ in a small neighborhood of any point except $z = 0$ in such a way that f is continuous, we cannot do so in the whole plane. The difficulty occurs in domains large enough to contain the origin because every time we encircle the origin and return to a point, the argument of z changes by $\pm 2\pi$ while the argument of $z^{\frac{1}{2}}$ changes only by $\pm \pi$, thus bringing us home to a new value. On the other hand, if we traverse any path which does *not* encircle the origin, then arg z and arg $z^{\frac{1}{2}}$ *do* return to the same value. The conclusion is inescapable that any attempt to define $z^{\frac{1}{2}}$ uniquely in the whole plane will produce a function which jumps, i.e., is discontinuous, somewhere along any path encircling the origin.

    This suggests the following artifice: By restricting arg z to the range $0 \leq \arg z < 2\pi$,* we simply deny z the right to pass around the origin. It is as if we had taken a pair of scissors and cut the complex plane along the positive real axis up to and including $z = 0$ (Figure 1.12). Now z is stymied when it tries to move around the origin because it is not allowed to leap

---

* Or to any other range such as $-\pi < \arg z \leq \pi$.

1.4. ALGEBRAIC FUNCTIONS OF A COMPLEX VARIABLE 27

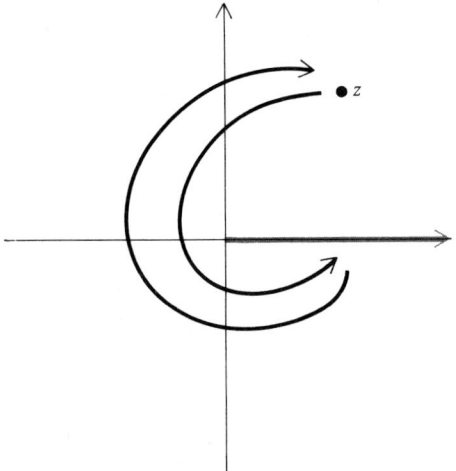

Figure 1.12

the cliff (the positive real axis had argument 0 when it left, argument 2π when it returned, which is quite a jump).

Once we have performed this cut and restricted arg z as above, we can choose one *branch* of the square root; either the branch

$$z^{\frac{1}{2}} = r^{\frac{1}{2}} \left(\cos \frac{\theta}{2} + i \sin \frac{\theta}{2} \right) \tag{1.10}$$

or the branch

$$z^{\frac{1}{2}} = r^{\frac{1}{2}} \left[\cos \left(\frac{\theta}{2} + \pi \right) + i \sin \left(\frac{\theta}{2} + \pi \right) \right]. \tag{1.11}$$

Whichever branch we choose is uniquely defined (and varies continuously) because the only way we can go from one branch to the other is by encircling the origin, which is forbidden.

Now, Equation (1.10) defines a single-valued function whose domain is the complex plane (as cut) and whose range is the upper half plane less the negative real axis, $0 \leq \arg z^{\frac{1}{2}} < \pi$. Equation (1.11) defines a single-valued function on the cut plane whose range is the lower half plane less the positive real axis, $\pi \leq \arg z^{\frac{1}{2}} < 2\pi$.

In this manner we shall often choose a single-valued, continuous *branch* of a multiple-valued function. Incidentally, the point which we cannot encircle without causing a jump and changing the value is called a *branch point* of the function, so $z = 0$ is a branch point of $z^{\frac{1}{2}}$; and the cut we perform to make possible the choice of the desired branch is called a *branch cut*. Notice, of course, that there is nothing privileged about the choice of the

28 INTRODUCTION TO COMPLEX NUMBERS AND FUNCTIONS

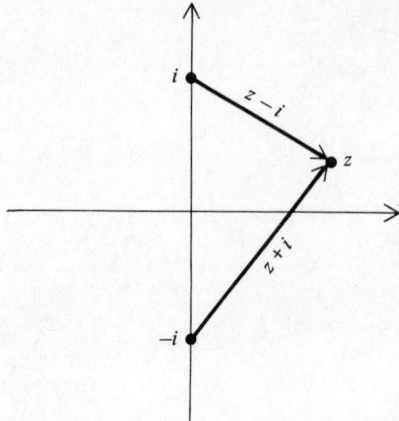

Figure 1.13

positive real axis for the cut here: Any ray or curve extending from the origin to infinity would have prevented our encircling the origin just as well.

Let us pursue another example in order to solidify our understanding of branch points:

EXAMPLE 6 Consider $w = (z^2 + 1)^{\frac{1}{2}} = (z + i)^{\frac{1}{2}}(z - i)^{\frac{1}{2}}$. Note first that for any value of z, $z + i$ is a complex number whose magnitude and direction is the same as a vector pointing from $-i$ to z, while $z - i$ corresponds to a vector pointing from i to z (Figure 1.13). If we consider these vectors positioned thus, it will be much easier to visualize what happens next. Let us begin with a point z_0 on the imaginary axis above i, and choose $\arg(z_0 + i) = \frac{1}{2}\pi$ and $\arg(z_0 - i) = \frac{1}{2}\pi$; then as we proceed counterclockwise along a curve encircling both i and $-i$, the argument of both $z + i$ and $z - i$

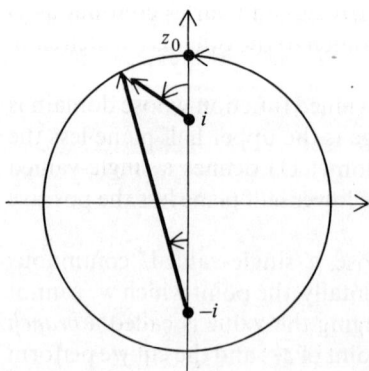

Figure 1.14

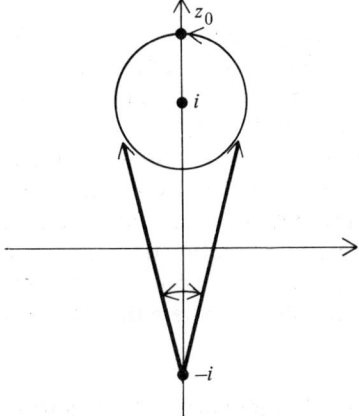

Figure 1.15

changes by 2π. The argument of w changes by half the sum of these two values, $\frac{1}{2}(2\pi) + \frac{1}{2}(2\pi) = 2\pi$, so that w remains unchanged (Figure 1.14). If we encircle just the point i (Figure 1.15), then $\arg(z - i)$ changes by 2π, while $\arg(z + i)$ changes by 0. arg w changes by half the sum again, which is π this time, so that w comes back to a different value. Therefore $z = i$ and, similarly, $z = -i$ are both branch points of w. We must encircle each point by a curve small enough to exclude the other in order to detect this, however. w can be redefined as a single-valued function by making two branch cuts, one up the imaginary axis from i, the other down from $-i$ (Figure 1.16), choosing $\arg z = 0$ for real z, and demanding that w be real for real values of z. Noting that as we move around $z = i$ in a small arc, $\arg(z + i)$ moves between its extremes of $\frac{3}{2}\pi$ and $-\frac{1}{2}\pi$, while $\arg(z - i)$ stays nearly at

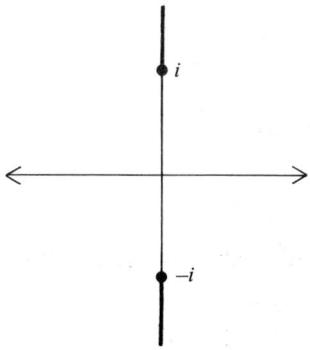

Figure 1.16

$-\frac{1}{2}\pi$, and similarly for the other branch point, arg w takes on all values between but not equal to $-\pi$ and π. $|w|$ takes on all possible values; hence, the domain of this single-valued branch of our function includes the entire plane except for the branch cuts illustrated in Figure 1.16, and the range is the entire w plane except for the negative real axis.

Note that it would also be possible in this example to make a single, *finite* branch cut along the imaginary axis joining the two points $z = i$ and $z = -i$. This effectively prohibits encircling one point without encircling the other, an action which does not change the value of the function, as we saw previously. In fact, the two infinite branch cuts proposed above can be considered as a single cut containing the "point at infinity" where the two lines "meet."

EXAMPLE 7 If $p(z)$ and $q(z)$ are two polynomials and if $q(z) \neq 0$, then their quotient

$$r(z) = \frac{p(z)}{q(z)}$$

is called a *rational function* of z. A rational function is defined for all values of z except for those values where $q(z) = 0$, and is single-valued. The range of a nontrivial rational function (one which does not reduce to a constant or zero) is the entire complex plane, a fact which we shall prove later in the course when we have the necessary apparatus.

EXAMPLE 8 A particularly interesting and important example of a rational function is a function of the form

$$w = \frac{az + b}{cz + d}.$$

Any nontrivial function of this form is called a *bilinear* or *Möbius* function. If $c = 0$, then w reduces to a linear function

$$w = \frac{a}{d}z + \frac{b}{d}.$$

If $a = 0$, then w is the reciprocal of a linear function

$$w = \frac{b}{cz + d}.$$

Clearly a and c could not both vanish, or w would reduce to the constant b/d. When neither a nor c is zero, we have

$$w = \frac{a}{c}\left(\frac{z + b/a}{z + d/c}\right) = \frac{a}{c}\left(1 + \frac{b/a - d/c}{z + d/c}\right) = \frac{a}{c} - \frac{1}{c}\left(\frac{ad - bc}{cz + d}\right).$$

1.4. Algebraic Functions of a Complex Variable

This is clearly a nontrivial function if and only if $ad - bc \neq 0$; no other condition is necessary. The domain of a bilinear function is the entire plane (except the point $z = -d/c$ when $c \neq 0$). The range is also the entire plane (except the point $w = a/c$ when $c \neq 0$), since a given w comes from

$$z = \frac{-dw + b}{cw - a}.$$

A function $f(z)$ itself defines a complex variable, namely the variable $w = f(z)$. w, like z, has a real and an imaginary part, so it can be represented in the form $w = u + iv$, where u and v are both real. As z varies over the domain of f, u, and v each vary over their own domains, which are the projections of the range of f on the real and imaginary axes, respectively. Hence u and v are also functions of the complex variable z; however, because they are real-valued functions, we normally represent them as functions of x and y, thus: $u = u(x, y)$, $v = v(x, y)$.

Because of this, f can be written as

$$f(z) = u(x, y) + iv(x, y),$$

a linear combination of two real-valued functions of two real variables. Therefore many of the properties which hold for real-valued functions carry over to complex-valued functions: This type of inheritance will save us much work in the future.

EXAMPLE 9 To find the real and imaginary parts of the function $f(z) = z^2$, we write z explicitly as $z = x + iy$ and obtain

$$f(z) = z^2 = (x + iy)^2 = (x + iy)(x + iy) = (x^2 - y^2) + i(2xy).$$

Hence in this case $u(x, y) = x^2 - y^2$, $v(x, y) = 2xy$.

EXAMPLE 10 A more difficult problem is to find the real and imaginary parts of

$$f(z) = z^{\frac{1}{2}} = r^{\frac{1}{2}}\left(\cos\frac{\theta}{2} + i\sin\frac{\theta}{2}\right)$$

in terms of x and y instead of the polar coordinates r and θ. Note that we are choosing the branch given by Equation (1.10) after performing the branch cut described in Example 5. First, since $r = \sqrt{x^2 + y^2}$, we have $r^{\frac{1}{2}} = (x^2 + y^2)^{\frac{1}{4}}$. Next, to find the value of $\cos(\theta/2)$ and $\sin(\theta/2)$, we recall the fact that

$$\cos\theta = \frac{x}{\sqrt{x^2 + y^2}} \qquad (0 \leq \theta < 2\pi)$$

32 INTRODUCTION TO COMPLEX NUMBERS AND FUNCTIONS

and the basic trigonometric identities

$$\cos\frac{\theta}{2} = \pm\sqrt{\frac{1}{2} + \frac{1}{2}\cos\theta}, \quad \sin\frac{\theta}{2} = \pm\sqrt{\frac{1}{2} - \frac{1}{2}\cos\theta}.$$

The signs must be chosen to accord with the quadrant containing $\theta/2$. Because $\theta/2$ is always in the first two quadrants, we can always choose the plus sign on the expression for $\sin(\theta/2)$. However, $\cos(\theta/2)$ is positive in the first quadrant and negative in the second, so its sign is the same as the sign of x. Hence

$$\cos\frac{\theta}{2} = \text{sgn } x \sqrt{\frac{1}{2} + \frac{1}{2}\frac{x}{\sqrt{x^2+y^2}}},^{*} \quad \sin\frac{\theta}{2} = \sqrt{\frac{1}{2} - \frac{1}{2}\frac{x}{\sqrt{x^2+y^2}}}.$$

Therefore, putting these into the expression for $z^{1/2}$, we obtain

$$f(z) = z^{1/2} = \frac{1}{\sqrt{2}}\text{sgn } x \sqrt{\sqrt{x^2+y^2}+x} + \frac{i}{\sqrt{2}}\sqrt{\sqrt{x^2+y^2}-x}.$$

In this case $u(x, y)$ and $v(x, y)$ are rather unwieldy. Normally it is much more convenient to deal with u and v as functions of the polar coordinates r and θ in a case like this.

The functions that we have studied in this section are all arrived at by the four arithmetic operations plus the extraction of roots; for this reason we have called them *algebraic functions*. The most important algebraic functions for our future studies are the polynomials; they are the building blocks from which, either by algebraic or by limiting operations, we shall construct the most important functions in complex analysis. Bear this in mind.

A. Exercises

1 Determine which of the following functions are single-valued; for those which are not, perform branch cuts which will make them single-valued. Then determine the range and domain of each function.

- **a** $5zw + 4z - 2w = i$
- **b** $5z^2 - 2iwz + 1 = 0$
- **c** $5z^2 - 2iw^2z + 1 = 0$
- **d** $w = (z+1)^{\frac{1}{3}}$
- **e** $w = |z|$
- **f** $w = (z^2 + 2iz - 1)^{-1}$
- **g** $w = (z^2 - 2iz - 1)^{\frac{1}{4}}$
- **h** $w = (z+1)^{\frac{1}{4}}(z-1)^{\frac{1}{3}}$
- **i** $w = \arg z$
- **j** $w = 1/z + (z-1)^2$

---

* The notation "sgn x," read "signum of x," is defined to be $+1$ if $x > 0$, -1 if $x < 0$, and 0 if $x = 0$.

2 In the following problems, determine the real and imaginary parts of the given functions in terms of x and y. In the case of multiple-valued functions, use the branch with smallest argument.

- **a** $1/z$
- **b** $z^2 - 2z + 1$
- **c** z^3
- **d** $z^3 - 3z$
- **e** $1/(1 + 2z^2)$
- **f** $(z - 2)/(z + 1)$
- **g** $(az + b)/(cz + d)\,;\, ad - bc \neq 0$
- **h** $1/z + (z - 1)^2$
- **•i** $z^{\frac{1}{3}}$
- **•j** $\sqrt{z^2 + 1}$

B. Problems

1 Show that the range of any quadratic polynomial $w = az^2 + bz + c\ (a \neq 0)$ is the entire complex plane.

2 Describe the range of a nontrivial rational function of the form

$$w = \frac{az^2 + bz + c}{dz^2 + cz + f},$$

where at least one of the two constants a and d is nonzero.

3 Let

$$p(z) = a_n z^n + a_{n-1} z^{n-1} + \cdots + a_1 z + a_0$$

with

$$n \geq 1,\quad a_n \neq 0,\quad n \geq 1.$$

Using the polar representation of the complex numbers involved,

$$z = r(\cos \theta + i \sin \theta),$$
$$a_k = \rho_k(\cos \phi_k + i \sin \phi_k) \quad (k = 0, 1, \ldots, n)$$

find expressions for the real and imaginary parts of $p(z)$ in terms of the polar coordinates r and θ.

C. Proofs

1 Assuming that every polynomial of degree $n \geq 1$ has at least one zero, prove that the range of a polynomial is the entire complex plane.

2 Under the same assumption, describe the range of a nontrivial rational function.

3 Prove, using the result of Problem 3 above, that neither the real nor the imaginary part of a polynomial $p(z)$ can vanish identically, assuming $a_n \neq 0$.

4 Extend the result of the previous proof to a nontrivial rational function.

• This denotes a problem of unusual difficulty.

1.5. TRANSCENDENTAL FUNCTIONS OF A COMPLEX VARIABLE

Up until now, other than giving a few hints of the marvels to come, we have dealt exclusively with algebraic operations on a complex variable and those functions which result therefrom. We are now prepared to turn ourselves to the study of the transcendental functions, of which the first and most paramount is the exponential function.

We have noticed the following properties of the expression $\cos\theta + i\sin\theta$, which are extraordinarily like those of a real exponential:

$$(\cos\theta + i\sin\theta)^n = \cos n\theta + i\sin n\theta,$$

$$(\cos\theta + i\sin\theta)^{-1} = \cos(-1)\theta + i\sin(-1)\theta,$$

$$(\cos\theta + i\sin\theta)^{1/n} = \cos\frac{\theta}{n} + i\sin\frac{\theta}{n} \text{ (one branch)}$$

$$(\cos\theta_1 + i\sin\theta_1)(\cos\theta_2 + i\sin\theta_2) = \cos(\theta_1 + \theta_2) + i\sin(\theta_1 + \theta_2).$$

If our mathematical antennae are working correctly, then we should have decided by now that it is possible to represent $\cos\theta + i\sin\theta$ as some number raised to some multiple of θ:

$$\cos\theta + i\sin\theta = \alpha^{\beta\theta},$$

where α and β are as yet unknown.

It turns out that $\alpha = e$ and $\beta = i$ are the proper choices for these two numbers. We do not yet have all the background to establish this beyond doubt; but even at this early point we shall see some extraordinarily compelling reasons for this choice; and later in this text we shall prove that it is the only possible choice.

First, recall that for real x_1 and x_2 we have

$$e^{x_1 + x_2} = e^{x_1}e^{x_2}.$$

However, if we are to define the exponential of a complex number e^z, it ought to have this property. In particular, then, we should have

$$e^z = e^{x+iy} = e^x e^{iy}.$$

We understand the e^x because this is a real exponential. The e^{iy} part is the troublesome one, because it is not at all clear how we can exponentiate a pure imaginary number. We would have to be extraordinarily unperceptive not to see that $\cos y + i\sin y$ is an extremely likely candidate for this, however.

One possible justification for this would be the purely formal manipu-

1.5. Transcendental Functions of a Complex Variable

lation of the "power series"

$$e^{iy} = \sum_{n=0}^{\infty} \frac{(iy)^n}{n!}$$

for e^{iy}.* (This is advisedly called "formal" because we do not yet have any information on the convergence of power series containing nonreal terms.) First we divide this series into its odd and even components,

$$e^{iy} = \sum_{k=0}^{\infty} \frac{(iy)^{2k}}{(2k)!} + \sum_{k=0}^{\infty} \frac{(iy)^{2k+1}}{(2k+1)!};$$

then, noting that $i^{2k} = (i^2)^k = (-1)^k$ and $i^{2k+1} = i^{2k}i = i(-1)^k$, we have

$$e^{iy} = \sum_{k=0}^{\infty} \frac{(-1)^k y^{2k}}{(2k)!} + i \sum_{k=0}^{\infty} \frac{(-1)^k y^{2k+1}}{(2k+1)!}.$$

The right-hand side of this expression is recognizable as $\cos y + i \sin y$. One can also perceive the eminent reasonableness of the definition

$$e^{iy} = \cos y + i \sin y$$

in another way. For then we would have to have also

$$e^{-iy} = \cos y - i \sin y.$$

By adding these two expressions, we have

$$e^{iy} + e^{-iy} = 2 \cos y.$$

How very extraordinary! Surely from some previous mathematics course we must know that

$$e^y + e^{-y} = 2 \cosh y.$$

Thus by defining e^{iy} as above, we might discover extremely fruitful relationships among the trigonometric and hyperbolic functions. Hence we do precisely this:

1.5.1. Definition $e^{iy} \triangleq \cos y + i \sin y$, for all real y. †

1.5.2. Definition $e^z = e^x (\cos y + i \sin y)$, for all complex $z = x + iy$.

---

* This comes from replacing μ in the expansion

$$e^\mu = 1 + \mu/1! + \mu^2/2! + \cdots$$

(from real analysis), purely formally, by iy.
† The notation "$\triangleq$" means "is defined to be."

1.5.3. Proposition $|e^z| = e^x$; $\arg e^z = y \pm 2k\pi$, $k = 0, 1, 2, \ldots$.

Proof $|e^z| = |e^x||\cos y + i \sin y| = e^x \sqrt{\cos^2 y + \sin^2 y} = e^x$.
Thus, if $e^z = r(\cos \theta + i \sin \theta)$, we have shown that $r = e^x$. It follows that $\cos \theta = \cos y$ and $\sin \theta = \sin y$, whence θ and y can differ at most by an integer multiple of 2π. QED

We also note that e^z is a single-valued function of the complex variable z, and that it is periodic with period $2\pi i$, because

$$e^{z+2\pi i} = e^z e^{2\pi i} = e^z(\cos 2\pi + i \sin 2\pi) = e^z.$$

An immediate consequence of our definition of the complex exponential is that the polar form of a complex variable $z = r(\cos \theta + i \sin \theta)$ can be shortened into the form $z = re^{i\theta}$. This important result is called "Euler's formula." [Indeed, the letter "e" itself is used as a tribute to the great mathematician Leonard Euler (1707–1783)]. In particular, if we take $z = -1$, then $r = |z| = 1$ and $\arg z = \pi$, so that

$$e^{\pi i} = -1, \quad \text{or} \quad e^{\pi i} + 1 = 0,$$

a compact formula linking the two transcendental constants e and π, the identity elements 0 and 1 of $\mathbb{R}$, and the imaginary unit i.

Logically, the next function to consider after the exponential is its inverse, the logarithm. For real quantities, if $y = e^x$, then $x = \ln y$;* that is, the natural logarithm of a (positive) real number is the power to which e must be raised to equal that number. Naturally we would like the same property to be true of the complex logarithm; likewise, we want the basic property $\log(xy) = \log x + \log y$ to hold true in the complex realm. Using these two properties, we define the logarithm w of $z = re^{i\theta}$ as follows:

$$w = \log z = \log re^{i\theta} = \ln r + \log e^{i\theta} = \ln r + i\theta,$$

so that we finally have

$$w = \ln|z| + i \arg z.$$

Therefore we see that the real part of the logarithm is the real number $\ln|z|$, and the imaginary part is the real number $\arg z$. Notice that only the logarithm of 0 is not defined; every *nonzero* complex number has infinitely many logarithms, since $\arg z$ is defined only to within an integer multiple of 2π.

Thus $\arg z$, and hence $\log z$, has a branch point at the origin, for every time z encircles the origin, $\arg z$ changes by $\pm 2\pi$ and the logarithm changes by $\pm 2\pi i$. As with any multiple-valued function, we often perform a branch cut and choose to deal with a single-valued branch of the logarithm. For

---

* We shall use the notation "ln" to represent the (single-valued) real natural logarithm of a positive real number; we shall use "log" to represent the complex logarithm.

example, we might remove the negative real axis and restrict ourselves to the choice $-\pi < \arg z \leq \pi$. Then the domain of $\log z$ is the entire z plane minus the origin; and since its real part $\ln|z|$ takes on all possible real values as z ranges over the complex plane, we conclude that the range of $\log z$ is the infinite strip in the w plane, where $-\pi < \operatorname{Im} w \leq \pi$ (Figure 1.17).

We have now succeeded in constructing the complex logarithm in such a way that, given any complex α, we have

$$\alpha = e^{\log \alpha}.$$

While the logarithm is multiple-valued, its values differ by multiples of $2\pi i$, and the exponential is periodic with this period, so the right-hand side above is uniquely defined. Now we can pursue the question of raising α to an arbitrary complex power β. We must have

$$\alpha^\beta = (e^{\log \alpha})^\beta = e^{\beta \log \alpha};$$

hence, by putting $\beta = \beta_1 + i\beta_2$ and using our definition of logarithm, we have

$$\alpha^\beta = \exp[(\beta_1 + i\beta_2)(\ln|\alpha| + i \arg \alpha)]$$

$$= \exp[\beta_1 \ln|\alpha| - \beta_2 \arg \alpha] \exp[i(\beta_2 \ln|\alpha| + \beta_1 \arg \alpha)].$$

Each time we encircle the origin, we will get a different value for α^β if $\beta_2 \neq 0$, since the real exponential is a monotone function and will assume a different value whenever $\arg \alpha$ changes by 2π. Even if $\beta_2 = 0$, the imaginary exponential will assume an infinite number of values unless for some integers k and l we have

$$\beta_1 \arg \alpha + 2l\pi = \beta_1(\arg \alpha + 2k\pi),$$

which implies that

$$l\pi = \beta_1 k\pi$$

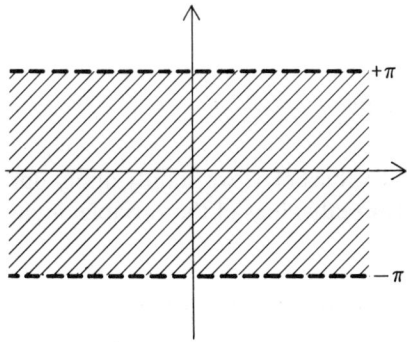

Figure 1.17

or $\beta_1 = l/k$, a rational number. Therefore, *in general*, a complex power of a complex number has infinitely many values. The expression α^β will have a finite number of values if and only if β is *real* and *rational*.

EXAMPLE 11 For instance, consider $\alpha = 1 + i, \beta = i$. Then

$$\alpha^\beta = (1 + i)^i = e^{i \log(1+i)} = e^{i(\ln\sqrt{2} + i[\pi/4 + 2n\pi])}, \quad n = 0, \pm 1, \pm 2, \ldots,$$

and we have

$$(1 + i)^i = e^{-(\pi/4 + 2n\pi)} [\cos(\ln\sqrt{2}) + i\sin(\ln\sqrt{2})], \quad n = 0, \pm 1, \pm 2, \ldots.$$

For each possible integer n we have a distinct value. Note that the ith powers of $1 + i$ have moduli which approach zero as $n \to \infty$ and which approach infinity as $n \to \infty$.

We are unaccustomed, from our experience with the reals, to thinking of a number raised to a power as being anything but uniquely defined. Yet this example shows that there are an infinite number of values for $(1 + i)^i$ and that the moduli of these values can be chosen arbitrarily small or arbitrarily large. Even if we consider the case when α is real and positive and β is real and nonzero, we have

$$\alpha^\beta = e^{\beta \log \alpha} = e^{\beta(\ln \alpha + 2n\pi i)} = e^{\beta \ln \alpha}(\cos 2n\pi\beta + i \sin 2n\pi\beta).$$

When n is zero we obtain the real positive value of α^β which we are accustomed to thinking of as the unique value. However, we now see that the inclusion of complex values allows an infinite number of possible values of α^β (if β is irrational), and that all of these values lie on the circle of radius $e^{\beta \ln \alpha}$ around the origin.

We can also use the complex exponentials to define most of the interesting transcendental functions. We have already seen that, because for real y it is true that

$$e^{iy} = \cos y + i \sin y, \quad e^{-iy} = \cos y - i \sin y,$$

we consequently have

$$\cos y = \frac{e^{iy} + e^{-iy}}{2}, \quad \sin y = \frac{e^{iy} - e^{-iy}}{2}.$$

Since this pair of formulas holds for real values of y, we simply apply our principle that any formula which holds for real values of the variable ought to hold for complex values, and obtain

1.5.4. Definition $\cos z \triangleq \frac{1}{2}(e^{iz} + e^{-iz}), \sin z = (1/2i)(e^{iz} - e^{-iz})$.

We now proceed as in the real case:

1.5.5. Definition

$$\tan z \triangleq \frac{\sin z}{\cos z}, \qquad \cot z \triangleq \frac{1}{\tan z},$$

$$\sec z \triangleq \frac{1}{\cos z}, \qquad \csc z \triangleq \frac{1}{\sin z}$$

To find the real and imaginary parts of $\cos z$, for instance, we simply expand the right-hand side:

$$\cos z = \tfrac{1}{2}(e^{i(x+iy)} + e^{-i(x+iy)}) = \tfrac{1}{2}(e^{-y}e^{ix} + e^{y}e^{-ix})$$
$$= \tfrac{1}{2}[e^{-y}(\cos x + i \sin x) + e^{y}(\cos x - i \sin x)]$$
$$= \tfrac{1}{2}(e^{y} + e^{-y}) \cos x - \tfrac{1}{2}i(e^{y} - e^{-y}) \sin x$$

and we therefore obtain

$$\cos z = \cosh y \cos x - i \sinh y \sin x.$$

By similar manipulations it is possible to show that

$$\sin z = \sin x \cosh y + i \cos x \sinh y.$$

Notice that for $y = 0$, the above formulas reduce to real identities, as they must. It is also clear that the complex sine and cosine are periodic with period 2π, since for instance

$$\sin(z + 2\pi) = \sin(x + 2\pi) \cosh y + i \cos(x + 2\pi) \sinh y$$
$$= \sin x \cosh y + i \cos x \sinh y = \sin z.$$

In fact, all the well-known trigonometric identities still hold in the complex plane, a fact which is left to the student to examine more closely in the problems.

In a manner quite analogous to the trigonometric functions, we can also define the complex hyperbolic functions:

1.5.6. Definition
$\cosh z \triangleq \tfrac{1}{2}(e^{z} + e^{-z}); \sinh z \triangleq \tfrac{1}{2}(e^{z} - e^{-z}).$

It is now possible for us to discern, in the field of complex numbers, the close ties between the trigonometric and hyperbolic functions, and to see some motivation for why they were named as they were. For example,

$$\sin(iz) = \frac{1}{2i}(e^{i(iz)} - e^{-i(iz)}) = \frac{1}{2i}(e^{-z} - e^{z}) = \frac{i}{2}(e^{z} - e^{-z}) = i \sinh z,$$

and, similarly

$$\sinh iz = \frac{1}{2}(e^{iz} - e^{-iz}) = i \sin z.$$

Similar relationships hold between the hyperbolic cosine and the trigonometric cosine, and are left to the student in the problems.

It is possible also to define inverses for all of these functions; we shall consider how to invert the function

$$z = \cosh w$$

in order to find a formula for its inverse

$$w = \cosh^{-1} z.$$

If we write

$$2z = e^w + e^{-w},$$

by multiplying both sides by e^w and gathering nonzero terms on one side, we have

$$e^{2w} - 2ze^w + 1 = 0,$$

which is a quadratic equation for e^w, the solution of which is

$$e^w = z + \sqrt{z^2 - 1}.$$

(Note that there are in general, two possible values for e^w.) We can now solve this equation for w by taking the logarithm of both sides:

$$w = \log(z + \sqrt{z^2 - 1})$$

Thus the inverse hyperbolic cosine is again an infinitely multiple-valued function, which is not surprising, since the hyperbolic cosine is, after all, periodic with period $2\pi i$. Furthermore, whenever $\sqrt{z^2 - 1} \neq 0$, there are in fact two infinite families of values for $\cosh^{-1} z$, one corresponding to each of the two values of the square root.

A. Exercises

1. Find expressions for each of the following:

   a. $\log i$
   b. $\log(1 + i)$
   c. i^i
   d. $(2 + i)^i$
   e. $3^{\sqrt{2}}$
   f. $(3 + 4i)^{i/2}$
   g. $(1 - 3i)^{1+i}$
   h. $(-2 + i)^{\sqrt{2}-i}$
   i. $[(1 + i)^{2+i}]^{2-i}$
   j. $(2 - i)^{\log(3+i)}$
   k. $e^{\log(6 + 7i)}$
   l. $[(1 + 2i)^i]^{i-1}$
   m. $[\log(3 - 2i)]^{-i}$
   n. $i^{\log i}$
   o. $\log(e^{1-3i})$
   p. $e^{2i} 2^i$

2. State which of the following are single-valued functions and why (use the results

of the proofs below). How would one choose a single-valued branch for each multiple-valued function?

  a $w = e^{2\sqrt{z}}$ **b** $w = \cos \sqrt{z}$
  c $w = \tan(z^2 + 1)$ **d** $\cos e^z + \sin^3 z$

  e $w = \dfrac{e^{z^4} + z + 1}{\sin z}$ **f** $w = e^{\cos 4z}$

  g $w = \sin z \cos^3 z^2$ **h** $w = \log \sin z$
  i $w = \left(\dfrac{az+b}{cz+d}\right)$, $ad - bc \neq 0$ **j** $w = e^z \sinh(z^{2/3})$

3 Find the real and imaginary parts of the following functions:

  a $w = \sinh z$ **b** $w = \cosh z$
  c $w = \tan z$ **d** $w = \cot z$
  e $w = \sec z$

B. Problems

1 Find expressions for $w = \arcsin z$ and $w = \arccos z$.

2 Prove the identity $\sin^2 z + \cos^2 z = 1$ from the definitions of the complex trigonometric functions.

3 Prove $\cos(z_1 + z_2) = (\cos z_1)(\cos z_2) - (\sin z_1)(\sin z_2)$.

4 Prove $\sin(z_1 + z_2) = (\sin z_1)(\cos z_2) + (\cos z_1)(\sin z_2)$.

5 Prove $\cos^2 \tfrac{1}{2}z = \tfrac{1}{2} + \tfrac{1}{2}\cos z$ and $\sin^2 \tfrac{1}{2}z = \tfrac{1}{2} - \tfrac{1}{2}\cos z$.

6 Find a formula for $\cosh(iz)$ in terms of $\cos z$ and one for $\cos(iz)$ in terms of $\cosh z$.

7 Show (in two ways) that $\cosh^2 z - \sinh^2 z = 1$.

8 Define the functions $\tanh z$ and $\coth z$ and discuss their domains and ranges. Determine their real and imaginary parts in reasonably simplified form.

C. Proofs

1 If f and g are each single-valued functions, prove that $f \circ g$ is, where $(f \circ g)(z) = f(g(z))$. Describe the range and domain.

2 If f and g are two single-valued functions, prove that $f \pm g$, fg, and f/g (where $g \neq 0$) are, and describe the range and domain of each.

Chapter **2**

Continuity, Differentiation, and Integration of Complex Functions

In the real calculus we learned of the close relationship between differentiation and integration—that they are, in a sense, inverse operations, and that the problem of finding the area under a curve given by a function is solved when one finds an anti-derivative (i.e., an indefinite integral) for the function. Of course there is no such facile geometric interpretation in the complex realm; however, the close relationship between differentiation and integration continues, and indeed is enhanced, by expanding our field of operations from the real line to the plane. The single condition of requiring a complex function to have a derivative induces a legion of equivalent conditions to appear, many of which involve integration. This will enable us to understand the relationship between integration and differentiation with greater depth and clarity than has been true in the past. Before we go into this, however, we shall have to study the usual preliminaries on limits and continuity. (Students already familiar with limits, continuity, and especially uniform continuity can skip Section 2.1 entirely.)

44 CONTINUITY, DIFFERENTIATION, AND INTEGRATION

2.1. CONTINUITY AND LIMITS OF FUNCTIONS OF A COMPLEX VARIABLE

Throughout this section we shall confine our discussion to single-valued functions or single-valued branches of multiple-valued functions, unless clear notification to the contrary is given.

From our studies of the real calculus (to which we shall often appeal), we should already have a good idea of what a continuous function is. (We first mentioned this topic back in Section 1.4.) Oftentimes we define continuity intuitively by considering the graph of the function under consideration. A function is continuous if its graph has no "jumps" or "breaks." Thus, the functions whose graphs are shown in Figure 2.1 are not continuous. Another way of looking at discontinuity is as follows: on the top graph, for instance, as the abscissa approaches x_1 from the left, the ordinate approaches y_1; as the abscissa approaches x_1 from the right, the ordinate approaches y_2. These two values are not equal, so if x is close to x_1 we do not know whether y is close to y_1 or y_2, but it certainly cannot be close to both values at once. In other words, at abscissa x_1, the function varies suddenly and violently; it "jumps." There is a hiatus in the graph. In the second graph, y approaches the same limit as x approaches x_0 from either right or left; but unfortunately this limit does not equal the value y_0 of the function at x_0.

We do not have ready at hand a helpful geometric concept like the graph in the case of a complex-valued function of a complex variable, since we would have to have a system of four coordinates to draw the graph [x, y, u, and v, where $f(z) = u + iv$], and this is simply not possible. But we get our hint on how to proceed from the fact that a continuous function is to have no sudden "jumps" or "breaks," just as in the real case. First of

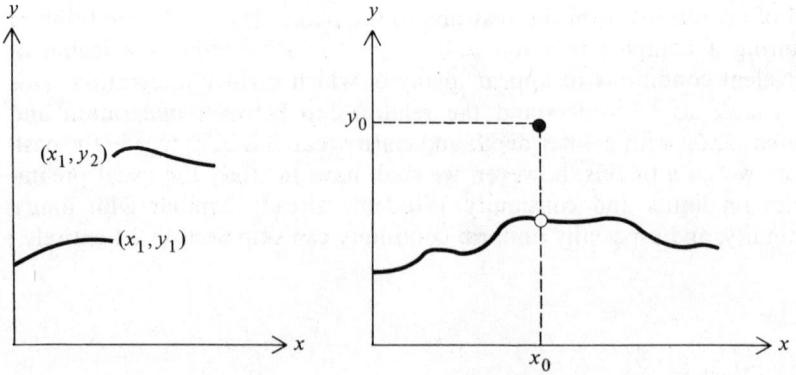

Figure 2.1

2.1. Continuity and Limits of Functions of a Complex Variable

all, for a function f to be continuous at a point z_0 we shall demand that $f(z_0) = w_0$ be defined; and secondly, that if z is "close" to z_0 in the domain of f, then $w = f(z)$ will be "close" to w_0 in the range of f. This latter requirement is imposed so that $f(z)$ will not "jump" at z_0.

We shall now have to make these rather hazy ideas more precise. First we need a better idea of what constitutes "closeness." Closeness involves being within a certain distance or radius of the given point. If we denote by $S_\varepsilon(z_0)$ the set $\{z : |z - z_0| < \varepsilon\}$, which is called the *open disk around z_0 of radius ε*, then whenever we say that a point can be made "sufficiently close" to z_0 we shall mean that it can be put inside $S_\varepsilon(z_0)$ for any ε.

We shall also use the concept of open disk to clarify the notion of domain. A set of points in the complex plane is said to be *open* if for every point z_0 in the set, an ε exists such that every point of $S_\varepsilon(z_0)$ is likewise in the set. In other words, an open set is one which contains an open disk around each of its points. The reader is urged to verify that the complex plane and an open disk are examples of open sets. On the other hand, the real line is not an open subset of $\mathbb{C}$. Further examples occur in the problems at the end of this section.

An open set is said to be disconnected if it consists of two or more nonempty *disjoint* open sets, that is, two or more open sets with no points in common. For example, the two open disks $S_1(2)$ and $S_1(2i)$ together constitute a disconnected open set (Figure 2.2). Obviously (and logically) a set which is not disconnected is said to be *connected*.

In topology and mathematical analysis, a *domain* is normally defined to be a connected open set. We shall follow this practice and shall further assume (unless notice is clearly given to the contrary) that the *domain of a function* is to be a domain in the topological sense, that is, a connected

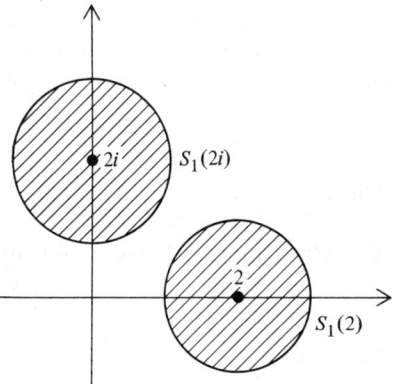

Figure 2.2

open set. A domain is said to be *arc-wise connected* if any two points in the domain can be connected by an arc (see Definition 2.3.1). It should be intuitively clear that a domain is arc-wise connected; in fact, however, this contention is not a trivial one, nor is its proof, which we shall therefore omit.

This brief introduction to open sets, open disks, and domains will serve to make our definitions more precise and the concepts more clear as we discuss continuity and limits. In particular, we are now ready to define the concept of continuity quite precisely:

2.1.1. Definition Let f be a given single-valued function with z_0 in its domain and $w_0 = f(z_0)$. Then f is said to be *continuous at* z_0 if, given $\varepsilon > 0$, there exists $\delta > 0$ (which may depend on both ε and z_0) such that whenever z lies in $S_\delta(z_0)$, $f(z)$ lies in $S_\varepsilon(w_0)$—i.e., $|z - z_0| < \delta$ implies that $|f(z) - f(z_0)| < \varepsilon$.

Although we are not able to graph the situation in all of its four dimensional complexity, we can consider the z plane and the w plane separately (Figure 2.3). Definition 2.1.1 says, geometrically, that if we are given the disk $S_\varepsilon(w_0)$ of radius ε around $f(z_0)$, then we can find a disk $S_\delta(z_0)$ of radius δ around z_0, such that the function f takes every point from inside $S_\delta(z_0)$ and places it inside $S_\varepsilon(w_0)$.

2.1.2. Definition $f(z)$ is said to be *continuous on a set* A if it is continuous at each $z \in A$.* If, furthermore, we can find a δ which depends only on the given ε and *not* on the particular point of A, then $f(z)$ is said to be *uniformly continuous* on A.

Let us work some examples to clarify these concepts.

EXAMPLE 1   The linear function $f(z) = az + b$, where a and b are constants and $a \neq 0$, is uniformly continuous on the whole plane. For, given $\varepsilon > 0$, if z_0 is any point of the plane and we choose z such that $|z - z_0| < \varepsilon/|a|$, we have
$$|f(z) - f(z_0)| = |az - az_0| = |a||z - z_0| < \varepsilon.$$
Here $\delta = \varepsilon/|a|$ and is clearly not dependent on z_0 at all.

EXAMPLE 2 $f(z) = z^2$ is continuous on the whole plane but not uniformly continuous. For, if $z_0 = 0$, then given $\varepsilon > 0$, if we choose $|z| < \sqrt{\varepsilon}$, we have $|f(z) - f(0)| = |z^2| = |z|^2 < \varepsilon$. If $|z_0| = M \neq 0$, then we have
$$|f(z) - f(z_0)| = |z^2 - z_0^2| = |z + z_0||z - z_0|.$$

---
* The notation "$z \in A$" means "z is a point in the set A."

2.1. Continuity and Limits of Functions of a Complex Variable

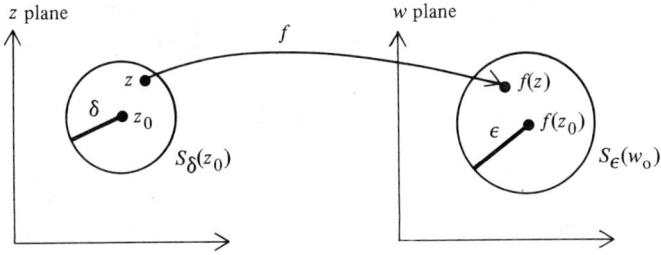

Figure 2.3

Now, given $\varepsilon > 0$, we wish to choose a $\delta > 0$ such that if $|z - z_0| < \delta$, then $|f(z) - f(z_0)| < \varepsilon$. From the above we see that

$$|f(z) - f(z_0)| < |z + z_0|\delta.$$

The quantity $|z + z_0|$ lies between $2M + \delta$ and $2M - \delta$ (Figure 2.4). Thus at the best we would have

$$|f(z) - f(z_0)| < (2M - \delta)\delta,$$

and at the worst,

$$|f(z) - f(z_0)| < (2M + \delta)\delta.$$

To make $|f(z) - f(z_0)| < \varepsilon$ for *all* values of z in $S_\delta(z_0)$, we must have at worst that

$$(2M + \delta)\delta = \delta^2 + 2M\delta = \varepsilon$$

or

$$\delta^2 + 2M\delta - \varepsilon = 0.$$

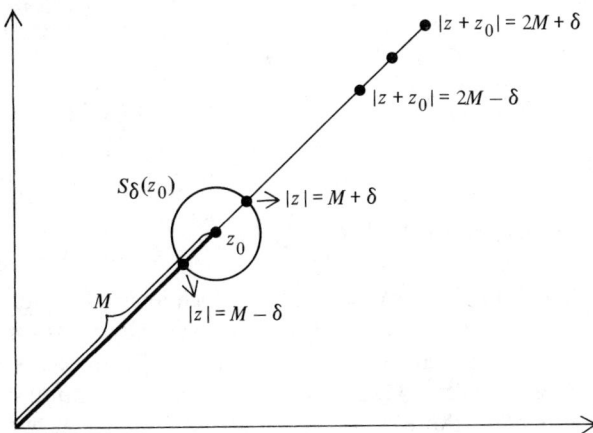

Figure 2.4

This quadratic expression in δ will be zero only when δ assumes one of the two values $-M \pm \sqrt{M^2 + \varepsilon}$; the negative value is not allowed, and hence we must choose

$$\delta = \sqrt{M^2 + \varepsilon} - M = \sqrt{|z_0|^2 + \varepsilon} - |z_0|.$$

In this case, δ clearly depends on both ε and z_0, and this value for δ is the best (i.e., largest) we can obtain. Note that $\delta(\varepsilon, z_0)$ must be taken smaller and smaller as z_0 increases in modulus—since the expression for δ approaches zero as $|z_0| \to \infty$. It is not just the dependence on z_0 that destroys uniform continuity, but the fact that δ is not bounded away from zero on the domain of f. This fact makes it impossible to choose one value of δ which will do for *all* z_0. Hence $f(z) = z^2$ is not uniformly continuous. However, $f(z)$ *is* uniformly continuous on any *bounded* subset of the plane, that is, on any set for which there exists a constant K such that $|z| \leq K$ for all z in the set (why?).

EXAMPLE 3 Let us now consider the single-valued branch of $f(z) = \sqrt{z}$ on the domain where $z \neq 0$ and $-\pi < \arg z < \pi$ (note that this domain is open) defined by taking the positive square root for real z and letting the argument of $\sqrt{z}$ vary continuously from $-\frac{1}{2}\pi$ to $\frac{1}{2}\pi$ as $\arg z$ goes from $-\pi$ to π. In other words, by taking $z = r(\cos\theta + i\sin\theta)(-\pi < \theta < \pi)$, we define

$$f(z) = \sqrt{z} = \sqrt{r}\left(\cos\frac{\theta}{2} + i\sin\frac{\theta}{2}\right), \quad (-\pi < \theta < \pi).$$

Then (assuming $|z_0| = M \neq 0$ since $z = 0$ is not in the domain),

$$|f(z) - f(z_0)| = |\sqrt{z} - \sqrt{z_0}| = \frac{|z - z_0|}{|\sqrt{z} + \sqrt{z_0}|} < \frac{\delta}{|\sqrt{z} + \sqrt{z_0}|},$$

and, since

$$|\sqrt{z} + \sqrt{z_0}| \geq \max(\sqrt{M} + \sqrt{M - \delta}, \sqrt{M} + \sqrt{M + \delta}) \geq \sqrt{M},$$

(see Figure 2.5), we have

$$|f(z) - f(z_0)| < \frac{\delta}{\sqrt{M}}.$$

This can be made less than ε if $\delta = \varepsilon\sqrt{M} = \varepsilon|z_0|^{1/2}$, which again depends on z_0. Thus $f(z) = \sqrt{z}$ is continuous for $z \neq 0$ but not uniformly so, since for a given value of ε, δ must be taken smaller and smaller as M decreases. This is also obvious because $S_\delta(z_0)$ can never contain $z = 0$, since the point 0 itself is not in the domain of f; there is no way that it could be. As we

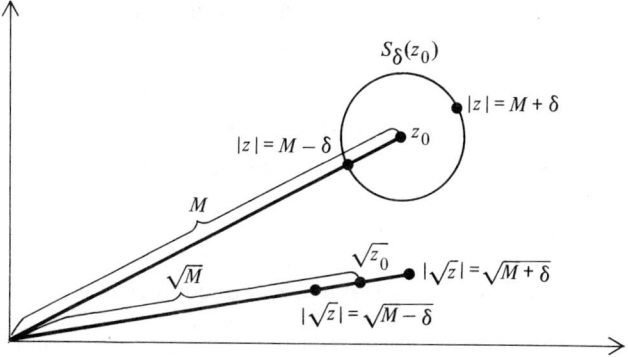

Figure 2.5

remarked in the last chapter, it is not possible to define $\sqrt{z}$ uniquely in any neighborhood of the origin in such a way that it is continuous there. Encircling the origin forces a jump. In our case, as we approach a point ζ on the negative real axis from above, $\sqrt{z}$ approaches $i\sqrt{-\zeta}$, while if we approach ζ from *below*, $\sqrt{z}$ approaches $-i\sqrt{-\zeta}$. Thus $\sqrt{z}$ fails to be continuous at any point on the negative real axis, which is why it is excluded from the domain.

In this example, δ becomes small as $z_0 \to 0$; but if we restricted our consideration to points bounded away from the origin and the negative real axis—say points z_0 for which $|z_0| \geq r > 0$ and such that z_0 is at least r units from the nearest point on the negative real axis—then the value $\delta = \varepsilon\sqrt{r}$ would always ensure that $|f(z) - f(z_0)| < \varepsilon$, since

$$\frac{\delta}{\sqrt{M}} = \frac{\varepsilon\sqrt{r}}{\sqrt{M}} \leq \varepsilon.$$

Thus again dependence on z_0 alone is not enough to prohibit uniform continuity. *If the dependence is such that it forces δ to go to zero in the domain of the function, then the continuity is nonuniform.*

An alternative way to define continuity for complex functions is in terms of real-valued functions, with which we are already acquainted. In fact, it should be clear to the reader that our definition is really basically the same as that for real-valued functions—the only difference being in the meaning of the absolute value notation "$|\cdot|$", which is simply a generalization of the real absolute value. (Much of the material in the rest of this section should be familiar to the student from real analysis; it can probably be skipped through rather rapidly.)

To see this, let us define continuity for a real-valued function $u = u(x, y)$ of two variables as follows*:

2.1.3. Definition $u = u(x, y)$ is said to be continuous at the point (x_0, y_0) in its domain if, given $\varepsilon > 0$, there exists an $\eta > 0$ and a $\delta > 0$ such that whenever $|x - x_0| < \eta$ and $|y - y_0| < \delta$, then $|u(x, y) - u(x_0, y_0)| < \varepsilon$.

We can then prove the following theorem:

2.1.4. Theorem The following two statements are equivalent:

a $f(z)$ is continuous at $z = z_0 = x_0 + iy_0$.
b The real and imaginary parts of f, $u(x, y)$ and $v(x, y)$, respectively, are continuous at (x_0, y_0).

To prove this theorem, and many of the other limit theorems, at this stage would involve detailed and burdensome calculations with epsilons and deltas. This is not difficult, but merely lengthy, and the ambitious student is strongly urged to supply such a proof.

2.1.5. Definition We shall say that a function $f(z)$ has a limit L as z approaches z_0, and write

$$\lim_{z \to z_0} f(z) = L$$

if and only if the real limit

$$\lim_{\substack{x \to x_0 \\ y \to y_0}} |f(x + iy) - L|$$

is zero.

A few comments on this definition will serve to erase any possible ambiguities. First, the point z_0 need not, of course, be in the domain of f; however, for any ε, the open disk $S_\varepsilon(z_0)$ must contain points in the domain of f, for this definition to be meaningful. In other words, it must be possible to evaluate f at points arbitrarily close to z_0. Second, the expression

$$\lim_{\substack{x \to x_0 \\ y \to y_0}} |f(x + iy) - L|,$$

which is a real limit in two variables, can equally well be written

$$\lim_{z \to z_0} |f(z) - L|,$$

---

* There are other ways of making this definition, of course, but they are inconsequential variations.

2.1. CONTINUITY AND LIMITS OF FUNCTIONS OF A COMPLEX VARIABLE

since it is clear that z gets close to z_0 if and only if x gets close to x_0 and y gets close to y_0.

Using this definition of the limit of a complex function in terms of a real limit, we are at liberty to borrow the results from the real calculus (which should already be known to the student), and thus eschew many of the complicated $\varepsilon - \delta$ proofs that might be necessitated by a more rigorous approach.

One of the most important principles we shall use is the "sandwich" principle, which states that if $|f(z)| \leq |g(z)|$ and if

$$\lim_{z \to z_0} |g(z)| = 0,$$

then

$$\lim_{z \to z_0} |f(z)| = 0$$

for $|f(z)|$ is "sandwiched" between 0 and $|g(z)|$; as $|g(z)|$ becomes small, so must $|f(z)|$.

We should now be able to see that it is possible to redefine continuity at a point as follows:

A function f is continuous at z_0 if and only if

a z_0 is in the domain of f;
b $\lim_{z \to z_0} f(z) = f(z_0)$.

Condition a is in our original definition; and condition b, by the definition of limit, says that the (real) quantity $|f(z) - f(z_0)|$ must get small whenever $|z - z_0|$ gets small. Thus both definitions do say the same thing.

Using the sandwich principle, it is a relatively easy task to prove all the usual limit theorems:

2.1.6. Theorem Let α be a constant, and let f and g be two functions with

$$\lim_{z \to z_0} f(z) = L_1, \qquad \lim_{z \to z_0} g(z) = L_2.$$

Then

a $\lim_{z \to z_0} \alpha f(z) = \alpha L_1$
b $\lim_{z \to z_0} f(z) + g(z) = L_1 + L_2$

c $\lim_{z \to z_0} f(z) g(z) = L_1 L_2$
d $\lim_{z \to z_0} \dfrac{f(z)}{g(z)} = \dfrac{L_1}{L_2}$ (if $L_2 \neq 0$).

Proof The proof of each of these four statements is an application of Definition 2.1.5, and (b), (c), and (d) involve the sandwich principle as well. Note in the last three proofs the use of the triangle inequality (Theorem

1.2.2) and in the last two, the insertion of an extra term. These techniques are standard, and should be studied carefully by the student until understood. Finally, note that all four proofs can easily be made rigorous by the use of $\varepsilon - \delta$ techniques.

a $|\alpha f(z) - \alpha L_1| = |\alpha| |f(z) - L_1| \to 0.$

b $|f(z) + g(z) - (L_1 + L_2)| = |f(z) - L_1 + g(z) - L_2|$
$\leq |f(z) - L_1| + |g(z) - L_2| \to 0.$

c $|f(z)g(z) - L_1 L_2| = |f(z)g(z) - L_1 g(z) + L_1 g(z) - L_1 L_2|$
$\leq |f(z) - L_1| |g(z)| + |L_1| |g(z) - L_2| \to 0.$

d $\left|\dfrac{f(z)}{g(z)} - \dfrac{L_1}{L_2}\right| = \left|\dfrac{f(z)L_2 - L_1 g(z)}{g(z) L_2}\right| = \left|\dfrac{f(z)L_2 - L_1 L_2 + L_1 L_2 - L_1 g(z)}{g(z) L_2}\right|$
$\leq \dfrac{1}{|g(z)|} |f(z) - L_1| + \left|\dfrac{L_1}{g(z) L_2}\right| |L_2 - g(z)| \to 0.$

Another limit theorem pertains to composite functions, that is, functions of functions:

2.1.7. Theorem Let f and g be two complex functions with
$$\lim_{z \to z_0} g(z) = L_1 \quad \text{and} \quad \lim_{w \to L_1} f(w) = L_2.$$
We also suppose that the range of g lies in the domain of f, so that $f(g(z))$ is defined for all z in the domain of g. Then
$$\lim_{z \to z_0} (f \circ g)(z) = L_2,$$
where $(f \circ g)(z) = f(g(z))$.

Proof As $w = g(z)$ assumes values close to L_1, $f(w)$ assumes values close to L_2, since each such w is in the domain of f. That is,
$$|(f \circ g)(z) - L_2| = |f(g(z)) - L_2| = |f(w) - L_2| \to 0. \quad \text{QED.}$$

Now that we have established the usual limit theorems, the parallel continuity theorems follow easily, since a statement regarding continuity is a statement that a limit exists *and* that a certain function is defined. And clearly, if f and g are defined at z_0, then so are their sum, product, and quotient [when $g(z_0) \neq 0$]:

2.1.8. Theorem Let f and g be two complex functions continuous at z_0, and let α be any constant. Then αf, $f + g$, fg, and f/g are continuous at z_0, the latter provided that $g(z_0) \neq 0$.

We also have a similar theorem regarding composite functions:

2.1.9. Theorem Let f and g be two complex functions with the range of g contained in the domain of f, with g continuous at z_0 and f continuous at $w_0 = g(z_0)$. Then $f \circ g$ is continuous at z_0.

Another theorem on continuity is the following corollary of Theorem 2.1.8:

2.1.10. Corollary If $f_1, f_2, \ldots, f_n$ are complex-valued functions which are continuous at $z = z_0$, then so is the function $f_1 + f_2 + \cdots + f_n$.

The proof is a simple exercise in mathematical induction and is left to the reader.

By putting these results all together, we have the very important conclusion that our prototype functions—the polynomials—are continuous.

EXAMPLE 4 First, $f(z) = z^n$ is continuous for any nonnegative integer value of n, because

$$|z^n - z_0^n| = |(z - z_0)(z^{n-1} + z^{n-2}z_0 + \cdots + zz_0^{n-2} + z_0^{n-1})|$$
$$= |z - z_0| |z^{n-1} + z^{n-2}z_0 + \cdots + zz_0^{n-2} + z_0^{n-1}|.$$

The first factor goes to zero while the second approaches the finite number $n|z_0|^{n-1}$; hence the whole quantity approaches zero. Now our statement is a foregone conclusion, since, by Theorem 2.1.8 and Example 4, $a_k z^k$ is continuous for any constant a_k, and, by Corollary 2.1.10, the sum

$$p(z) = a_n z^n + a_{n-1} z^{n-1} + \cdots + a_1 z + a_0$$

must also be continuous for any value of z.

Note that a function may have a limit at a point without being continuous there. For example, for the function

$$f(z) = \frac{z^2 + z - 2}{z - 1},$$

we have

$$\lim_{z \to 1} f(z) = 3:$$

But $f(z)$ is not defined at $z = 1$, since it assumes there the indeterminate form $0/0$; i.e., $w = 3$ does not lie in the range of f, nor does $z = 1$ lie in the domain of f.

The function $f(z) = (z^2 + z - 2)/(z - 1)$ is admittedly a rather artificial case, since if $z \neq 1$, then $f(z) = z + 2$; so $f(z)$ is actually linear and hence uniformly continuous everywhere except at $z = 1$; if, in fact, we extend the definition of f so that $f(1) = 3$, then $f(z)$ becomes continuous everywhere. The singularity of $f(z)$ at $z = 1$ in a case like this is called a *removable singularity*, since it can be "removed" and the function can be made continuous by simply defining the function at the singular point to equal the limiting value. We shall study this concept in more detail in a future chapter.

Another example of a function which is not continuous everywhere is $f(z) = \arg z$, where we take $-\pi < \arg z \leq \pi$. $\arg z$ is not defined for $z = 0$, and as z approaches a point on the negative real axis from above, $\arg z$ approaches π, a different value from that attained if z approached the same point from below the axis. For a function to be continuous at a point, then, the limit must be the same for every possible way of approaching a point. In the real calculus, we have only two directions; we have an infinite number of them in the complex plane. Not only can we approach a point in the complex plane from any direction, but also we can come into it along a spiral or in many more complicated ways than that.

EXAMPLE 5   To see the difficulties which can arise in the complex plane, consider the function $e^{1/z}$ near $z = 0$. Of course, this function is not defined at $z = 0$; however, everywhere else it is

$$e^{1/z} = e^{(1/r)(\cos\theta - i\sin\theta)} = e^{(1/r)\cos\theta}\left[\cos\left(\frac{1}{r}\sin\theta\right) - i\sin\left(\frac{1}{r}\sin\theta\right)\right].$$

If we let $z \to 0$ along the ray $\theta = 0$, we have

$$e^{1/z} = e^{1/r}$$

which goes to infinity as $r \to 0$. On the other hand, if $z \to 0$ along the ray $\theta = \frac{1}{2}\pi$, we have

$$e^{1/z} = \cos\frac{1}{r} - i\sin\frac{1}{r} = e^{-i/r},$$

which has no limit, but oscillates endlessly around the unit circle as $r \to 0$. Thus the behavior of this function as $z \to 0$ depends on the direction through which we allow z to approach 0. An even more unusual thing happens if we allow z to approach 0 along the set of points

$$(r_n, \theta_n) = \left(\frac{2}{(2n+1)\pi}, \frac{(2n+1)\pi}{2}\right) \quad (n = 1, 2, 3, \ldots),$$

because $\sin\theta_n = (-1)^n$, $\cos\theta_n = 0$, and hence $\cos[(1/r_n)\sin\theta_n] = 0$, $\sin[(1/r_n)\sin\theta_n] = 1$, so that $e^{1/z_n} = -i$. Hence $e^{1/z}$ equals $-i$ at every point along this path. It is possible to choose sets of points (r_n, θ_n) along which $e^{1/z}$ approaches other limits as well, a conclusion which is left to the reader to verify.

2.1. Continuity and Limits of Functions of a Complex Variable

This example was not intended merely to show the reader that $e^{1/z}$ is *really bad* near $z = 0$, although that is a pertinent fact; rather, the purpose was to show that a reasonably straightforward function can have quite different limits as one approaches a point in different ways. It is exceedingly important to realize that this freedom of movement which we have in the complex plane—two dimensions versus one in the real case—is not completely illusory. It does indeed introduce unexpected and nontrivial complications. The reader should bear this in mind as we struggle with two other limiting concepts, the derivative and the integral, in the sections ahead.

A. Exercises

1 Examine the continuity of the following functions at the points indicated:
  a $e^z, e^{1/z}, z^n e^{1/z}$ at $z = 0$; n is a positive integer.
  b $\sin z, \sin(1/z), z\sin(1/z)$, at $z = 0$.
  c $\dfrac{z^3 - 4z^2 + z + 2}{z^2 - 3z + 2}$ at $z = 1, 2,$ and 3.

2 Examine the uniform continuity of the following functions in the sets indicated. Find the modulus of continuity [that is, $\delta = \delta(\varepsilon)$], where one exists.
  a $z^2 - 3z + 2$ for $|z| \leq 5$; in the right half plane.
  b $(z^2 - 3z + 2)^{-1}$ for $3/2 \leq \text{Re}(z) \leq 3$; for $2 \leq \text{Im}(z) \leq 3$.
  c The function in 1(c) for $|z| \geq 3$.

3 Tell where each of the following functions is continuous. Quote the appropriate theorem in proof of your contention. Do any have removable singularities?

  a $|z|^2$
  b $\log z \, (-\pi < \arg z < \pi, z \neq 0)$
  c $\dfrac{z^2 - 2z - 3}{z + 1}$
  d $\dfrac{\sin z}{z + 1}$
  e $\dfrac{z + 1}{z^2 + z + 1}$
  f $e^{z^2 - z + 3}$
  g $\sin\left(\dfrac{z + 1}{z - 1}\right)$
  h $\dfrac{e^z}{z}$
  i $\text{Re}\,\dfrac{z}{z^2 + 4}$
  j $z^2 e^{\bar{z} - 4}$
  k $\dfrac{1}{e^z}$
  l $\dfrac{1}{(z^2 + \bar{z}^2 + 1)}$

4 Find the following limits.
  a $\displaystyle\lim_{z \to 1+i} \dfrac{z^2 - z + 1 - i}{z^2 - 2z + 2}$

b $\lim_{z \to i} \dfrac{z^2 + 1}{z^4 + 1}$

c $\lim_{z \to 3-i} \dfrac{z - 1}{z - 2}$

d $\lim_{z \to -1+i} \dfrac{z^3 - 3z^2 + 4z + 4}{z^2 - 2z + 1}$

e $\lim_{z \to 0} \dfrac{\sqrt{z+1} - 1}{z}$ (using the branch for which $\sqrt{1} = 1$)

f $\lim_{z \to 1} \dfrac{(1/z) - z}{z - 1}$

B. Problems

1 Prove that the following sets are open:

   a An open disk
   b The entire plane
   c $\{z : 0 < \operatorname{Im} z < 1\}$
   d $\{z : \alpha < \arg z < \beta\}$ $(0 < \alpha < \beta < \tfrac{1}{2}\pi)$.

2 Show that it is possible to cause $e^{1/z}$ to approach other limits than those discussed in the text. In fact, show that for any complex number a, it is possible to choose z_n $(n = 1, 2, 3, \ldots)$ in such a way that as $z_n \to 0$, we have $e^{1/z_n} \to a$.

3 Prove that the following sets are not open.

   a The real axis
   b $\{z : |z - a| \leq r\}$
   c $\{z : 0 \leq \operatorname{Im} z < 1\}$
   d A single point

C. Proofs

1 Let f and g be two complex functions which are both continuous at $z = z_0$. Then the function fg is continuous at $z = z_0$.

2 Let f and g be two complex functions, with $f(z)$ continuous at $z = z_0$ and g continuous at the value $f(z_0)$. Then show that the composite function $g \circ f$, defined by

$$(g \circ f)(z) = g[f(z)],$$

is continuous at $z = z_0$.

3 Suppose $f(z)$ is continuous at $z = z_0$, and that there is a positive constant m such that $|f(z)| \geq m$ for all z within a given radius η of z_0. Then $1/f$ is continuous at $z = z_0$.

2.2. DIFFERENTIABLE AND ANALYTIC FUNCTIONS

As has often been the case in our study of complex functions, so also in considering the derivative we shall find that there is no such facile geometric interpretation of a derivative as the slope of a tangent line, as there is in the real case. However, a derivative (even a complex-valued one) still does represent a rate of change; and we simply define derivative precisely as in the real case:

2.2.1. Definition Let f be a continuous single-valued function defined on a domain D of the complex plane. Then f is said to be *differentiable at the point* $z \in D$ if

$$\lim_{\Delta z \to 0} \frac{f(z + \Delta z) - f(z)}{\Delta z}$$

exists and is finite; when the limit exists it is denoted by $f'(z)$ or $(df/dz)(z)$.

Beware, however, of the similarity between this definition and the one for the real case! As when we discussed complex limits, Δz may approach zero along any path whatever; for $f'(z)$ to exist, the limit must exist and be the same for any possible approach.* It will be instructive for us first to consider a function which is nowhere differentiable, then a function which is everywhere differentiable.

EXAMPLE 6 The real-valued function $f(x) = |x|$ is differentiable everywhere except at $x = 0$, where the graph of f has a "corner." However, the complex function $f(z) = |z|$ is *nowhere* differentiable. To show that this is the case, let us take $z = re^{i\theta}$ ($r \neq 0$) and $\Delta z = \Delta r e^{i\phi}$, where ϕ is fixed. Then

$$\frac{f(z + \Delta z) - f(z)}{\Delta z} = \frac{|re^{i\theta} + \Delta r e^{i\phi}| - |re^{i\theta}|}{\Delta r e^{i\phi}} = \frac{|r + \Delta r e^{i(\phi - \theta)}| - r}{\Delta r} e^{-i\phi}.$$

If we take as our direction of approach $\phi = \theta$, then the limit of this expression as $\Delta r \to 0$ is $e^{-i\theta}$. On the other hand, if we take $\phi = \theta + \tfrac{1}{2}\pi$, then the expression becomes

$$e^{-i(\theta + \pi/2)} \frac{|r + \Delta r e^{i\pi/2}| - r}{\Delta r} = -ie^{-i\theta} \frac{\sqrt{r^2 + \Delta r^2} - r}{\Delta r}$$

$$= -ie^{-i\theta} \frac{\Delta r}{\sqrt{r^2 + \Delta r^2} + r},$$

---

* Note that if two approaches give different limits, one can combine the two to yield a new approach for which *no* limit exists.

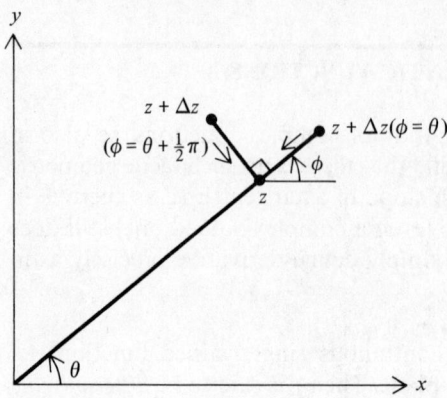

Figure 2.6

which obviously has the limit 0 (Figure 2.6). Thus at each nonzero point, the limit of the difference quotient

$$\frac{f(z+\Delta z)-f(z)}{\Delta z}$$

depends on the angle ϕ through which Δz approaches zero; hence the limit does not exist. The limit does not exist at $z = 0$ either because, for example, if $\Delta z = \Delta r$ then the limit is $+1$, while if $\Delta z = -\Delta r$ then the limit is -1. Therefore $f(z) = |z|$ is nowhere differentiable.

EXAMPLE 7 On the other hand, $f(z) = z^n$ (where n is a nonnegative integer) is everywhere differentiable and its derivative is (as one might expect) nz^{n-1}. For,

a $n = 0$: Then $f(z) = 1$ and

$$\frac{f(z+\Delta z)-f(z)}{\Delta z} = \frac{1-1}{\Delta z} = 0.$$

Thus $f'(z) \equiv 0$.

b $n > 0$: Then

$$\frac{f(z+\Delta z)-f(z)}{\Delta z} = \frac{(z+\Delta z)^n - z^n}{\Delta z}.$$

Now by the binomial theorem (Section 1.3, Proof 6) we have that

$$(z+\Delta z)^n = z^n + nz^{n-1}\Delta z + O(\Delta z^2);*$$

---

* The symbol "$O(\Delta z^2)$," read "big O of Δz^2," means that what follows goes to zero at least as fast as Δz^2, that is, the remaining terms each have a factor of Δz to the power two or higher.

hence

$$\frac{f(z+\Delta z)-f(z)}{\Delta z} = \frac{nz^{n-1}\Delta z + O(\Delta z^2)}{\Delta z} = nz^{n-1} + O(\Delta z);$$

and the limit of this quantity is clearly nz^{n-1} as $\Delta z \to 0$. [Note that $O(\Delta z^2)/\Delta z = O(\Delta z)$.]

A restatement of the definition of the fact that f is differentiable at the point z would be that for sufficiently small values of Δz, we have

$$\frac{f(z+\Delta z)-f(z)}{\Delta z} = f'(z) + o(1),*$$

or

$$f(z+\Delta z) = f(z) + f'(z)\Delta z + o(\Delta z).*$$

[Note that $\Delta z \, o(1) = o(\Delta z)$.]

By using this interpretation of differentiability, it is relatively simple to prove most of the usual results about derivatives.

2.2.2. Theorem Let f and g be differentiable at z, and let α be a constant. Then αf, $f+g$, fg, and f/g are all differentiable at z [the latter provided that $g(z) \neq 0$], and their derivatives are

(1) $\alpha f'(z),$ (2) $f'(z) + g'(z),$ (3) $f'(z)g(z) + f(z)g'(z),$

$$(4) \quad \frac{g(z)f'(z) - g'(z)f(z)}{[g(z)]^2},$$

respectively.

Proof We shall leave the first two to the reader. The proofs of the other two are as follows:

    a The product rule.

$$\frac{f(z+\Delta z)g(z+\Delta z) - f(z)g(z)}{\Delta z}$$

$$= \frac{[f(z) + f'(z)\Delta z + o(\Delta z)][g(z) + g'(z)\Delta z + o(\Delta z)] - f(z)g(z)}{\Delta z}$$

$$= \frac{f(z)g(z) + f'(z)g(z)\Delta z + f(z)g'(z)\Delta z + o(\Delta z) - f(z)g(z)}{\Delta z}$$

$$= f'(z)g(z) + f(z)g'(z) + o(1),$$

which does have the desired limit as $\Delta z \to 0$.

---

* The symbol "o(1)" is read "little o of 1," "o(Δz)" is read "little o of Δz." In general, for any $k = 0, 1, 2, 3, \ldots$, o(Δz^k) represents an expression which goes to zero *faster* than Δz^k, i.e., $\lim_{\Delta z \to 0} o(\Delta z^k)/\Delta z^k = 0$. Thus $\lim_{\Delta z \to 0} o(1) = \lim_{\Delta z \to 0} o(\Delta z^0)/\Delta z^0 = 0$.

b The quotient rule.

$$\left(\frac{f(z+\Delta z)}{g(z+\Delta z)} - \frac{f(z)}{g(z)}\right) \bigg/ \Delta z = \frac{f(z+\Delta z)g(z) - f(z)g(z+\Delta z)}{\Delta z g(z)g(z+\Delta z)}$$

$$= \frac{[f(z) + f'(z)\Delta z + o(\Delta z)]g(z) - f(z)[g(z) + g'(z)\Delta z + o(\Delta z)]}{\Delta z g(z)g(z+\Delta z)}$$

$$= \frac{g(z)f'(z) - g'(z)f(z) + o(1)}{g(z)g(z+\Delta z)}$$

which approaches the predicted limit as $\Delta z \to 0$. QED.

As an immediate corollary of the above we have the following.

2.2.3. Corollary If $f_1, f_2, \ldots, f_n$ are differentiable at z, then so is $f_1 + f_2 + \cdots + f_n$ and its derivative there is $f_1'(z) + f_2'(z) + \cdots + f_n'(z)$. The proof, a simple exercise in induction, is left to the student.

2.2.4. Corollary A polynomial $a_n z^n + a_{n-1} z^{n-1} + \cdots + a_1 z + a_0$ is everywhere differentiable, and its derivative is the polynomial $na_n z^{n-1} + (n-1)a_{n-1} z^{n-2} + \cdots + a_1$.

The proof is immediate; it is a consequence of Example 7, Theorem 2.2.2, and Corollary 2.2.3.

One of the most important results of the differential calculus, in the complex case as in the real case, is the formula for differentiating the function of a function, the "chain rule":

2.2.5. Theorem Let g be differentiable at z, and f be differentiable at $w = g(z)$. Then the composite function $f \circ g$ is differentiable at z and

$$(f \circ g)'(z) = f'(w)g'(z) = f'[g(z)]g'(z) = (f' \circ g)(z)g'(z)$$

Proof

$$\frac{(f \circ g)(z + \Delta z) - (f \circ g)(z)}{\Delta z} = \frac{f[g(z + \Delta z)] - f[g(z)]}{\Delta z}$$

$$= \frac{f[g(z) + g'(z)\Delta z + o(\Delta z)] - f[g(z)]}{\Delta z}$$

$$= \frac{f[g(z)] + f'[g(z)][g'(z)\Delta z + o(\Delta z)] + o(\Delta z) - f[g(z)]}{\Delta z}$$

$$= f'[g(z)]g'(z) + o(1),$$

which clearly has the desired limit as $\Delta z \to 0$.*

---

* In this calculation, terms which go to zero faster than $o(\Delta z)$ have been dropped.

The chain rule, as in the real case, enables us to differentiate such complicated expressions as $(z^2 - 2)^{32}$ without having to carry out the expansion: Indeed, a straightforward use of the chain rule shows that the derivative of this expression is $32(z^2 - 2)^{31}(2z)$.

Unfortunately, so far we have learned only how to differentiate polynomial expressions. Deriving formulas for the derivatives of other functions which we have encountered is not an easy process: The reader may wish to attempt to find the derivative of e^z based on the definition alone to see just how difficult this can be. The next theorem, however, enables us to use known formulas from the real calculus, thereby greatly simplifying the process of finding the derivatives.

2.2.6. Theorem Let $f = u + iv$ be differentiable at the point $z = x + iy$. Then u and v are differentiable at (x, y), and they satisfy the equations

$$\frac{\partial u}{\partial x} = \frac{\partial v}{\partial y}, \quad \frac{\partial u}{\partial y} = -\frac{\partial v}{\partial x}. \tag{2.1}$$

at that point. [Equations (2.1) are called the *Cauchy-Riemann equations*.] Conversely, if u and v possess continuous first derivatives at (x, y) and satisfy the Cauchy-Riemann equations there, then f is differentiable at $z = x + iy$.

Proof First, let us assume that f is differentiable. This means that the limit of the expression

$$\frac{f(z + \Delta z) - f(z)}{\Delta z}$$

must exist and be the same regardless of the direction through which $\Delta z = \Delta x + i\Delta y$ approaches zero. In particular, if this direction is horizontal, then $\Delta z = \Delta x$ and

$$\frac{f(z + \Delta z) - f(z)}{\Delta x} = \frac{u(x + \Delta x, y) - u(x, y)}{\Delta x} + i\frac{v(x + \Delta x, y) - v(x, y)}{\Delta x}.$$

Since $f'(z)$ exists, the limit of both the real and imaginary parts exist and we have

$$f'(z) = \frac{\partial u}{\partial x}(x, y) + i\frac{\partial v}{\partial x}(x, y). \tag{2.2}$$

On the other hand, if the direction of approach is vertical, then $\Delta z = i\Delta y$

and
$$\frac{f(z+\Delta z)-f(z)}{\Delta z} = \frac{v(x, y+\Delta y)-v(x,y)}{\Delta y} - i\frac{u(x, y+\Delta y)-u(x,y)}{\Delta y}$$

so that in the limit we have

$$f'(z) = \frac{\partial v}{\partial y}(x,y) - i\frac{\partial u}{\partial y}(x,y). \tag{2.3}$$

Both expressions for $f'(z)$ must be the same, and when we equate real and imaginary parts we obtain the Cauchy-Riemann equations, as promised.

To prove the converse of this theorem, we note that the continuous differentiability of u and v at the point (x, y) implies that

$$u(x+\Delta x, y+\Delta y) - u(x,y) = u_x(x,y)\Delta x + u_y(x,y)\Delta y + o(|\Delta z|),$$
$$v(x+\Delta x, y+\Delta y) - v(x,y) = v_x(x,y)\Delta x + v_y(x,y)\Delta y + o(|\Delta z|),$$

where $|\Delta z| = \sqrt{\Delta x^2 + \Delta y^2}$. By using these two expressions we obtain

$$\begin{aligned} f(z+\Delta z) - f(z) &= u(x+\Delta x, y+\Delta y) - u(x,y) \\ &\quad + i[v(x+\Delta x, y+\Delta y) - v(x,y)] \\ &= (u_x\Delta x + u_y\Delta y) + i(v_x\Delta x + v_y\Delta y) + o(|\Delta z|). \end{aligned}$$

By using the Cauchy-Riemann equations, we obtain

$$\begin{aligned} f(z+\Delta z) - f(z) &= (u_x\Delta x - v_x\Delta y) + i(v_x\Delta x + u_x\Delta y) + o(|\Delta z|) \\ &= u_x(\Delta x + i\Delta y) + iv_x(\Delta x + i\Delta y) + o(|\Delta z|) \\ &= (u_x + iv_x)(\Delta x + i\Delta y) + o(|\Delta z|) \\ &= (u_x + iv_x)\Delta z + o(|\Delta z|) \end{aligned}$$

and, hence,

$$\frac{f(z+\Delta z)-f(z)}{\Delta z} = u_x + iv_x + o(1),$$

which, in the limit as $\Delta z \to 0$, gives

$$f'(z) = u_x + iv_x,$$

the same result as we had before; the Cauchy-Riemann equations also imply

$$f'(z) = v_y - iu_y$$

as before. QED.

2.2. DIFFERENTIABLE AND ANALYTIC FUNCTIONS

The usefulness of the Cauchy-Riemann equations is that once we have written a function f in terms of its real and imaginary parts u and v, then we can calculate the first partials of u and v (if they exist); if they are continuous and satisfy the Cauchy-Riemann equations, then f is differentiable and its derivative is given by either formula (2.2) or (2.3).

EXAMPLE 8   To find the derivative of $f(z) = e^z$, we write

$$f(z) = e^x \cos y + ie^x \sin y.$$

Then

$$u_x = v_y = e^x \cos y, \qquad u_y = -v_x = -e^x \sin y.$$

The Cauchy-Riemann equations being satisfied, it follows that $f(z)$ is differentiable and that

$$f'(z) = e^x \cos y + ie^x \sin y = e^z.$$

By the chain rule, if $p(z)$ is any polynomial, then the function

$$f(z) = e^{p(z)}$$

has as its derivative

$$f'(z) = e^{p(z)} p'(z).$$

Thus in particular the derivative of $e^{\alpha z}$ is $\alpha e^{\alpha z}$ for any complex constant α.

EXAMPLE 9   Let us test to see whether the single-valued branch of $f(z) = \log z = \ln|z| + i \arg z$ (for $-\pi < \arg z < \pi$, $z \neq 0$) possesses a derivative. First we note that for $u = \ln|z| = \frac{1}{2}\ln(x^2 + y^2)$, we have

$$u_x = \frac{x}{x^2 + y^2}, \qquad u_y = \frac{y}{x^2 + y^2}.$$

The function $v = \arg z = \arctan(y/x)$, where the value of the arctangent is taken between $-\pi$ and π and is chosen according to the quadrant in which (x, y) lies and to be $\frac{1}{2}\pi$ when $x = 0$, $y > 0$ and $-\frac{1}{2}\pi$ when $x = 0$, $y < 0$, has as its partial derivatives

$$v_x = \frac{1}{1 + (y^2/x^2)}\left(-\frac{y}{x^2}\right) = -\frac{y}{x^2 + y^2},$$

$$v_y = \frac{1}{1 + (y^2/x^2)}\left(\frac{1}{x}\right) = \frac{x}{x^2 + y^2}.$$

Note that even though $\arg z$ is a multiple-valued function, its partial derivatives are well-defined single-valued functions, since the various values of $\arg z$ differ only by a constant. Once again the Cauchy-Riemann equations

64 CONTINUITY, DIFFERENTIATION, AND INTEGRATION

are satisfied, and hence $f(z) = \log z$ is differentiable in the cut plane $-\pi < \arg z < \pi, z \neq 0$ and its derivative is

$$f'(z) = \frac{x}{x^2 + y^2} - i\frac{y}{x^2 + y^2} = \frac{x - iy}{x^2 + y^2} = \frac{\bar{z}}{z\bar{z}} = \frac{1}{z}.$$

EXAMPLE 10 If we consider the function $f(z) = z^\alpha$, where α is any complex constant and where we consider f only in some domain where we can define a single-valued branch of z^α (a cut plane, for example), then we can write

$$f(z) = e^{\alpha \log z}$$

and find the derivative by the chain rule:

$$f'(z) = e^{\alpha \log z} \cdot \alpha \cdot \frac{1}{z} = \frac{\alpha z^\alpha}{z} = \alpha z^{\alpha - 1}.$$

Again a single-valued branch of this derivative must be chosen, consistently with the choice of the branch of z^α.

In the exercises, we shall ask the student to calculate the derivatives of many other standard functions using the techniques of these examples.

Example 7 showed us that the derivative of $f \equiv 1$ is zero, and Theorem 2.2.2 implies that the derivative of any constant is zero. It is now possible to use the Cauchy-Riemann equations plus facts from the real calculus to show the converse:

EXAMPLE 11 Suppose f is a differentiable function and that $f' \equiv 0$ in a domain D. Then f is a constant. For

$$f' \equiv u_x + iv_x \equiv v_y - iu_y \equiv 0,$$

implying that $u_x \equiv u_y \equiv v_x \equiv v_y \equiv 0$. But then we know from the real calculus that u and v are independent of both x and y; hence u and v are constants and, therefore, so is f.

All of the differentiable functions we have studied so far turn out to have first derivations which are continuous. There is no *a priori* reason why this should always be so. Functions with *continuous* first derivatives are very important to our future studies; accordingly we give them a special name:

2.2.7. Definition Let f be a single-valued function defined and possessing a continuous first derivative in a domain D. Then f is said to be *analytic* in D. If D is the entire complex plane, then f is said to be *entire*.

2.2. Differentiable and Analytic Functions

It was not until 1900 that Goursat discovered that the word "continuous" in the above definition is superfluous, since the very act of requiring a complex function to have a derivative in a domain *forces* that derivative to be continuous. We shall see the proof of this in the final section of this chapter.

It is now possible for us to show that the complex exponential function is uniquely defined by its properties and by the requirement that it be analytic. We do this as follows:

EXAMPLE 12   Let $f(z)$ be a nonconstant function such that (a) $f(z + a) = f(z) f(a)$ for all complex z and a, and (b) $f'(0)$ exists and equals α. Then it turns out that $\alpha \neq 0$ and $f(z) = e^{\alpha z}$. To prove this, first, note that (a) implies that either $f(0) = 0$ or $f(0) = 1$ (put $0 = a = z$). If $f(0) = 0$, then (a) also implies $f(z) \equiv 0$. But $f(z)$ is nonconstant; hence $f(0) = 1$ is the only tenable conclusion. Next, form the difference quotient and apply this fact plus (b):

$$\frac{f(z + \Delta z) - f(z)}{\Delta z} = \frac{f(z) f(\Delta z) - f(z) f(0)}{\Delta z} = f(z) \frac{f(\Delta z) - f(0)}{\Delta z}.$$

By hypothesis, $f'(0)$ exists; therefore, because of the above equation, $f'(z)$ exists for all z and, in fact,

$$f'(z) = f(z) f'(0) = \alpha f(z).$$

This shows that $\alpha \neq 0$, for if it were zero, $f'(z)$ would be zero for all z, making f a constant, contrary to the hypothesis.

Now let us consider the function g defined by $g(z) = f(z) e^{-\alpha z}$. By taking the derivative of g, we have (by Theorem 2.2.2, the product rule) that

$$g'(z) = f'(z) e^{-\alpha z} - \alpha f(z) e^{-\alpha z} = \alpha f(z) e^{-\alpha z} - \alpha f(z) e^{-\alpha z}$$

and, therefore,

$$g'(z) \equiv 0.$$

It follows from Example 10 that $g(z)$ is a constant, and since

$$g(0) = f(0) e^0 = 1,$$

we have

$$g(z) = f(z) e^{-\alpha z} \equiv 1$$

and, therefore, $f(z) = e^{\alpha z}$. Hence the complex exponential is uniquely defined: It is the *only* nonconstant complex-valued function differentiable at $z = 0$ that possesses the basic property (a) of an exponential.

As a final application of the methods of this section, we can prove the complex version of a very helpful limit theorem:

CONTINUITY, DIFFERENTIATION, AND INTEGRATION

2.2.8. Theorem (*L'Hôpital's Rule*) Suppose that
$$\lim_{z \to z_0} f(z) = \lim_{z \to z_0} g(z) = 0.$$

If f and g are both differentiable at z_0 and if
$$\lim_{z \to z_0} g'(z)$$
exists and is (finite and) nonzero, then
$$\lim_{z \to z_0} \frac{f(z)}{g(z)} = \frac{f'(z_0)}{g'(z_0)}.$$

Proof For values of z sufficiently close to z_0, we have
$$\frac{f(z)}{g(z)} = \frac{f(z_0) + f'(z_0)(z - z_0) + o(1)(z - z_0)}{g(z_0) + g'(z_0)(z - z_0) + o(1)(z - z_0)} = \frac{f'(z_0) + o(1)}{g'(z_0) + o(1)},$$
whence the conclusion follows.

As in the real case, L'Hôpital's rule can be generalized to the case in which the derivatives of f and g are zero as well, in which case the limit will be $f''(z_0)/g''(z_0)$, assuming that these derivatives exist and that $g''(z_0) \neq 0$. This generalization will be left to the reader.

EXAMPLE 13

$$\lim_{z \to 0} \frac{e^z - 1}{z} = \lim_{z \to 0} \frac{e^z}{1} = e^0 = 1.$$

A. Exercises

1 Differentiate the following, using the rules derived in this chapter:

a $\left(z + \dfrac{1}{z}\right)^2$

b $\dfrac{z^3 + 4z^2 - 2z + 3}{z^2 - 3z + 2}$

c $(z^3 - 27)^3 (z^2 + 1)$

d $(z^3 + 10z - 1)^{2/3}$

e $e^{7\sqrt{z-3}}$

f $\log(e^{2z} + z^2)$

g $z \log z - 1$

h $\dfrac{(z + 1)^{1/2}}{z^2 + 2}$

i $\exp \dfrac{(z^2 + 2)}{(z^2 - 1)}$

j $\dfrac{az + b}{cz + d} \; (ad - bc \neq 0)$

2 Aided by the results of the problems below, find derivatives of the following:

- **a** $\sin(z^2 + 3)$
- **b** $\cos^2(z^3 - 3z^2 + 4z - 1)$
- **c** $\sinh[\log(z^2 + 2)]$
- **d** $\tan z$
- **e** $\tanh z$
- **f** $1 + \cosh^2(z^3 - 1)$
- **g** $(\frac{1}{2} + \frac{1}{2}\cos 2z)^{\frac{1}{2}}$
- **h** $\log(\cos z)$
- **i** $\sin\left(\dfrac{z-1}{z+2}\right)$
- **j** $[\cos(z + \sqrt{z})]^{\frac{1}{3}}$

3 Calculate the following limits by the use of L'Hôpital's rule:

- **a** $\lim\limits_{z \to 0} \dfrac{\sin z}{z}$ [HINT: $(d/dz)\sin z = \cos z.$]

- **b** $\lim\limits_{z \to 0} \dfrac{\cos z - 1}{\sin z}$ [HINT: $(d/dz)(\cos z) = -\sin z.$]

- **c** $\lim\limits_{z \to 1} \dfrac{z - 1}{\log z}$ [HINT: $(d/dz)(\log z) = 1/z.$]

- **d** $\lim\limits_{z \to i} \dfrac{\sinh \pi z}{z^4 - 1}$ [HINT: $(d/dz)\sinh \pi z = \pi \cosh \pi z.$]

- **e** $\lim\limits_{z \to 0} \dfrac{z - \sin z}{z^2}$

- **f** $\lim\limits_{z \to 0} \dfrac{\cos^2 z - 1}{z^2}$ [HINT: $(d/dz)\cos^2 z = -2\cos z \sin z.$]

B. Problems

1 Find the derivatives of the following functions two different ways; first, by use of the Cauchy-Riemann equations, and second, by using the known derivative of e^z and various differentiation rules:

- **a** $\sin z$
- **b** $\cos z$
- **c** $\sinh z$
- **d** $\cosh z$

2 Attempt to calculate the derivative of e^z by the definition alone. Carry out the calculation as far as you can using every possible fact from the real calculus.

C. Proofs

1 If f is differentiable at z_0, then it is continuous at z_0.
2 If α is a constant and f is differentiable at z, then so is αf and $(\alpha f)'(z) = \alpha f'(z)$.
3 If f and g are differentiable at z, then so is $f + g$ and $(f + g)'(z) = f'(z) + g'(z)$.

4 *Logarithmic differentiation.* If $f_1, f_2, \ldots, f_n$ are differentiable at z then so is $f = f_1 f_2 \cdots f_n$ and

$$f'(z) = f_1(z) f_2(z) \cdots f_n(z) \left(\frac{f_1'(z)}{f_1(z)} + \frac{f_2'(z)}{f_2(z)} + \cdots + \frac{f_n'(z)}{f_n(z)} \right),$$

provided that $f_k(z) \neq 0$ for any k.

5 Generalize L'Hôpital's rule to show that if $f(z_0)$, $f'(z_0)$, $g(z_0)$, $g'(z_0)$ are all zero and $g''(z_0) \neq 0$, then

$$\lim_{z \to z_0} \frac{f(z)}{g(z)} = \frac{f''(z_0)}{g''(z_0)}.$$

2.3. INTEGRATION OF FUNCTIONS OF A COMPLEX VARIABLE

In defining the integral of a real-valued function $f(x)$ over an interval $a \leq x \leq b$, one normally considers all finite sums of the form

$$\sum_{i=1}^{n} f(\tilde{x}_i)(x_i - x_{i-1}),$$

where $x_0 = a$, $x_n = b$, and $x_{i-1} < x_i$ for all i, and where $\tilde{x}_i$ always satisfies $x_{i-1} \leq \tilde{x}_i \leq x_i$. If all such sums have the same limit I as n approaches ∞ and as the maximum of the $x_i - x_{i-1}$ approaches zero, then we call this value the integral of $f(x)$ from a to b and write

$$I = \int_a^b f(x) \, dx.$$

This may seem to be a pretty difficult process; the student may have to study it many times before he really understands what is going on. However, it is very worthwhile, because this type of integration is simpler than the integration of a complex-valued function, and yet, at the same time, it is the prototype for this integration.

Let us attempt to understand why complex integration is more difficult. In the real case, when we write

$$\int_a^b f(x) \, dx = \lim \sum_{i=1}^{n} f(\tilde{x}_i)(x_i - x_{i-1}),$$

with the prescriptions as above, it is clear that we have divided the real interval from a to b up into a number of subintervals $(x_0, x_1), (x_1, x_2), \ldots, (x_{n-1}, x_n)$; then we have evaluated f at some point in each interval and formed the sum as indicated. We then approach the value of the integral

2.3. INTEGRATION OF FUNCTIONS OF A COMPLEX VARIABLE

more and more closely as the number of intervals increases and their maximum length decreases.

Now, however, suppose that we wish to do the same thing with a complex-valued function; that is, given $f = f(z)$ defined on a certain domain in the complex plane, and given two points a and b in that domain, suppose we attempt to form a sum as in the real case:

$$\sum_{i=1}^{n} f(\tilde{z}_i)(z_i - z_{i-1}). \tag{2.4}$$

Other than $z_0 = a$ and $z_n = b$, there is little else in this sum that makes sense. Among complex numbers, there is no relationship like the relationship "$<$" among real numbers. Thus it makes no sense to say that $z_{i-1} < z_i$ or that $\tilde{z}_i$ is between z_{i-1} and z_i.

To get around this, one might contemplate joining a and b with a straight line (assuming that this line lies entirely in the domain of f), dividing it into intervals by means of a set of points $z_0, z_1, \ldots, z_n$ such that z_{i+1} is always closer to $z_n = b$ than is z_i, and choosing $\tilde{z}_i$ somewhere on the segment between z_{i-1} and z_i. Then, as the number of intervals increases and their maximum length decreases, if this sum approaches a limit, we call it the integral of f from a to b.

There is only one difficulty. In the complex plane, we could equally well have chosen some other curve joining a to b, for example, an arc of a circle (Figure 2.7). If we divide the circular arc into smaller arcs by means of points $z'_1, z'_2, \ldots$ approaching $z'_m = b$, then we can certainly consider all the sums

$$\sum_{i=1}^{m} f(\tilde{z}'_i)(z'_i - z'_{i-1}),$$

where $\tilde{z}'_i$ always lies on the arc connecting z'_{i-1} to z'_i, and ask if this collection of sums has a limit. And we can do this for any arc whatsoever which

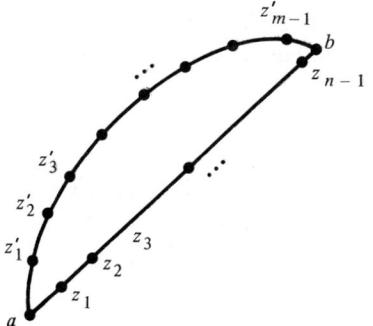

Figure 2.7

joins a and b. Thus, for each arc joining a and b, we have to take the limit of all such sums, and there is simply no reason at all to expect that the limit will be the same on two different arcs, or even that it exists on either or both arcs.

In order to clarify this discussion further and end up with a reasonable definition of the integral, then we shall have to be a bit more precise about the notion of an arc joining two points a and b in the complex plane. For the purposes of our study, we shall ask that all arcs be specified by parametric equations, that is, in the form $x = x(t)$, $y = y(t)$, or in complex notation

$$z = z(t) = x(t) + iy(t);$$

that, as t runs between two values t_0 and t_1, z should run continuously between a and b without "jumps," "gaps," and "breaks," and without crossing itself; and that all of our arcs possess a tangent except perhaps at a finite number of corners. *This is certainly not the most general definition of an arc*; but it will suffice for all the applications which we shall consider. Phrased rigorously we have the following:

2.3.1. Definition A *piecewise smooth arc* (or simply *arc*, or *contour*) C is a complex-valued function of a real variable, $z(t) = x(t) + iy(t)$, defined for $t_0 \leq t \leq t_1$, with the following properties:

1. Both $x(t)$ and $y(t)$ are continuous functions of t;
2. If $z(t') = z(t'')$, then $t' = t''$;
3. $z'(t) = x'(t) + iy'(t)$ exists, is continuous, and is nonzero except at most at a finite number of points between t_0 and t_1.

Condition (1) ensures that the arc will be a continuous one, in the usual sense; (2), that the arc will not cross itself; and (3), that the arc will possess a continuously turning tangent at all but at most a finite number of points.

To see that this latter sentence is true, we simply consider the complex number $z(t + \Delta t) - z(t)$. This is a chord joining the two points $z(t)$ and $z(t + \Delta t)$ (Figure 2.8). Divided by the real number Δt, it becomes

$$\frac{z(t + \Delta t) - z(t)}{\Delta t} = \frac{x(t + \Delta t) - x(t)}{\Delta t} + i\frac{y(t + \Delta t) - y(t)}{\Delta t},$$

which is still a vector pointing in the direction of the indicated chord. Now if our curve possesses a tangent at $z(t)$, then as $\Delta t \to 0$, this chord more and more closely approaches the tangent direction. Hence its limit, if it exists and is nonzero, which is

$$x'(t) + iy'(t),$$

is a tangent vector to the curve at the point $z(t)$.

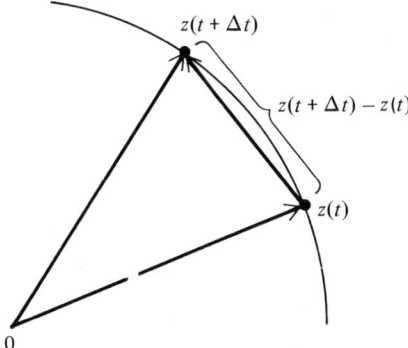

Figure 2.8

Many times we shall restrict our attention, as in what follows, to *smooth arcs*. A smooth arc joining two points a and b is an arc which possesses a continuously turning tangent at *every* point. No generality is lost in this assumption since every (piecewise smooth) arc consists of a finite number of smooth arcs joined end to end, and any concept defined or theorem proved for a smooth arc is hence true for a piecewise smooth arc.

Let us suppose, then, that $f(z)$ is a continuous function defined in a domain D, and that C is a smooth arc in D joining a and b (Figure 2.9). We wish to define the integral of $f(z)$ on the arc C from a to b. Let C be given by the parametric equation $z(t) = x(t) + iy(t)$. As our first step, we divide C into n parts by means of the points

$$z_0 = a, z_1, z_2, \ldots, z_n = b,$$

where $\qquad z_i = z(t_i) \quad \text{and} \quad t_i < t_{i+1}, \qquad i = 0, 1, 2, \ldots, n-1.$

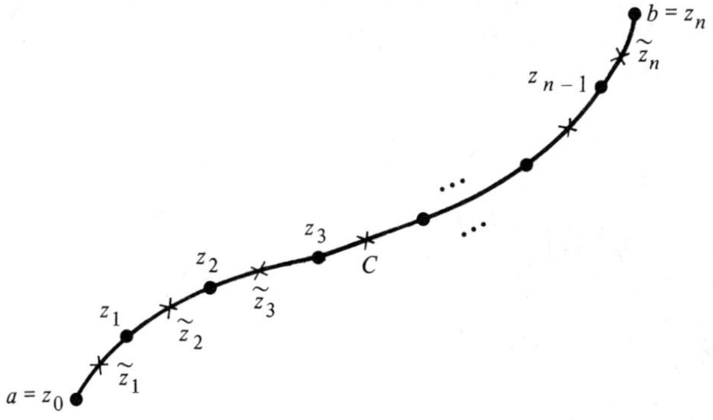

Figure 2.9

72 CONTINUITY, DIFFERENTIATION, AND INTEGRATION

(Note that though the z_i's cannot be ordered by "<", the t_i's can.) On each of these arcs of C we choose a point, labelling the point from the kth arc $\tilde{z}_k = z(\tilde{t}_k)$, where $t_{k-1} \leq \tilde{t}_k \leq t_k$. Then we consider the sum

$$\sum_{k=1}^{n} f(\tilde{z}_k)(z_k - z_{k-1}).$$

This looks quite similar to the sum which we use when defining a real integral; the analogy is more striking if we put $z_k - z_{k-1} = \Delta z_k$ and write the sum as

$$\sum_{k=1}^{n} f(\tilde{z}_k) \Delta z_k.$$

The primary difference is that this sum is, of course, a complex number, since both $f(\tilde{z}_k)$ and Δz_k are complex. If, as we let n go to infinity in such a way that the maximum $|\Delta z_k|$ goes to zero, this sum has a limit which is the same regardless of the particular choice of subdivisions, then we write the limit as

$$\int_C f(z)\,dz = \lim_{\substack{n \to \infty \\ \max|\Delta z_k| \to 0}} \sum_{k=1}^{n} f(\tilde{z}_k)\Delta z_k.$$

Thus we must answer the question as to whether this limit exists, supposing that $f(z)$ is continuous and $z(t)$ is continuously differentiable. The answer is yes, as we can see by reducing this complex integral to two real ones; the process begins as follows:

$$\sum_{k=1}^{n} f(\tilde{z}_k)(z_k - z_{k-1}) = \sum_{k=1}^{n} f(z(\tilde{t}_k))[z(t_k) - z(t_{k-1})]$$

$$= \sum_{k=1}^{n} f(z(\tilde{t}_k)) \frac{z(t_k) - z(t_{k-1})}{t_k - t_{k-1}} (t_k - t_{k-1}).$$

Now, by continuity and from the fact that C is a smooth arc, $\max |z_k - z_{k-1}| \to 0$ implies that $\max(t_k - t_{k-1}) \to 0$, so that it suffices to take the limit as $\max(t_k - t_{k-1}) = \max \Delta t_k$ goes to zero (why?). By writing $f(z) = u(z) + iv(z)$ and $z = x + iy$, we have

$$\sum_{k=1}^{n} f(\tilde{z}_k)(z_k - z_{k-1})$$

$$= \sum_{k=1}^{n} [u(z(\tilde{t}_k)) + iv(z(\tilde{t}_k))] \left(\frac{x(t_k) - x(t_{k-1})}{\Delta t_k} + i \frac{y(t_k) - y(t_{k-1})}{\Delta t_k} \right) \Delta t_k$$

$$= \sum_{k=1}^{n} \left(u(z(\tilde{t}_k)) \frac{\Delta x_k}{\Delta t_k} - v(z(\tilde{t}_k)) \frac{\Delta y_k}{\Delta t_k} \right) \Delta t_k$$

$$+ i \left(u(z(\tilde{t}_k)) \frac{\Delta y_k}{\Delta t_k} + v(z(\tilde{t}_k)) \frac{\Delta x_k}{\Delta t_k} \right) \Delta t_k.$$

2.3. INTEGRATION OF FUNCTIONS OF A COMPLEX VARIABLE 73

As max Δt_k goes to zero and n goes to infinity, it can be shown* that the limit exists and

$$\int_C f(z)\,dz = \int_{t_0}^{t_n} \left(u(z(t))\frac{dx}{dt}(t) - v(z(t))\frac{dy}{dt}(t) \right) dt$$
$$+ i \int_{t_0}^{t_n} \left(u(z(t))\frac{dy}{dt}(t) + v(z(t))\frac{dx}{dt}(t) \right) dt.$$

Thus the integral of a continuous complex function along a smooth arc is evaluated simply by substituting the parametrization of the arc $z(t) = x(t) + iy(t)$ for $z = x + iy$, and then evaluating the two resulting real integrals.† To evaluate $\int_C f(z)\,dz$ when C is not smooth, we recall that in this case C is the union of a finite number of smooth arcs, $C = C_1 \cup C_2 \cup \cdots \cup C_m$, and we define

$$\int_C f(z)\,dz = \int_{C_1} f(z)\,dz + \int_{C_2} f(z)\,dz + \cdots + \int_{C_m} f(z)\,dz.$$

We shall usually ignore the distinction between smooth and nonsmooth arcs in the future, unless the distinction has some particular importance at the time.

Although hitherto we have considered only arcs which do not cross themselves, it is certainly useful to consider arcs for which the initial point and the final point coincide, that is, $a = b$. Such an arc is called a *simple closed contour*. For example, the unit circle, as parametrized by

$$z(t) = \cos 2\pi t + i \sin 2\pi t \qquad (0 \leq t \leq 1)$$

is a simple closed contour.

One final bit of notation: If C is a simple closed contour, we write

$$\oint_C f(z)\,dz$$

for the integral around C from start to finish, provided that the integration proceeds in the positive (counterclockwise) direction, that direction in which the interior of C is always on the left. The reader should note that what we have just said is not as simple and intuitive as it may seem. (The fact that a simple closed contour C divides the plane into two open sets with no points in common—one, called the interior of C [or $I(C)$] and the other, called the exterior of C [or $E(C)$]—and the fact that one cannot

---

* See Appendix A for a proof of the existence of the integral of a real continuous function on a finite interval.

† It can also be shown that, under the conditions we have assumed, the value of the integral depends only on f and C, and not on the particular parametrization of C.

move from one to the other without crossing C, is one of the most profound theorems of topology, called the *Jordan curve theorem*. Because the proof of this theorem is lengthy and difficult, we shall omit it and take the result for granted.) An alternative way to define the positive direction around a simple closed contour is to state that this is the direction in which the tangent rotates through $+2\pi$ radians as the curve is traversed from start to finish. However, we should remark that the orientation of a simple closed curve (in general) is not a simple concept to define rigorously.

Let us now work some examples:

EXAMPLE 14 Suppose $f(z) = \bar{z}$, and let C be given by $t^2 - it$, $0 \leq t \leq 1$. Then

$$\int_C \bar{z}\, dz = \int_0^1 (x - iy)(x' + iy')\, dt = \int_0^1 (t^2 + it)(2t - i)\, dt$$

$$= \int_0^1 [(2t^3 + t) + it^2]\, dt = \left[\left(\frac{1}{2}\right)t^4 + \left(\frac{1}{2}\right)t^2\right]_0^1 + \frac{it^3}{3}\bigg|_0^1 = 1 + \frac{i}{3}.$$

Note how i, being a constant, slips through the integral sign just like any real constant.

EXAMPLE 15 Let $f(z) = 3x + 2iy$, and let C be the triangle with hypotenuse $C_1: y = 2x + 3$ for $0 \leq x \leq 1$ and legs $C_2: x = 1$ for $3 \leq y \leq 5$ and $C_3: y = 3$ for $0 \leq x \leq 1$. On C_1, we use $x = t$ as parameter; then $x'(t) = 1$ and $y'(t) = 2$; on C_2, $y = t$ is the parameter with $x'(t) = 0$ and $y'(t) = 1$; on C_3, $x = t$ is the parameter again with $x'(t) = 1$ and $y'(t) = 0$. Thus

$$\int_C f(z)\, dz$$

$$= \int_1^0 [3t + 2i(2t + 3)](1 + 2i)\, dt + \int_3^5 (3 + 2it)\, i\, dt + \int_0^1 (3t + 6i)\, dt$$

$$= \int_1^0 [(-5 + 10i)t + (-12 + 6i)]\, dt + \int_3^5 (3i - 2t)\, dt + \int_0^1 (3t + 6i)\, dt$$

$$= (-5 + 10i)\frac{t^2}{2}\bigg|_1^0 + (-12 + 6i)t\bigg|_1^0 + (3it - t^2)\bigg|_3^5 + \left(\frac{3t^2}{2} + 6it\right)\bigg|_0^1$$

$$= (5 - 10i)\tfrac{1}{2} + 12 - 6i + 15i - 25 - 9i + 9 + \tfrac{3}{2} + 6i$$

$$= (\tfrac{5}{2} + 12 - 25 + 9 + \tfrac{3}{2}) + (-5 - 6 + 15 - 9 + 6)i = i.$$

The reader can verify that this integration proceeds around C in the positive direction.

2.3. INTEGRATION OF FUNCTIONS OF A COMPLEX VARIABLE

EXAMPLE 16 Suppose $f(z) = 1/z$, and let C be a circle around the origin given by $z(\theta) = r(\cos \theta + i \sin \theta)$, $0 \le \theta \le 2\pi$. Then on the circle, $1/z = 1/r(\cos \theta - i \sin \theta)$. Thus

$$\oint_C \frac{dz}{z} = \oint_0^{2\pi} \frac{1}{r}(\cos \theta - i \sin \theta) r(-\sin \theta + i \cos \theta) d\theta = i \int_0^{2\pi} d\theta = 2\pi i.$$

If, on the other hand, $f(z) = z^n = r^n(\cos n\theta + i \sin n\theta)$, where n is any integer not equal to -1,

$$\oint_C z^n \, dz = \int_0^{2\pi} r^n(\cos n\theta + i \sin n\theta) r(-\sin \theta + i \cos \theta) d\theta$$

$$= ir^{n+1} \int_0^{2\pi} [\cos(n+1)\theta + i \sin(n+1)\theta] \, d\theta = 0.$$

A couple of computational rules have been assumed in the above examples, which would be appropriate to mention at this point, along with a number of other simple rules:

2.3.2. Theorem Let f and g be defined and continuous in a domain D, let α be a constant, and let C^+ be any arc in D joining two points β and γ oriented in that direction, with C^- the same curve with the opposite orientation. Let δ be any other point on C^+ with C_1 the arc from β to δ, C_2 the arc from δ to γ. Then the following formulas hold:

1. $\quad \int_{C^+} \alpha f(z) \, dz = \alpha \int_{C^+} f(z) \, dz,$

2. $\quad \int_{C^+} [f(z) + g(z)] \, dz = \int_{C^+} f(z) \, dz + \int_{C^+} g(z) \, dz,$

3. $\quad \int_{C^+} f(z) \, dz = - \int_{C^-} f(z) \, dz,$

4. $\quad \int_{C^+} f(z) \, dz = \int_{C_1} f(z) \, dz + \int_{C_2} f(z) \, dz.$

The formulas are simple consequences of the definition of the integral and the proofs are left to the reader.

At this point the student may be curious as to whether integration has any relation to differentiation as in the real calculus, that is, whether the integral of a function has anything to do with the existence of an "anti-

76 CONTINUITY, DIFFERENTIATION, AND INTEGRATION

derivative;" we shall investigate this question carefully, always referring (as is our wont) to the known results of the real calculus.

2.3.3. Definition Suppose $f = u + iv$ is a continuous single-valued function on a domain D and that there exists a single-valued function $F = U + iV$ on D such that $F' = f$. Then F (which is analytic) is called an *antiderivative* of f.

As in the real calculus, we can find numerous examples of antiderivatives simply by reversing our familiar differentiation formulas. For instance, $z^{n+1}/(n+1)$ is an antiderivative of z^n; $\sin z$ is an antiderivative of $\cos z$. Then the fundamental theorem carries over intact from the real calculus:

2.3.4. Theorem (*The Fundamental Theorem of Calculus*) Let f be a continuous, single-valued function in a domain D and suppose that F is an antiderivative of f in D. Then if C is any arc in D whose initial point is α and whose final point is β, we have

$$\int_C f(z)\,dz = F(\beta) - F(\alpha).$$

Proof Let $z(t) = x(t) + iy(t)$, $a \leq t \leq b$, be the parametric representation of C. Then because $F'(z) = f(z)$, we have

$$f = u + iv = U_x + iV_x = V_y - iU_y.$$

By writing the integral in terms of t, we have

$$\int_C f(z)\,dz = \int_a^b f(z(t))\,z'(t)\,dt = \int_a^b (U_x + iV_x)(x' + iy')\,dt$$

$$= \int_a^b (U_x x' - V_x y')\,dt + i \int_a^b (V_x x' + U_x y')\,dt$$

$$= \int_a^b (U_x x' + U_y y')\,dt + i \int_a^b (V_x x' + V_y y')\,dt.$$

The latter step is justified since F is analytic and hence U and V satisfy the Cauchy–Riemann equations. But now because of the chain rule for real functions of two variables,

$$U_x x' + U_y y' = \frac{dU}{dt}, \qquad V_x x' + V_y y' = \frac{dV}{dt},$$

we have

$$\int_C f(z)\,dz = \int_a^b \frac{dU}{dt}\,dt + i \int_a^b \frac{dV}{dt}\,dt,$$

which, by the fundamental theorem of the real calculus, yields

$$\int_C f(z)\,dz = U(x(t), y(t))\Big|_a^b + iV(x(t), y(t))\Big|_a^b$$

$$= (U + iV)\Big|_a^b = F(\beta) - F(\alpha). \quad \text{QED}$$

Thus if a function of a complex variable possesses an antiderivative, then the evaluation of its integral is a straightforward process: Simply evaluate the antiderivative at the two endpoints and subtract. In fact, we have the following:

2.3.5. Corollary If f possesses an antiderivative F in D and if α and β are any two points in D, then the integral of f from α to β is independent of the path from α to β (as long as it lies in D).

Obviously! For, given any arc C in D joining α and β, we have shown

$$\int_\alpha^\beta f(z)\,dz = F(\beta) - F(\alpha),$$

which is unambiguous since F is single-valued.

And now we even have the following theorem:

2.3.6. Theorem If f possesses an antiderivative in D and if C is any simple closed contour in D, then

$$\oint_C f(z)\,dz = 0.$$

Proof Let α and β be any two points on C and let C_1 be one of the arcs of C between α and β, with C_2 representing the other arc oriented from β to α. Suppose that the orientation of C_1 and C_2 coincides with the positive orientation of C (Figure 2.10). Then

$$\oint_C f(z)\,dz = \int_{C_1} f(z)\,dz + \int_{C_2} f(z)\,dz = F(\beta) - F(\alpha) + F(\alpha) - F(\beta) = 0$$

QED

A converse of Theorem 2.3.6 is also true, and we shall turn to this in a moment; but first let us consider some examples.

EXAMPLE 17 Consider the function $f(z) = z^n$ for n any integer except -1. Then $f(z)$ has a single-valued antiderivative in the domain D consisting of all z (if $n \geq 0$) or of all nonzero z (if $n < -1$), namely $z^{n+1}/(n+1)$. Hence

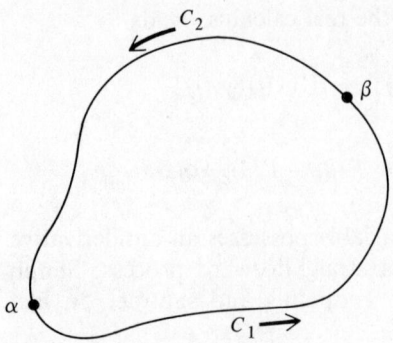

Figure 2.10

if α and β are any two points (neither of which is zero if $n < 0$), we have

$$\int_\alpha^\beta z^n \, dz = \frac{\beta^{n+1} - \alpha^{n+1}}{n+1}.$$

If C is any simple closed contour in the plane (not passing through $z = 0$ if $n < 0$), then

$$\oint_C z^n \, dz = 0.$$

By putting this together with Theorem 2.3.2, we have the following:

2.3.7. Theorem If $p(z) = a_n z^n + a_{n-1} z^{n-1} + \cdots + a_1 z + a_0$ is any complex polynomial, then

1    $\int_\alpha^\beta p(z) \, dz$ is independent of the path joining α to β; and

2    $\oint_C p(z) \, dz = 0$ for any simple closed contour.

This theorem is true of more functions than just polynomials. In fact, it is true, as we have seen, of any function having an antiderivative. It would be nice if we could characterize all functions which possess an antiderivative —and we shall do that in very simple terms in the next section.

In Example 17 we specifically neglected the case $n = -1$ because, of course, $1/z$ does not have a single-valued antiderivative in any domain containing the origin; its antiderivative is a branch of the logarithm, which can only be single-valued in a cut plane. We can, however, define a single-valued branch of the logarithm of much more general functions than $1/z$:

EXAMPLE 18 Let f be continuously differentiable and nonzero in a domain D. We shall assume that D is *simply connected*, which means that no simple closed curve in D encloses points not in D. Let z_0 be any fixed point in D, let z be an arbitrary point in D, and suppose $\log f(z_0)$ is any fixed value of the logarithm of the complex number $f(z_0)$. Then the function defined by

$$L(z) = \int_{z_0}^{z} \frac{f'(\zeta)}{f(\zeta)} d\zeta + \log f(z_0)$$

is a single-valued logarithm of $f(z)$ in D. First, to show that L is a logarithm of f, we consider the function g defined by $g(z) = f(z) e^{-L(z)}$. Then

$$g'(z) = f' e^{-L(z)} - f e^{-L(z)} L'(z) = e^{-L(z)} \left(f' - f \frac{f'}{f} \right) \equiv 0.$$

Therefore $g(z)$ is a constant (Example 10), and because

$$g(z_0) = f(z_0) e^{-L(z_0)} = f(z_0) e^{-\log f(z_0)} = \frac{f(z_0)}{f(z_0)} = 1,$$

this constant is 1; therefore

$$f(z) = e^{L(z)},$$

and thus by definition L is a logarithm of f.

To see that L is single-valued, note that we can write L as

$$\begin{aligned} L(z) &= \int_{z_0}^{z} \frac{f'(\zeta)}{f(\zeta)} d\zeta + \log f(z_0) \\ &= \log f(z) - \log f(z_0) + \log f(z_0) \\ &= \log f(z) = \ln |f(z)| + i \arg f(z). \end{aligned}$$

It is not clear that $\arg f(z)$ is necessarily well-defined (i.e., single-valued) in D, even given that we have unambiguously chosen a fixed value for $\arg f(z_0)$. However, if two values of $L(z)$ differ, it must be by a constant, since L has a single-valued derivative (namely f'/f); and further, this constant must be an integer multiple of $2\pi i$, since we always have

$$f(z) = e^{L(z)}$$

and the exponential is periodic with period $2\pi i$. But the argument of f can change by an even multiple of π only if the variable $w = f(z)$ encircles the origin in the w plane one or more times as z encircles some point z_1. This would demand that either $f(z_1) = 0$ or $\lim_{z \to z_1} 1/f(z_1) = 0$, i.e., that $f(z_1)$ be a branchpoint of the logarithm. But our hypotheses prevent this; $f(z)$ is nonzero and analytic in D, and no point z_1 where $f(z_1)$ is zero or where f has a singular point can be either in D (by hypothesis), or encircled by

80 CONTINUITY, DIFFERENTIATION, AND INTEGRATION

a curve in D, since D is simply connected. Therefore $\arg f(z)$ is single-valued in D and $L(z)$ is also single-valued.

As an application of this example let us consider Example 12 from another point of view. There we had given that f was a nonconstant function for which (a) $f(z + a) = f(z) f(a)$ for all z and a; and (b) $f'(0)$ exists and equals α. We then derived from this that $\alpha \neq 0$ and that

$$f'(z) = \alpha f(z)$$

for all z; hence

$$\alpha = \frac{f'(z)}{f(z)}.$$

Therefore by the preceding example,

$$L(z) = \int_{z_0}^{z} \frac{f'(\zeta)}{f(\zeta)} d\zeta + \log f(z_0) = \int_{z_0}^{z} \alpha \, d\zeta + \log f(z_0)$$

is a single-valued logarithm of f. Simply take $z_0 = 0$ and $\log f(z_0) = \log 1 = 0$, and we have that

$$L(z) = \alpha z$$

is a single-valued logarithm of $f(z)$. Hence by definition $f(z) = e^{\alpha z}$.   QED

Sometimes we shall be more interested in estimating integrals than in evaluating them. Before we go into this, let us review the idea of arc length of an arc C, and relate it to the complex parametric representation of the arc $z(t) = x(t) + iy(t)$, $a \leq t \leq b$. Assuming C is a smooth arc, let us divide it into n sub-arcs by means of the sequence of points $z_0 = a, z_1, z_2, z_3, \ldots, z_{n-1}, z_n = b$ as in the past, where $z_i = z(t_i)$ and $t_i < t_{i+1}$ for all i. Then the broken line consisting of the straight line segments joining each z_i to z_{i+1} (Figure 2.11) is an approximation to C, and its length is obviously

$$\sum_{i=1}^{n} |z_i - z_{i-1}| = \sum_{i=1}^{n} \sqrt{(x_i - x_{i-1})^2 + (y_i - y_{i-1})^2}.$$

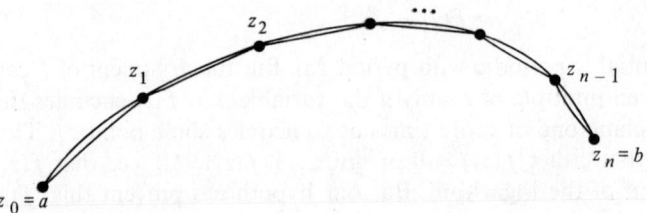

Figure 2.11

2.3. INTEGRATION OF FUNCTIONS OF A COMPLEX VARIABLE

If we represent $z_i - z_{i-1}$ by Δz_i, $x_i - x_{i-1}$ by Δx_i, and $y_i - y_{i-1}$ by Δy_i, we have

$$\sum_{i=1}^{n} |z_i - z_{i-1}| = \sum_{i=1}^{n} |\Delta z_i| = \sum_{i=1}^{n} \sqrt{\Delta x_i^2 + \Delta y_i^2} = \sum_{i=1}^{n} \sqrt{\frac{\Delta x_i^2}{\Delta t_i} + \frac{\Delta y_i^2}{\Delta t_i}} \Delta t_i.$$

Clearly, if this quantity approaches a limit L as $n \to \infty$ and max $\Delta t_i \to 0$, for all possible choices of the z_i's, then this limit L would be a reasonable thing to *define* to be the length of C. This is made all the more reasonable by the fact that if C is a straight line, then any subdivision of C—and hence the limit—gives us exactly the length of C as defined by the Pythagorean theorem.

As a matter of fact, as $n \to \infty$ and max $t_i \to 0$ the last sum approaches

$$\int_a^b \sqrt{x'(t)^2 + y'(t)^2}\, dt;$$

hence we *define* the length of a smooth curve C, joining a and b and parametrized by $z(t) = x(t) + iy(t)$, to be the quantity

$$\int_C |dz| = \int_a^b \sqrt{x'(t)^2 + y'(t)^2}\, dt.$$

(It can be shown that the length of an arc does not depend on the particular parametrization used.)

Bearing this in mind, the next theorem is the most important and basic one for estimating the value of complex integrals.

2.3.8. Theorem Let f be given and continuous on its domain D, and let C be an arc in D of length L. Let M be the maximum of $|f(z)|$ on C, i.e., $|f(z)| \leq M$ for $z \in C$. Then

$$\left| \int_C f(z)\, dz \right| \leq \int_C |f(z)|\, |dz| \leq ML.$$

Proof $\left| \int_C f(z)\, dz \right|$ is a limit of sums of the form

$$\left| \sum_{n=1}^{n} f(\tilde{z}_k) \left(\frac{\Delta x_k}{\Delta t_k} + \frac{i \Delta y_k}{\Delta t_k} \right) \Delta t_k \right| \leq \sum_{k=1}^{n} |f(z_k)| \left| \frac{\Delta x_k}{\Delta t_k} + \frac{i \Delta y_k}{\Delta t_k} \right| \Delta t_k$$

$$\leq M \sum_{k=1}^{n} \sqrt{\left(\frac{\Delta x_k}{\Delta t_k}\right)^2 + \left(\frac{\Delta y_k}{\Delta t_k}\right)^2} \Delta t_k.$$

The triangle inequality can be applied because the sum is finite. Now, by taking the limit of this string of inequalities as $n \to \infty$ and max $\Delta t_k \to 0$,

82 CONTINUITY, DIFFERENTIATION, AND INTEGRATION

we obtain

$$\left| \int_C f(z) \, dz \right| \leq \int_C |f(z)| \left| \frac{dx}{dt} + i \frac{dy}{dt} \right| dt \leq M \int_{t_0}^{t_n} \sqrt{\left(\frac{dx}{dt}\right)^2 + \left(\frac{dy}{dt}\right)^2} \, dt.$$

We recognize the last integral as being the formula for the length of C, thus giving us

$$\left| \int_C f(z) \, dz \right| \leq \int_C |f(z)| \, |dz| \leq ML,$$

completing the proof.

And now we can keep our promise to prove the converse of Theorem 2.3.6:

2.3.9. Theorem Let f be defined and continuous in a domain D, and suppose that the integral of f between any two points of D is independent of the path chosen. Then f possesses an antiderivative in D.

Proof Let z_0 be any fixed point in D. Then the function

$$F(z) \triangleq \int_{z_0}^{z} f(\zeta) \, d\zeta$$

is well-defined and single-valued, since the integral from z_0 to z is independent of path. Now F is an antiderivative of f, for

$$F(z + \Delta z) - F(z) = \int_{z_0}^{z + \Delta z} f(\zeta) \, d\zeta - \int_{z_0}^{z} f(\zeta) \, d\zeta.$$

Because the last two integrals are independent of path, we can choose the path from z_0 to $z + \Delta z$ to coincide with the path from z_0 to z in the second integral, then to proceed along a straight line from z to $z + \Delta z$ (which is always possible if Δz is small enough, since every point z in a domain can

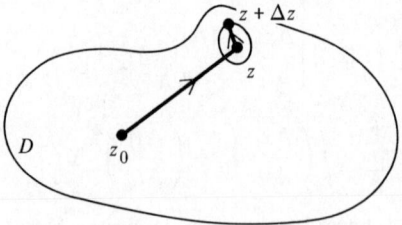

Figure 2.12

be enclosed by a small open disk in the domain; Figure 2.12). Thus

$$F(z + \Delta z) - F(z) = \int_z^{z+\Delta z} f(\zeta)\, d\zeta.$$

Because f is continuous, we can choose $|\Delta z|$ so small that $|f(z) - f(\zeta)| < \varepsilon$ for all ζ along the segment from z to $z + \Delta z$, for any given ε. Hence if we write $\eta(\zeta) = f(\zeta) - f(z)$, we have $|\eta(\zeta)| \leq \varepsilon$ along this segment. Furthermore, $\eta(\zeta) \to 0$ as $|\zeta - z| \to 0$ and hence as $|\Delta z| \to 0$, since $|\zeta - z| \leq |\Delta z|$. Thus

$$F(z + \Delta z) - F(z) = \int_z^{z+\Delta z} f(\zeta)\, d\zeta = \int_z^{z+\Delta z} [f(z) + \eta(\zeta)]\, d\zeta$$

$$= f(z) \int_z^{z+\Delta z} d\zeta + \int_z^{z+\Delta z} \eta(\zeta)\, d\zeta$$

$$= f(z)\, \Delta z + \int_z^{z+\Delta z} \eta(\zeta)\, d\zeta.$$

Now by Theorem 2.3.8,

$$\left| \int_z^{z+\Delta z} \eta(\zeta)\, d\zeta \right| \leq \varepsilon |\Delta z|.$$

Because ε is $o(1)$ with respect to Δz, that is, it goes to zero as Δz goes to zero, it follows that

$$F(z + \Delta z) - F(z) = f(z)\, \Delta z + |\Delta z|\, o(1),$$

which implies that $F'(z)$ exists and equals $f(z)$. Hence F is indeed an antiderivative of f. QED

We shall now apply the results of this and the previous sections to study the gamma function of Euler, a particular analytic function of considerable importance. The gamma function is defined by an improper integral:

$$\Gamma(z) = \int_0^\infty t^{z-1} e^{-t}\, dt.$$

The parameter can be real or complex; t is a real variable. In order to be certain that $\Gamma(z)$ is well-defined, we first have to determine the values of z for which the improper integral converges, that is, the values of z for which

$$\lim_{\substack{\delta \to 0 \\ \Delta \to \infty}} \int_\delta^\Delta t^{z-1} e^{-t}\, dt$$

exists.

84 CONTINUITY, DIFFERENTIATION, AND INTEGRATION

First, let us consider
$$\int_\delta^1 t^{z-1} e^{-t}\, dt \qquad (\delta < 1).$$

We know that
$$t^{z-1} = e^{(z-1)\log t} = e^{(x-1)\log t + iy\log t}$$

and
$$|t^{z-1}| = |e^{(x-1)\log t}||e^{iy\log t}| = e^{(x-1)\log t} = t^{x-1},$$

so, if $x > 0$,
$$\left| \int_\delta^1 t^{z-1} e^{-t}\, dt \right| \leq \int_\delta^1 t^{x-1} e^{-t}\, dt \leq \int_\delta^1 t^{x-1}\, dt = \frac{t^x}{x}\bigg|_\delta^1 = \frac{1-\delta^x}{x}.$$

As $\delta \to 0$, this quantity has a finite limit provided that $x > 0$, i.e., Re$(z) > 0$. Clearly the limit is infinite if $x < 0$. If $x = 0$, the integral is $-\ln \delta$, which also has an infinite limit.

Next we examine
$$\int_1^\Delta t^{z-1} e^{-t}\, dt \qquad (\Delta > 1).$$

We have
$$\left| \int_1^\Delta t^{z-1} e^{-t}\, dt \right| \leq \int_1^\Delta t^{x-1} e^{-t}\, dt;$$

if $x \leq 1$, then
$$\int_1^\Delta t^{x-1} e^{-t}\, dt \leq \int_1^\Delta e^{-t}\, dt = -(e^{-\Delta} - e^{-1}),$$

which has limit e^{-1} as $\Delta \to \infty$.

On the other hand, if $x > 1$ and if N is an integer not less than $x - 1$, we have
$$\int_1^\Delta t^{x-1} e^{-t}\, dt \leq \int_1^\Delta t^N e^{-t}\, dt = e^{-\Delta} p(\Delta) - e^{-1} p(1),$$

where p is a certain polynomial of degree N. Since $e^{-\Delta} p(\Delta) \to 0$ as $\Delta \to \infty$, the improper integral exists for $x > 1$.

Therefore, since at the upper limit the integral converges for all x and at the lower limit it converges for $x > 0$, we know that the complex function
$$\Gamma(z) = \int_0^\infty t^{z-1} e^{-t}\, dt$$

is well-defined for all z with Re$(z) > 0$, i.e., in the right half plane.

By constructing the difference quotient $[\Gamma(z + \Delta z) - \Gamma(z)]/\Delta z$, using the definition of $\Gamma(z)$ as

$$\lim_{\substack{\delta \to 0 \\ \Delta \to \infty}} \int_{\delta}^{\Delta} t^{z-1} e^{-t}\, dt,$$

and then applying arguments somewhat more involved than the above, it is possible to show that

$$\Gamma'(z) = \int_0^\infty t^{z-1} (\ln t)\, e^{-t}\, dt$$

for $\operatorname{Re}(z) > 0$, so that $\Gamma(z)$ is analytic in the right half plane.

One important feature of the gamma function is that it represents a generalization of the factorial. For, first,

$$\Gamma(1) = \int_0^\infty e^{-t}\, dt = 1;$$

next, integrating by parts,

$$\Gamma(z + 1) = \int_0^\infty t^z e^{-t}\, dt = -e^{-t} t^z \Big|_0^\infty + z \int_0^\infty t^{z-1} e^{-t}\, dt = z \int_0^\infty t^{z-1} e^{-t}\, dt,$$

and hence $\Gamma(z + 1) = z\Gamma(z)$. From these two facts it is easy to prove by induction that $\Gamma(n + 1) = n!$, where n is any positive integer. One can also show that $\Gamma(z + k) = (z + k - 1)(z + k - 2) \cdots z\Gamma(z)$, so that it is possible to compute all the values of the gamma function provided that its values are known in the strip $0 < \operatorname{Re}(z) \leq 1$. In fact, many books of tables contain only these values of the gamma function.

A. Exercises

1 Evaluate $\int_C f(z)\, dz$, where f and C are as indicated:

   a $f(z) = x - iy^2$; C is the portion of $xy = 1$ between $\tfrac{1}{3} + 3i$ and $2 + 1/2i$.
   b Same f; C the straight line between $\tfrac{1}{3} + 3i$ and $2 + 1/2i$.
   c $f(z) = x^2$; C the unit circle.
   d $f(z) = x^2 + y^2 - 2ixy$; C the line from $z = 0$ to $z = 2 + i$.
   e $f(z) = x^2 - y^2 + 2ixy$; C the unit circle.
   f $f(z) = 2x - 3iy$; C the ellipse $z(t) = \cos t + 2i \sin t$, $0 \leq t \leq 2\pi$.
   g Same f; C the parabola $y = 2x^2$ from $x = 0$ to $x = 2$.
   h $f(z) = (3x - 2y + 1) + i(2x + 3y - 5)$; C as in Exercise 1(g).

2 Evaluate the following integrals:

   a $\displaystyle\int_1^{\pi i} e^{2z}\, dz$

86 CONTINUITY, DIFFERENTIATION, AND INTEGRATION

b $\quad \displaystyle\int_{1+i}^{-3-i} (z^2 - 3z + 2)\, dz$

✓ c $\quad \displaystyle\int_0^{1+5i} ze^{z^2}\, dz$

✓ d $\quad \displaystyle\oint_{|z-2|=1} z^{-3}\, dz$

✓ e $\quad \displaystyle\int_{2i}^{2+2i} (z^2 + z^{-2})\, dz$

f $\quad \displaystyle\int_{-2-2i}^{1+i} z \cos 2z\, dz$

✓ g $\quad \displaystyle\oint_{|z|=1} \frac{z+1}{(z-2)^3}\, dz$

h $\quad \displaystyle\int_i^{-1} z^2(3z^3 - 12)^5\, dz$

i $\quad \displaystyle\int_1^{3i} \log z\, dz$ (using the branch of the log which is real for z real).

j $\quad \displaystyle\int_i^{i+2} (z+1)^{1/2}\, dz$ (using the branch of the square root which is positive for z real and greater than -1.)

B. Problems

✓ 1   Show by two methods that $\displaystyle\oint_{|z|=1} e^{iz}\, dz = 0$.

    Use this result to evaluate $\displaystyle\oint_{|z|=1} \cos z\, dz,\ \oint_{|z|=1} \sin z\, dz$.

2   Evaluate the integral in Exercise 2(c) by parametrizing the straight line between the two points and substituting for z.

3   Evaluate the same integral by parametrizing the path from 0 to 1, then from 1 to $1 + 5i$.

4   Evaluate $\Gamma(\tfrac{1}{2})$, where $\Gamma(z)$ is the Euler gamma function.

$$\text{HINT:} \quad \int_0^\infty e^{-x^2}\, dx = \frac{\sqrt{\pi}}{2}$$

5   Prove that $\Gamma(z) = 2\displaystyle\int_0^\infty t^{2z-1} e^{-t^2}\, dt,\ \operatorname{Re}(z) > 0$.

6 If m, n, and a are real positive constants, show that
$$\int_0^\infty t^m e^{-at^n} \, dt = \frac{1}{n} a^{-(m+1)/n} \Gamma\left(\frac{m+1}{n}\right),$$

7 Use the previous result to evaluate the following:

  a $\displaystyle\int_0^\infty x^3 e^{-2x} \, dx$ b $\displaystyle\int_0^\infty x^4 e^{-3x^2} \, dx$ c $\displaystyle\int_0^\infty y^2 e^{-7y^2} \, dy.$

C. Proofs

1 Prove that $\displaystyle\int_C \alpha f(z) \, dz = \alpha \int_C f(z) \, dz.$

2 Prove that $\displaystyle\int_C [f(z) + g(z)] \, dz = \int_C f(z) \, dz + \int_C g(z) \, dz.$

3 Prove that $\displaystyle\int_C [\alpha_1 f_1(z) + \alpha_2 f_2(z) + \cdots + \alpha_n f_n(z)] \, dz$
$$= \alpha_1 \int_C f_1(z) \, dz + \alpha_2 \int_C f_2(z) \, dz + \cdots + \alpha_n \int_C f_n(z) \, dz.$$

2.4. INTEGRAL THEOREMS AND HIGHER DERIVATIVES

We saw in the last section that if f is a continuous, single-valued function possessing a single-valued antiderivative in a domain D, and if C is any simple closed contour in C, then
$$\oint_C f(z) \, dz = 0.$$
In particular, we saw that this was always true for polynomials, since they possess antiderivatives in the whole plane. It is now our task to characterize all functions which possess antiderivatives; not surprisingly, such functions turn out to be very similar to polynomials, that is, they can be approximated by polynomials, so in a sense they are "almost" polynomials. This will be the crowning result of the next chapter, together with the conclusion that all such functions are analytic and conversely. Let us first prove the converse of the fact stated at the beginning of this paragraph.

2.4.1. Theorem Let f be a single-valued continuous function defined in a domain D, and suppose that $\oint_C f(z) \, dz = 0$ for each simple closed contour C in D. Then f has an antiderivative in D.

88 CONTINUITY, DIFFERENTIATION, AND INTEGRATION

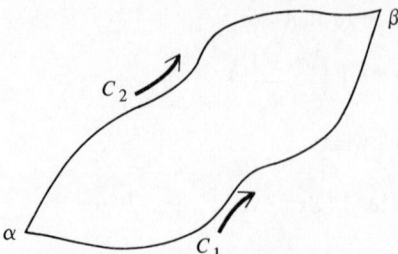

Figure 2.13

Proof We shall show that the integral of f between two points is independent of path; the contention will then follow from Theorem 2.3.9. Let α and β be two points in D and let C_1 and C_2 be two smooth (or even piecewise smooth) arcs joining α to β and oriented in that direction (Figure 2.13). For simplicity we assume that C_1 and C_2 have no points in common except for α and β. Then the contour C consisting of C_1^+ and C_2^- is a simple closed contour in D, and hence

$$\oint_C f(z)\,dz = 0.$$

But

$$\oint_C f(z)\,dz = \int_{C_1^+} f(z)\,dz + \int_{C_2^-} f(z)\,dz = \int_{C_1} f(z)\,dz - \int_{C_2} f(z)\,dz.$$

Therefore

$$\int_{C_1} f(z)\,dz = \int_{C_2} f(z)\,dz$$

and the contention is proved. This proof can be extended easily to the case in which C_1 and C_2 coincide along a portion of their length, or intersect each other a finite number of times (see the proofs at the end of the section). We shall not be concerned with any more pathological cases which might occur.

Our next step shall be to show that analytic functions possess antiderivatives. We shall first need the following:

2.4.2. Lemma *(Green's Theorem)* Suppose $P(x, y)$ and $Q(x, y)$ are two real-valued functions defined inside and on a simple closed contour

2.4. Integral Theorems and Higher Derivatives

C and possessing continuous first partial derivatives there; then*

$$\oint_C (P\,dx + Q\,dy) = \int\int_{I(C)} \left(\frac{\partial Q}{\partial x} - \frac{\partial P}{\partial y} \right) dx\,dy.$$

The proof of this theorem is in Appendix A.

Now we are ready for one of our most significant results:

2.4.3. Theorem *(Cauchy's Theorem)* Suppose f is defined and analytic in a domain D, and C is any simple closed contour in D whose interior is also in D; then

$$\oint_C f(z)\,dz = 0;$$

thus, every analytic function possesses an antiderivative.

Proof By Theorem 2.2.6, the real and imaginary parts of f satisfy the Cauchy-Riemann equations. Because u_x, u_y, v_x and v_y are continuous (by the definition of analyticity), we can apply Green's theorem to obtain

$$\oint_C f(z)\,dz = \oint_C (u + iv)(dx + i\,dy) = \oint_C (u\,dx - v\,dy) + i\oint_C (v\,dx + u\,dy)$$

$$= \int\int_{I(C)} \left(\frac{\partial u}{\partial y} + \frac{\partial v}{\partial x} \right) dx\,dy + i\int\int_{I(C)} \left(\frac{\partial v}{\partial y} - \frac{\partial u}{\partial x} \right) dx\,dy,$$

and these last two integrals vanish because of the Cauchy-Riemann equations.

Cauchy's Theorem has as a consequence (for example) that if $R(z)$ is any rational function, i.e., a quotient of two polynomials $p(z)/q(z)$, and if C is any simple closed contour not containing a zero of $q(z)$ inside or on it, then

$$\oint_C R(z)\,dz = 0.$$

For, a rational function is analytic anywhere the denominator does not vanish. Furthermore, any rational function $p(z)/q(z)$ can be transformed by division into the sum of a polynomial and a rational function whose numerator is of lower degree than the denominator; and such a rational function can be decomposed by partial fractions techniques into terms of

---

* $I(C)$ represents the plane domain interior to C.

the form $\alpha(z - \beta)^{-k}$, k being a positive integer.* For example,

$$\frac{z}{(z - 1)^2 (z + i)} = \frac{A}{z + i} + \frac{B}{z - 1} + \frac{C}{(z - 1)^2}.$$

A is determined by multiplying both sides by $z + i$ and setting $z = -i$:

$$\frac{-i}{(-i - 1)^2} = \frac{-i}{2i} = -\frac{1}{2} = A;$$

C is determined by multiplying both sides by $(z - 1)^2$,

$$\frac{z}{z + i} = 1 - \frac{i}{z + i} = \frac{A}{z + i}(z - 1)^2 + B(z - 1) + C, \quad (2.5)$$

then setting $z = 1$:

$$C = 1 - \frac{i}{1 + i} = 1 - \frac{i}{2}(1 - i) = \frac{1 - i}{2};$$

B is then determined first by differentiating (2.5) with respect to z,

$$\frac{i}{(z + i)^2} = B + O(z - 1),\dagger$$

and then setting $z = 1$:

$$\frac{i}{(1 + i)^2} = B = \frac{i}{2i} = \frac{1}{2}.$$

Provided that none of the zeros of the denominators lie on a simple closed contour C, the integrals around C of all the terms of the form $\alpha(z - \beta)^{-k}$ for $k \neq 1$ will be zero, since as we have seen, all such terms have a single-valued antiderivative. However, we have yet to attack the case of $k = 1$; this is the subject of our next lemma.

2.4.4. Lemma Let C be a simple closed curve and let z be a point not on C. Then

$$\oint_C \frac{d\zeta}{\zeta - z} = \begin{cases} 2\pi i & \text{if } z \in I(C) \\ 0 & \text{if } z \in E(C) \end{cases}.$$

Proof First let us consider the case when z is inside C (Figure 2.14). We draw a circle Σ around z whose radius δ is small enough so that Σ is

---

* We can consider this a purely algebraic fact and calculate as in the example given here. However, in Section 3.3 we shall prove that such a decomposition always exists—a fact which may, at first glance, seem obvious.
† The second term still contains a factor $(z - 1)$ after differentiating, which will vanish when $z = 1$.

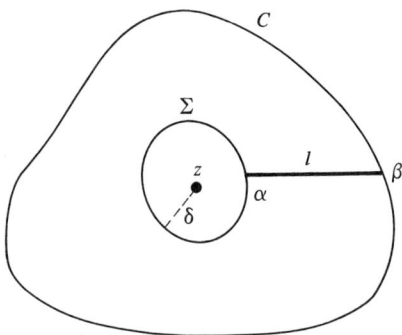

Figure 2.14

contained completely inside C; then we connect Σ with C by an arc l, which intersects Σ at the point $z = \alpha$ and C at the point $z = \beta$. If we traverse the resulting curve first from α to β along l, then around C in the positive direction, then along l from β to α, and finally around Σ in the negative direction, we have in effect traversed a simple closed curve K (Figure 2.15) inside which $1/(\zeta - z)$ is analytic. Hence

$$\oint_K \frac{d\zeta}{\zeta - z} = \int_{l_{\alpha\beta}} \frac{d\zeta}{\zeta - z} + \oint_{C^+} \frac{d\zeta}{\zeta - z} + \int_{l_{\beta\alpha}} \frac{d\zeta}{\zeta - z} + \oint_{\Sigma^-} \frac{d\zeta}{\zeta - z} = 0.$$

The two integrals back and forth along l cancel one another, and hence we have

$$\oint_{C^+} \frac{d\zeta}{\zeta - z} + \oint_{\Sigma^-} \frac{d\zeta}{\zeta - z} = 0,$$

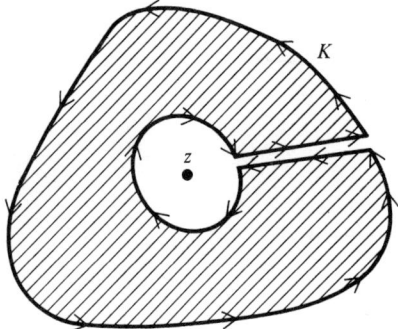

Figure 2.15

or, finally,

$$\oint_{C^+} \frac{d\zeta}{\zeta - z} = \oint_{\Sigma^+} \frac{d\zeta}{\zeta - z}.$$

The latter integral is simple enough to evaluate; we use the parametrization $\zeta - z = \delta e^{i\theta}$ ($0 \leq \theta \leq 2\pi$), so that $d\zeta = i\delta e^{i\theta}\,d\theta$; and we have

$$\oint_{\Sigma^+} \frac{d\zeta}{\zeta - z} = \int_0^{2\pi} \frac{i\delta e^{i\theta}}{\delta e^{i\theta}} d\theta = i\int_0^{2\pi} d\theta = 2\pi i,$$

as claimed.

In the case when z is not in the interior of C, one can form a branch cut from z to ∞ which nowhere intersects C, and hence define a single-valued antiderivative of $(\zeta - z)^{-1}$ over the whole of C, whence

$$\oint_C \frac{d\zeta}{\zeta - z} = 0,$$

by Theorem 2.3.6. A couple of cases and their branch cuts are shown in Figure 2.16. We shall not attempt a rigorous proof or spend any time worrying about unusual cases. In most practical examples C will be a circle or some similarly nice curve.

By putting this theorem together with previous results, we obtain the following.

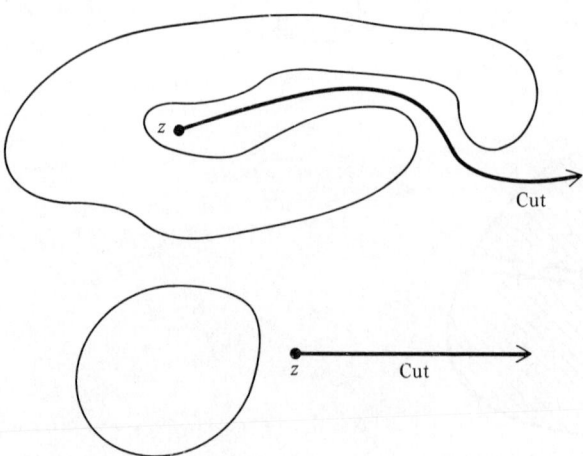

Figure 2.16

2.4. Integral Theorems and Higher Derivatives

2.4.5. Corollary Let C be a simple closed contour, let z be a point not on C, and let k be any integer. Then

$$\oint_C (\zeta - z)^k \, d\zeta = \begin{cases} 2\pi i & \text{if } k = -1, \ z \in I(C); \\ 0 & \text{otherwise.} \end{cases}$$

We can use this corollary to evaluate the integral of any rational function around any simple closed contour not passing through a zero of the denominator.

EXAMPLE 19 For example, consider the function $z/[(z-1)^2 (z+i)]$, which we examined just prior to Lemma 2.4.4. We can, for example, integrate this function around the circle $|z - 1| = 1$ as follows:

$$\oint_{|z-1|=1} \frac{z \, dz}{(z-1)^2 (z+i)^2} = -\frac{1}{2} \oint \frac{dz}{z+i} + \frac{1}{2} \oint \frac{dz}{z-1} + \frac{(1-i)}{2} \oint \frac{dz}{(z-1)^2}$$

$$= -\frac{1}{2} \cdot 0 + \frac{1}{2} \cdot 2\pi i + \frac{(1-i)}{2} \cdot 0 = \pi i.$$

The first integral vanishes because $-i$ is not in the circle $|z - 1| = 1$; the last, because $(z-1)^{-2}$ has a single-valued antiderivative everywhere except at $z = 1$.

EXAMPLE 20 Let us evaluate

$$\oint_C \frac{z^2 + z - 1}{z^3 + 3z^2 - z - 3} \, dz,$$

where C is the circle $|z| = 2$. Then, by using partial fractions, we have

$$\frac{z^2 + z - 1}{z^3 + 3z^2 - z - 3} = \frac{z^2 + z - 1}{(z-1)(z+1)(z+3)}$$

$$= \frac{1}{8(z-1)} + \frac{1}{4(z+1)} + \frac{5}{8(z+3)}$$

and so

$$\oint_C \frac{z^2 + z - 1}{z^3 + 3z^2 - z - 3} \, dz = \frac{1}{8} \oint_C \frac{dz}{z-1} + \frac{1}{4} \oint_C \frac{dz}{z+1} + \frac{5}{8} \oint_C \frac{dz}{z+3}$$

$$= 2\pi i \left(\frac{1}{8} + \frac{1}{4} \right),$$

so that the answer is $\frac{3}{4} \pi i$. The last integral vanishes because the point $z = -3$ is outside the circle $|z| = 2$.

94 CONTINUITY, DIFFERENTIATION, AND INTEGRATION

We are now ready to prove one of the most important and useful consequences of Cauchy's theorem:

2.4.6. Theorem *(The Cauchy Integral Formula)** Let f be analytic in a domain which includes a simple closed contour C and its interior, and let $z \in I(C)$. Then

$$f(z) = \frac{1}{2\pi i} \oint_C \frac{f(\zeta)}{\zeta - z} d\zeta.$$

Proof First note that we have already proved this theorem in the special case $f \equiv 1$; this was Lemma 2.4.4. Now we shall attempt to prove basically the same thing for the case of a nonconstant analytic function. We note that just as in Lemma 2.4.4, we have

$$\oint_C \frac{f(\zeta)}{\zeta - z} d\zeta = \oint_\Sigma \frac{f(\zeta)}{\zeta - z} d\zeta,$$

where Σ is a circle centered at z and contained in the interior of C.

Because f is continuous at z (see Section 2.2, Proof 1), one can choose δ, the radius of Σ, so small that $|f(\zeta) - f(z)| < \varepsilon$ for all ζ on Σ. Then our procedure will be to show that the difference

$$\left| \frac{1}{2\pi i} \oint_\Sigma \frac{f(\zeta)}{\zeta - z} d\zeta - f(z) \right|$$

is less than ε, and hence must be zero, since ε can be taken as small as desired. Now

$$f(z) = \frac{f(z)}{2\pi i} \oint_\Sigma \frac{d\zeta}{\zeta - z} = \frac{1}{2\pi i} \oint_\Sigma \frac{f(z) \, d\zeta}{\zeta - z},$$

since $f(z)$ does not depend on the variable of integration ζ and can be looked at as a constant; and we have

$$\left| \frac{1}{2\pi i} \oint_\Sigma \frac{f(\zeta)}{\zeta - z} d\zeta - f(z) \right| = \frac{1}{2\pi} \left| \oint_\Sigma \frac{f(\zeta) - f(z)}{\zeta - z} d\zeta \right|$$

$$\leq \frac{1}{2\pi} \max_{\zeta \in \Sigma} \left| \frac{f(\zeta) - f(z)}{\zeta - z} \right| 2\pi\delta.$$

On Σ, $|f(\zeta) - f(z)| < \varepsilon$ and $|\zeta - z| = \delta$. Hence

$$\left| \frac{1}{2\pi i} \oint_\Sigma \frac{f(\zeta)}{\zeta - z} d\zeta - f(z) \right| \leq \frac{1}{2\pi} \cdot \frac{\varepsilon}{\delta} \cdot 2\pi\delta = \varepsilon. \quad \text{QED.}$$

---

* Note the difference between Cauchy's *theorem* and Cauchy's *formula*.

2.4. INTEGRAL THEOREMS AND HIGHER DERIVATIVES

The Cauchy Integral formula allows us to evaluate many integrals which involve something more complicated than rational functions:

EXAMPLE 21 Consider

$$\oint_C \frac{\cos \pi z}{z - 1} dz,$$

where C is the circle $|z - 1| = 1$. Then according to Theorem 2.4.6, this integral will be $2\pi i$ times the value of $\cos \pi z$ at $z = 1$, which is $\cos \pi = -1$; thus

$$\oint_C \frac{\cos \pi z}{z - 1} dz = -2\pi i.$$

As with many of the properties of complex functions which we have studied thus far, the "Cauchy formula" property goes both ways:

2.4.7. Theorem Let f be a function defined and continuous inside and on a simple closed contour C. If f satisfies the Cauchy integral formula for every $z \in I(C)$, that is, if

$$f(z) = \frac{1}{2\pi i} \oint_C \frac{f(\zeta)}{\zeta - z} d\zeta, \qquad z \in I(C), \qquad (2.6)$$

then f possesses derivatives of all orders at z, and

$$f^{(k)}(z) = \frac{k!}{2\pi i} \oint_C \frac{f(\zeta)}{(\zeta - z)^{k+1}} d\zeta.$$

In particular, if (2.6) is true for every $z \in I(C)$, then f is analytic in $I(C)$.

Proof If $k = 1$, the formula we wish to establish is

$$f'(z) = \frac{1}{2\pi i} \oint_C \frac{f(\zeta)}{(\zeta - z)^2} d\zeta. \qquad (2.7)$$

Note that if we knew that it was legal to differentiate with respect to z under the integral sign, then this is exactly what we would get.

To prove the above formula, we calculate

$$\frac{f(z + \Delta z) - f(z)}{\Delta z} = \frac{1}{2\pi i \Delta z} \oint_C \frac{f(\zeta)}{\zeta - z - \Delta z} d\zeta - \frac{1}{2\pi i \Delta z} \oint_C \frac{f(\zeta)}{\zeta - z} d\zeta$$

$$= \frac{1}{2\pi i} \oint_C \frac{f(\zeta)}{(\zeta - z - \Delta z)(\zeta - z)} d\zeta,$$

96 CONTINUITY, DIFFERENTIATION, AND INTEGRATION

which looks as if it ought to approach the right limit as $\Delta z \to 0$. To see that this is so, we take the difference between Equation 2.7 and this and estimate:

$$\left| \frac{1}{2\pi i} \oint_C \frac{f(\zeta)}{(\zeta - z)^2} d\zeta - \frac{1}{2\pi i} \oint_C \frac{f(\zeta)}{(\zeta - z - \Delta z)(\zeta - z)} d\zeta \right|$$

$$= \frac{1}{2\pi} \left| \oint_C \frac{(\zeta - z - \Delta z) f(\zeta) - (\zeta - z) f(\zeta)}{(\zeta - z - \Delta z)(\zeta - z)} d\zeta \right|$$

$$\leq \frac{1}{2\pi} \frac{M}{\min |\zeta - z - \Delta z| |\zeta - z|^2} L |\Delta z|,$$

where M is the maximum of $|f(\zeta)|$ on C and L is the length of C; the denominator is bounded away from zero for sufficiently small $|\Delta z|$, and hence the whole expression goes to zero as $\Delta z \to 0$.

To complete the proof by induction we would need to show that if the conclusion is true for $k = n$, then it is true for $k = n + 1$. This aspect of the proof is quite similar to the preceding calculations and is left to the reader as an exercise.

Theorem 2.4.7 has the following as an immediate consequence.

2.4.8. Corollary An analytic function in a domain D is infinitely differentiable in D.

Thus, imposing on a complex function the requirement that it have a continuous derivative forces it to have derivatives of all orders. As we shall see in the next section, one need not even require that the first derivative be continuous.

Theorem 2.4.7 can be used to help in the evaluation of integrals also:

EXAMPLE 22 Evaluate

$$\oint_{|z|=3} \frac{z^2 - 2z + 3}{(z - 1)(z + 2)^3} dz.$$

By partial fractions,

$$\frac{1}{(z - 1)(z + 2)^3} = \frac{1}{27(z - 1)} - \frac{1}{27(z + 2)} - \frac{1}{9(z + 2)^2} - \frac{1}{3(z + 2)^3},$$

2.4. Integral Theorems and Higher Derivatives

so that

$$\oint_{|z|=3} \frac{z^2 - 2z + 3}{(z-1)(z+2)^3} dz = \frac{1}{27} \oint_{|z|=3} \frac{(z^2 - 2z + 3)\,dz}{z-1}$$

$$- \frac{1}{27} \oint_{|z|=3} \frac{(z^2 - 2z + 3)\,dz}{z+2}$$

$$- \frac{1}{9} \oint_{|z|=3} \frac{(z^2 - 2z + 3)}{(z+2)^2} dz - \frac{1}{3} \oint_{|z|=3} \frac{(z^2 - 2z + 3)\,dz}{(z+2)^3}$$

$$= 2\pi i \left(\frac{1}{27}(z^2 - 2z + 3) \bigg|_{z=1} - \frac{1}{27}(z^2 - 2z + 3) \bigg|_{z=-2} \right.$$

$$\left. - \frac{1}{9}(2z - 2) \bigg|_{z=-2} - \frac{1}{3} \frac{(2)}{2!} \bigg|_{z=-2} \right)$$

$$= 2\pi i \left(\frac{2}{27} - \frac{11}{27} + \frac{6}{9} - \frac{1}{3} \right) = \frac{2\pi i}{27}(2 - 11 + 18 - 9) = 0.$$

(It is purely an accident that this result came out to be zero. The next example shows that this need not always be the case.)

EXAMPLE 23 Evaluate

$$\oint_{|z|=3} \frac{z \cos \pi z \, dz}{(z-1)(z+2)^2}.$$

By using partial fractions, we have

$$\frac{1}{(z-1)(z+2)^2} = \frac{1}{9(z-1)} + \frac{-1}{3(z+2)^2} + \frac{1}{9(z+2)},$$

so that the integral becomes

$$\frac{1}{9} \oint_{|z|=3} \frac{z \cos \pi z \, dz}{z-1} - \frac{1}{9} \oint_{|z|=3} \frac{z \cos \pi z \, dz}{z+2} - \frac{1}{3} \oint_{|z|=3} \frac{z \cos \pi z \, dz}{(z+2)^2}.$$

Since $z \cos \pi z$ is analytic, all integrals can be evaluated by means of the Cauchy integral formula and Theorem 2.4.7, yielding

$$2\pi i \left[\tfrac{1}{9} z \cos \pi z \big|_{z=1} - \tfrac{1}{9} z \cos \pi z \big|_{z=-2} - \tfrac{1}{3}(\cos \pi z - z\pi \sin \pi z)_{z=-2} \right]$$

$$= 2\pi i (-\tfrac{1}{9} + \tfrac{2}{9} - \tfrac{1}{3}) = -\tfrac{4}{9}\pi i.$$

98 CONTINUITY, DIFFERENTIATION, AND INTEGRATION

This section will not be complete until we tie up the package by demonstrating that Cauchy's Theorem also possesses a converse:

2.4.9. Theorem *(Morera's Theorem)* Suppose that f is defined and continuous in a domain D and that for every simple closed contour C in D,

$$\oint_C f(z)\,dz = 0;$$

then f is analytic in D.

Proof By Theorem 2.4.1, f has an antiderivative F in D. Thus $F'(z) = f(z)$ for all z in D; and, since f is continuous, F is analytic. But then by Theorem 2.4.7, F possesses all derivatives. Hence, $F'' = f'$ exists, and likewise $F''' = f''$; hence f' is continuous (Section 2.2, Proof 1). Therefore f, possessing a continuous derivative, is analytic. QED.

A. Exercises

1 Evaluate the following integrals:

a $\oint_{|z|=2} \dfrac{(z^2+1)\,dz}{(z-3)(z^2-1)}$

b $\oint_{|z-1|=2} \dfrac{(iz-1)\,dz}{z^2+2iz-2}$

c $\oint_{|z|=4} \dfrac{z^3\,dz}{(z-3)^3(z-2)^2}$

d $\oint_{|z-i|=1} \dfrac{dz}{z^3+z^2+z+1}$

e $\oint_{|z|=4} \dfrac{(z^2-i)\,dz}{z^3-z^2-5z-3}$

f $\oint_{|z|=2} \dfrac{(z^2-i)\,dz}{z^3-z^2-5z-3}$

2 Evaluate the following:

a $\dfrac{1}{2\pi i}\oint_C \dfrac{e^z\,dz}{z-1}$, where C is $\{z:|z|=\tfrac{1}{2}\}$; where C is $\{z:|z|=2\}$.

b $\dfrac{1}{2\pi i}\oint_C \dfrac{e^{az}\,dz}{z^2+1}$, if a is real and positive and C is $\{z:|z|=2\}$.

c $\dfrac{1}{2\pi i}\oint_C \dfrac{e^{iz}\,dz}{z^2+2}$, where C is $\{z:|z|=2\}$.

d $\dfrac{1}{2\pi i}\oint_C \dfrac{\sin 3z\,dz}{z-1/2}$, where C is $\{z:|z|=2\}$.

3 Evaluate the following integrals around circles centered at the origin, with radius as indicated:

a $\oint \dfrac{e^z\,dz}{z^4}$, $r=1$.

b $\dfrac{1}{2\pi i}\oint \dfrac{e^{az}\,dz}{(z^2+1)^2}$, $r=3$.

c $\oint \dfrac{\cos \pi z \, dz}{(z-1)^2}$, $r = 3$.

d $\dfrac{1}{2\pi i} \oint \dfrac{\sin^2 \pi z \, dz}{z^3}$, $r = 1$.

e $\oint \dfrac{(z^3 + 1) \, dz}{(z-1)^3 (z-2)^2}$, $r = \dfrac{3}{2}$.

f $\oint \dfrac{z^4 + 3z^3 - 2z + 1}{(z+1)^3} \, dz$, $r = 2$.

g $\oint \dfrac{z^2 e^z \, dz}{(z-1)^2 (z-2)}$, $r = 3$.

h $\oint \dfrac{(z^3 + 2z^2 + 1) \cos \pi z \, dz}{(z-1)^2}$, $r = 2$.

4 Express each of the following polynomials as new polynomials in powers of $z - 2$. (Use the result of Problem 3 below.)

a $z^3 - 2z^2 + 4z - 5i$
b $5z^6 + 2z^5 - 17z^4 + 3z^3 - 2z^2 + 1$
c $4(z+1)^3 - 2(z+1)^2 + (3+i)(z+1) + 4 - i$.

B. Problems

1 Show that Green's theorem is true for $\oint_C [(y^2 - 2xy) \, dx + (x^3 + 2x^2 y) \, dy]$, where C is the rectangle with vertices at $2 \pm i$, $-2 \pm i$.

2 Show by using Green's theorem that the area enclosed by a simple closed contour C is given by $\tfrac{1}{2} \oint_C (x \, dy - y \, dx)$.

3 Show that if $p(z) = a_n z^n + a_{n-1} z^{n-1} + \cdots + a_1 z + a_0$, then for any complex z_0,

$$p(z) = p(z_0) + (z - z_0) p'(z_0) + \dfrac{(z-z_0)^2}{2!} p''(z_0) + \cdots + \dfrac{(z-z_0)^n}{n!} p^{(n)}(z_0).$$

C. Proofs

1 Sketch a proof of Theorem 2.4.1 in the case when C_1 and C_2 share an arc, and when C_1 and C_2 intersect in a finite number of points.

2 Complete the proof of Theorem 2.4.7 by induction, assuming the theorem true for $k = n$, and using this to prove its truth for $k = n + 1$.

3 If z is any complex number not equal to 1, then prove by induction that

$$\dfrac{1}{1-z} = 1 + z + z^2 + \cdots + z^n + z^{n+1} \left(\dfrac{1}{1-z} \right).$$

2.5. THE MAXIMUM MODULUS THEOREM, HARMONIC FUNCTIONS, AND GOURSAT'S THEOREM

In this section we shall discover how rare indeed are analytic functions, for we shall prove more theorems which even further circumscribe their properties. We shall also carry out some applications of the results of this

100 CONTINUITY, DIFFERENTIATION, AND INTEGRATION

chapter; then, finally, we shall prove the long-awaited result that a single-valued function differentiable in any domain is analytic there.

Our first theorem is one which says, in effect, that the absolute value of a nonconstant analytic function is always larger somewhere on a simple closed curve in its domain than it is anywhere inside the curve.

2.5.1. Theorem (*The Maximum Modulus Theorem*) Let $f(z)$ be nonconstant and analytic in a domain D, and let C be a simple closed contour in D whose interior is also contained in D. Then if $|f(z)| \leq M$ for some $M > 0$ and for all $z \in C \cup I(C)$, we have $|f(z)| < M$ for any $z \in I(C)$.

Proof First, let us assume C is a circle, $\{z : |z - z_0| = r\}$. Then since f is analytic,

$$f(z_0) = \frac{1}{2\pi i} \oint_C \frac{f(z)\,dz}{z - z_0} = \frac{1}{2\pi i} \oint_0^{2\pi} \frac{f(z_0 + re^{i\theta}) ire^{i\theta}\,d\theta}{re^{i\theta}},$$

and hence

$$f(z_0) = \frac{1}{2\pi r} \oint_0^{2\pi} f(z_0 + re^{i\theta}) r\,d\theta.$$

In other words, at the center of the circle the value of f is the average of the values of f over the circumference of the circle. If we call this average value A, then since $|f(z_0 + re^{i\theta})| \leq M$ we have

$$|f(z_0)| = |A| \leq \frac{1}{2\pi r} M 2\pi r = M,$$

so that $|A| \leq M$. It is our purpose to show that $|A| < M$, i.e., that the value at the center of a circle can never equal the maximum unless the function is a constant. We shall do this by assuming that $|A| = M$, and showing that this implies that $|f(z)| \equiv M$ throughout the entire disk, which would force f to be a constant in the disk. Hence let us assume $|A| = M$. This will force $|f| = M$ on the entire circle, for if for some θ_0 between 0 and 2π we have $|f(z_0 + re^{i\theta_0})| = M - \delta < M$, then because of continuity there would exist an η so that $|f(z_0 + re^{i\theta})| < M - \frac{1}{2}\delta$ for $\theta_0 - \eta < \theta < \theta_0 + \eta$. Then we would have

$$|A| = |f(z_0)| \leq \frac{1}{2\pi}\left\{\int_0^{\theta_0 - \eta} + \int_{\theta_0 - \eta}^{\theta_0 + \eta} + \int_{\theta_0 - \eta}^{2\pi}\right\} |f(z + re^{i\theta})|\,d\theta$$

$$\leq \frac{1}{2\pi} M(\theta_0 - \eta) + \left[M - \left(\frac{1}{2}\right)\delta\right] 2\eta + M(2\pi - \theta_0 - \eta)$$

$$= \frac{1}{2\pi}\left\{(2\pi - 2\eta)M + \left[M - \left(\frac{1}{2}\right)\delta\right] 2\eta\right\} = \frac{1}{2\pi}(2\pi M - \delta\eta) < M,$$

2.5. THE MAXIMUM MODULUS THEOREM

which is a contradiction. Thus, if $|A| = M$ we must have $|f(z_0 + re^{i\theta})| = M$ around the whole circumference of C; and by the same argument, $|f(z_0 + \rho e^{i\theta})| = M$ for all $\rho < r$. This implies $|f(z)| \equiv M$ throughout $I(C)$; that is, if $f(z) = u + iv$, then

$$u^2 + v^2 \equiv M^2,$$

and, differentiating this expression with respect to x and y,

$$2uu_x + 2vv_x = 2uu_y + 2vv_y = 0.$$

Applying the Cauchy-Riemann equations,

$$uv_y + vv_x = 0, \qquad vv_y - uv_x = 0.$$

If we multiply the first equation by u and the second by v and add, we obtain $(u^2 + v^2)v_y = M^2 v_y = 0$. Therefore $v_y = 0$. On the other hand, if we multiply the first equation by v and the second by u and subtract, we obtain $v_x = 0$. Therefore v must be a constant in the disk. But if v is a constant, then so is u, by the Cauchy-Riemann equations.

Consequently, we have shown that if $|A| = M$, then $f(z) = u + iv$ is a constant, which is a contradiction. Therefore the only tenable hypothesis is that $|A| < M$.

Now, if C is *any* curve in D with z_0 in $I(C)$ such that $|f(z_0)| = M$, then by the above argument $f(z)$ is a constant in any circle around z_0 lying inside C; and if z' is any other point inside C, then z_0 and z' can be connected to each other by a string of small circles inside C, each containing the center of the next, so that f is constant in each circle, so $f(z')$ is also equal to this same constant. Thus, if $|f(z)| = M$ for *any* $z_0 \in I(C)$, then $f(z)$ is constant throughout $I(C)$, a contradiction. We conclude that $|f(z)| < M$ for every $z \in I(C)$. QED.

EXAMPLE 24 The function $f(z) = e^z$ has as its modulus

$$|f(z)| = e^x.$$

If C is any simple closed contour, then $|f(z)|$ will assume its maximum on C at the point or points with the largest x coordinate, and will be strictly less than that value anywhere in $I(C)$ because e^x is a strictly monotone function. If C is a circle around the origin of radius R, then

$$\max_{z \in C} |f(z)| = e^R,$$

and hence

$$|f(z)| < e^R$$

for all z such that $|z| < R$. Note that the *minimum* value of $|f(z)|$ also occurs on C, namely e^{-R}, at the point $z = -R$. (See Section 2.5, Proof 1.) Thus, just because $|f(z)|$ assumes its maximum at some point on C does not mean that it assumes this value at *every* point of C. The *maximum* modulus grows as C grows, but at points of C where $|f(z)|$ is not its maximum, anything can happen.

EXAMPLE 25 Consider the modulus of $f(z) = \cos z$ on a circle of radius R around the origin. For real values of z, the modulus varies between zero and one since $0 \leq |\cos x| \leq 1$. However, if $z = iR$, then

$$|\cos z| = |\cos iR| = \cosh R$$

and, of course, $\cosh R$ is a strictly monotone increasing function of R.

We know that a nonconstant analytic function attains its maximum modulus in a bounded closed set only on the boundary of the set. Liouville's theorem tells us even more, when the function is entire; that its maximum modulus actually goes to infinity as $|z|$ goes to infinity, at least in some direction or along some path. Certainly we have seen this to be true in the two preceding examples.

2.5.2. Theorem (*Liouville's Theorem*) If $f(z)$ is a bounded entire function, i.e., if there exists a finite positive constant M so that $|f(z)| \leq M$ for all values of z, then $f(z)$ is a constant.

Proof By Theorem 2.4.7,

$$f'(z) = \frac{1}{2\pi i} \oint_{|\zeta - z| = r} \frac{f(\zeta)\, d\zeta}{(\zeta - z)^2}$$

for all values of z and r. Hence

$$|f'(z)| \leq \frac{1}{2\pi} \int_{|\zeta - z| = r} \frac{|f(\zeta)||d\zeta|}{|\zeta - z|^2} \leq \frac{M 2\pi r}{2\pi r^2} = \frac{M}{r}.$$

Given $\varepsilon > 0$, if we choose any $r > M/\varepsilon$, we have $|f'(z)| < \varepsilon$. This implies that $|f'(z)| = 0$, for all z, since ε is arbitrary. Therefore $f' \equiv 0$, and hence $f(z)$ is a constant. QED.

From the maximum modulus theorem and from Liouville's theorem we can conclude that, if $f(z)$ is nonconstant and entire and $|f(z)| \leq M(r)$ on the circumference of a circle of radius r around the origin, then $M(r)$ is a monotone strictly increasing function of r and $\lim_{r \to \infty} M(r) = \infty$. These two theorems give us a large amount of information about entire functions. For example, it is not hard to prove that if $f(z)$ is entire and $|f(z)|/|z|^k \leq M < \infty$ for all z outside a sufficiently large circle around the origin, then $f(z)$ is a polynomial of degree k or less. This proof will be left for the questions at the end of this section. Another application of Liouville's theorem is the following proof of the fundamental theorem of algebra: That every polynomial has at least one complex zero is well known. However, it is an unfortunate fact that the first proof most undergraduates see is the following, which is akin to shooting a mosquito to death with an elephant gun. There are more elementary proofs.

2.5.3. Corollary *(The Fundamental Theorem of Algebra)* Let $p(z) = a_n z^n + \cdots + a_1 z + a_0$ $(a_n \neq 0)$ be a polynomial of degree $n \geq 1$. Then $p(z)$ has at least one root, i.e., there exists at least one point z_0 in the complex plane such that $p(z_0) = 0$.

Proof $p(z)$ is an entire function. If it had no roots, then $1/p(z)$ would also be an entire function, since the only place it could fail to be analytic would be where the denominator vanishes. Furthermore, as $|z| \to \infty$, $|1/p(z)| \to 0$. Thus for all values of z, $|1/p(z)|$ must be bounded. Therefore, by Liouville's theorem, $1/p(z)$ is a constant. But this contradicts the hypothesis, for a polynomial of degree 1 or higher is manifestly not a constant. Therefore we conclude that $p(z)$ has at least one root.

From the fundamental theorem of algebra one can conclude that a polynomial of degree n has precisely n roots, by removing a linear factor $(z - z_0)$ corresponding to the root and examining what is left, using induction. (Of course it is possible that some of these roots may be equal.)

It turns out that theorems akin to the maximum modulus theorem also hold for the real and imaginary parts of an analytic function. First, before we show this, we shall need to consider the following question: Given a real-valued function $u(x, y)$, is there a function $v(x, y)$ so that $f(z) = u(x, y) + iv(x, y)$ is analytic? To do this we would have to set up the Cauchy-Riemann equations with the known function u, and solve for v. However, it may turn out that no such function exists. Something, memory if nothing else, should tell us that not every u can be the real part of an analytic function. In fact, the next theorem tells us what condition u must satisfy before a v can be found:

2.5.4. Theorem If $f(z) = u(x, y) + iv(x, y)$ is analytic in a domain D, then u and v satisfy Laplace's equation in D, i.e.,

$$\frac{\partial^2 u}{\partial x^2} + \frac{\partial^2 u}{\partial y^2} = 0 \quad \text{and} \quad \frac{\partial^2 v}{\partial x^2} + \frac{\partial^2 v}{\partial y^2} = 0.$$

Proof Since $f(z)$ is analytic, it possesses all possible derivatives in D, and so u and v possess all possible partial derivatives. Furthermore, in D

104 CONTINUITY, DIFFERENTIATION, AND INTEGRATION

the Cauchy-Riemann equations hold:

$$\frac{\partial u}{\partial x} = \frac{\partial v}{\partial y}, \qquad \frac{\partial u}{\partial y} = -\frac{\partial v}{\partial x}.$$

Differentiating the first equation with respect to x and the second with respect to y, we obtain

$$\frac{\partial^2 u}{\partial x^2} = \frac{\partial^2 v}{\partial x\, \partial y}, \qquad \frac{\partial^2 u}{\partial y^2} = -\frac{\partial^2 v}{\partial y\, \partial x}.$$

Now $\partial^2 v/\partial x\, \partial y = \partial^2 v/\partial y\, \partial x$ since all derivatives of v exist and are continuous; hence, if we add the two preceding equations we obtain

$$\frac{\partial^2 u}{\partial x^2} + \frac{\partial^2 u}{\partial y^2} = 0.$$

The equation for v is obtained in a similar fashion.

Functions which satisfy Laplace's equation are called *harmonic functions*. A pair of harmonic functions $u(x, y)$ and $v(x, y)$ which satisfy the Cauchy-Riemann equations are called *conjugate harmonic functions*. Note that this is a different use of the word "conjugate" from what we have encountered before, but no confusion should arise, as it will always be clear from context which usage is to be understood.

EXAMPLE 26   Suppose $u(x, y) = 3x - 2xy$. Let us verify that $u(x, y)$ is harmonic and find a harmonic conjugate $v(x, y)$ for u, so that $f(z) = u(x, y) + iv(x, y)$ is analytic. Clearly u is harmonic, since it is linear in both x and y, so that $\partial^2 u/\partial x^2 = 0$ and $\partial^2 u/\partial y^2 = 0$. Next, u and v must satisfy the Cauchy-Riemann equations, so that

$$\frac{\partial v}{\partial y} = \frac{\partial u}{\partial x} = 3 - 2y, \qquad \frac{\partial v}{\partial x} = -\frac{\partial u}{\partial y} = 2x.$$

If this pair of equations has a solution v, then $u + iv$ must be analytic, as required. Integrating the first equation with respect to y, holding x constant, and the second with respect to x, holding y constant, we obtain

$$v(x, y) = 3y - y^2 + c_1(x) \quad \text{and} \quad v(x, y) = x^2 + c_2(y),$$

where c_1 and c_2 are arbitrary functions of x and y, respectively. Since these two expressions for v must be equal, we have $3y - y^2 + c_1(x) = x^2 + c_2(y)$ or $3y - y^2 - c_2(y) = x^2 - c_1(x)$. x and y are independent variables; the only way a function of y can equal a function of x is when both are equal to a constant, $3y - y^2 - c_2(y) = x^2 - c_1(x) = -c$, giving us that $c_2(y) = 3y - y^2 + c$ or $c_1(x) = x^2 + c$. Therefore, $f(z) = 3x - 2xy + i(x^2 + 3y - y^2 + c)$. This example illustrates the fact that given the real part of an analytic function, the imaginary part can be obtained to within an arbitrary additive constant by integrating the Cauchy-Riemann equations. This example does not constitute a proof of this fact, of course. This is an important result which we shall assume without proof, since the calculations are straightforward in all cases which will interest us.

EXAMPLE 27 $u(x, y) = 6x^2 + y$ is not a harmonic function, as $\partial^2 u/\partial x^2 + \partial^2 u/\partial y^2 = 12 \neq 0$. Not being harmonic, u cannot be the real part of an analytic function.

In the problems we shall consider an alternative method for determining an analytic function given its real part u. In the meantime, it is now possible to prove a corollary of the maximum modulus theorem which applies to harmonic functions:

2.5.5. Corollary *(The Maximum Principle for Harmonic Functions)* If $u(x, y)$ is a nonconstant harmonic function in a domain D and C is a simple closed curve in D with $u(x, y) \leq M$ throughout $C \cup I(C)$, then $u(x, y) < M$ for $(x, y) \in I(C)$.

Proof Let $f(z) = u(x, y) + iv(x, y)$ be an analytic function on $C \cup I(C)$ with u its real part. By Theorem 2.2.5, the composite function $e^{f(z)}$ is also analytic on $C \cup I(C)$. But $e^{f(z)} = e^{u+iv} = e^{u(x,y)}(\cos v + i \sin v)$. By the maximum modulus theorem (Theorem 2.5.1), since $|e^{f(z)}| = e^{u(x,y)} \leq e^M$ for $z \in C \cup I(C)$, we must have $|e^{f(z)}| < e^M$ for $z \in I(C)$. Hence $e^{u(x,y)} < e^M$ in $I(C)$, and since the real exponential is strictly monotone increasing, we have $u(x, y) < M$ in $I(C)$, completing the proof.

Similar theorems hold relating to the minimum modulus of an analytic function and to a minimum principle for harmonic functions; their proofs are straightforward applications of Theorem 2.5.1 and Corollary 2.5.5, and are left to the reader in the questions at the end of this section.

One can interpret the maximum principle for harmonic functions as follows: The frying pan is always hottest where it is closest to the fire. This is because solutions of Laplace's equation can be interpreted physically as the temperatures in a body which has reached a steady state or equilibrium (the flow of heat is such that the temperature at each point is constant with respect to time).*

---

* For discussion and further information see, for instance, Carslaw and Jaeger, *Conduction of Heat in Solids*, 2nd ed., Oxford, London, 1959.

106 CONTINUITY, DIFFERENTIATION, AND INTEGRATION

EXAMPLE 28 Consider the harmonic function $u(x, y) = x^2 - y^2$, which is the real part of the analytic function $f(z) = z^2$. On a circle of radius R around the origin, it is clear that the maximum of $u(x, y)$ always occurs where $y = 0$, that is, that max $u(x, y) = R^2$. It is also clear that u is strictly less than R^2 everywhere inside the circle.

Our final task in this section will be to show that a function differentiable at every point in a domain is analytic there. First we shall need the following:

2.5.6. Lemma Let f be single-valued and differentiable in a domain D and let T be an arbitrary triangle in D. Then

$$\oint_T f(z)\,dz = 0.$$

Proof We subdivide T into four congruent subtriangles t_1, t_2, t_3, t_4 by connecting the midpoints of the sides of T (Figure 2.17). Then

$$I = \oint_T f(z)\,dz = \sum_{i=1}^{4} \oint_{t_i} f(z)\,dz,$$

because, since f is single-valued, the integrals back and forth along the common sides of the t_i's cancel. By the triangle inequality,

$$|I| \leq \sum_{i=1}^{4} \left| \oint_{t_i} f(z)\,dz \right|.$$

Let us denote by T_1 the t_i for which

$$\left| \oint_{t_i} f(z)\,dz \right|$$

is a maximum, and let

$$I_1 = \oint_{T_1} f(z)\,dz.$$

Then we have

$$|I| \leq 4|I_1|.$$

In a similar manner we can divide T_1 into four subtriangles, denote by I_2 the value of the integral of f around the one giving the largest modulus (call this one T_2), and obtain

$$|I| \leq 4|I_1| \leq 4^2|I_2|,$$

and, if we continue the process by induction, we eventually obtain

$$|I| \leq 4^n|I_n|.$$

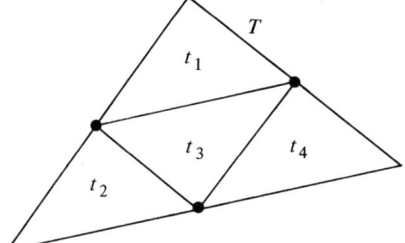

Figure 2.17

The areas and perimeters of the sequence of triangles T_n are decreasing to zero, and each triangle is inside the previous one; hence there exists a *unique* point z_0 which is common to all the triangles.* Since f is differentiable at z_0, we have that

$$f(z) = f(z_0) + f'(z_0)(z - z_0) + |z - z_0| \, o(1)$$

for z sufficiently close to z_0; and for n sufficiently large,

$$I_n = \oint_{T_n} f(z)\,dz = \oint_{T_n} f(z_0)\,dz + \oint_{T_n} f'(z_0)(z - z_0)\,dz$$
$$+ \oint_{T_n} |z - z_0|\, o(1)\,dz$$
$$= f(z_0) \oint_{T_n} dz + f'(z_0) \oint_{T_n} (z - z_0)\,dz + \oint_{T_n} |z - z_0|\, o(1)\,dz;$$

because 1 and $(z - z_0)$ both possess antiderivatives, their integrals around the simple closed contour T_n vanish, and we have

$$I_n = \oint_{T_n} f(z)\,dz = \oint_{T_n} |z - z_0|\, o(1)\,dz,$$

which can be estimated as follows:

$$|I_n| = \left| \oint_{T_n} f(z)\,dz \right| \leq o(1) \max |z - z_0| \, l_n,$$

where l_n is the perimeter of T_n; furthermore, $|z - z_0| < l_n$, also, so

$$|I_n| = \left| \oint_{T_n} f(z)\,dz \right| \leq l_n^2 o(1) \leq \frac{l^2}{4^n} o(1);$$

---

* This is a nontrivial fact which can be proved by constructing the real and imaginary parts of z_0 much as in the proof of Theorem 1 in Appendix A.

108 CONTINUITY, DIFFERENTIATION, AND INTEGRATION

the last inequality holds since if l is the perimeter of the original triangle, then by construction, we always have $l_{n+1} = l_n/2$ for all n; hence $l_n = l/2^n$. Going back to our original integral I, then

$$|I| \leq 4^n |I_n| \leq l^2 o(1);$$

the latter term can be made arbitrarily small as $n \to \infty$, since $o(1)$ goes to zero as $z \to z_0$. Hence

$$|I| = 0. \quad \text{QED}.$$

It is now a simple matter to prove our crucial result:

2.5.7. Theorem *(Goursat's Theorem)* Let f be single-valued in a domain D. If f is differentiable in D, then f is analytic in D.

Proof First, let us suppose that D is the interior of a circle. Let z_0 be any point in D, and let us define the (single-valued) function

$$F(z) = \int_{z_0}^{z} f(\zeta) \, d\zeta.$$

The integration is taken along the straight line segment joining z_0 and z, which does lie inside D since D is convex. Then we write

$$F(z + \Delta z) - F(z) = \int_{z_0}^{z+\Delta z} f(\zeta) \, d\zeta - \int_{z_0}^{z} f(\zeta) \, d\zeta,$$

where the first integral is over the straight line joining z_0 to $z + \Delta z$. Now the integral around the triangle with vertices $z_0, z, z + \Delta z$ is zero by Lemma 2.5.6 (Figure 2.18), i.e.,

$$\int_{z_0}^{z} f(\zeta) \, d\zeta + \int_{z}^{z+\Delta z} f(\zeta) \, d\zeta + \int_{z+\Delta z}^{z_0} f(\zeta) \, d\zeta = 0,$$

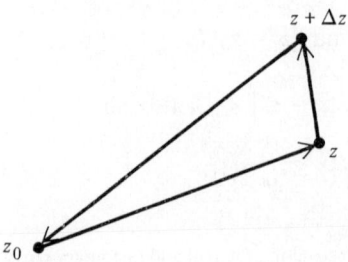

Figure 2.18

so

$$\int_{z_0}^{z+\Delta z} f(\zeta)\,d\zeta - \int_{z_0}^{z} f(\zeta)\,d\zeta = \int_{z}^{z+\Delta z} f(\zeta)\,d\zeta.$$

Thus

$$F(z+\Delta z) - F(z) = \int_{z}^{z+\Delta z} f(\zeta)\,d\zeta,$$

and it follows as in Theorem 2.3.9 that F is an antiderivative of f. Therefore F is analytic and hence by Theorem 2.4.7 so is its derivative f. Since f is analytic in each open disk of an arbitrary domain, and single-valued by hypothesis, it is analytic in the whole domain, as claimed.

A. Exercises

1 Check the following functions for harmonicity, and if they are harmonic, find their harmonic conjugate functions:

a $xe^x \cos y - xe^x \sin y$ **b** $x^3 - 3xy^2$

c $x^2 - y^2 + 2x$ **d** $\dfrac{x}{(x^2+y^2)} (x+iy \ne 0)$

e $ax + by$ (a, b constants)   **f** $e^{2x}(\cos^2 y - \sin^2 y)$
g $\sinh x \cos y$ **h** $e^x \cos y \cosh y \sin x$
i $5x^3 - 7xy^2$ **j** $e^{3x^2 - 3y^2 - x} \cos(6xy - y)$
k $e^{x^2 - y^2} \cos 2xy$ **l** $3x^2 - 2x + 3y^2 + 1$

2 Examine each of the following harmonic functions on the circle $|z| = 1$; show that they assume their maxima *and* minima on the boundary; and find the values:

a $e^x \cos y$ **b** $3(x^2 - y^2) - 2x$
c $x^3 - 3xy^2$ **d** $ax + by$ (a, b constants)

B. Problems

1 Let $f(z) = u(x, y) + iv(x, y)$ be any function of z. Show that

$$f(z) = u\left(\frac{z}{2}, \frac{z}{2i}\right) + iv\left(\frac{z}{2}, \frac{z}{2i}\right);$$

that is, x can be replaced by $z/2$ and y by $z/2i$ without changing the value of f.

2 Assume that the function f in Problem 1 is analytic. Find an expression for $f'(z)$ from the formula in Problem 1.

3 Using the Cauchy-Riemann equations, write the expression for f' derived in Problem 2 in terms of u alone.

4 Integrate the expression in Problem 3 and obtain that

$$f(z) = 2u\left(\frac{z}{2}, \frac{z}{2i}\right) + C,$$

where, if 0 is in the domain of f, $C = -u(0,0) + iv(0,0)$.

5 Problem 4 gives a method for constructing $f(z)$ given its real part $u(x, y)$. For example, if $u(x, y) = x^2 - y^2$, then

$$f(z) = 2u\left(\frac{z}{2}, \frac{z}{2i}\right) + C = 2\left(\frac{z^2}{4} + \frac{z^2}{4}\right) + C = z^2 + C.$$

Use this method to find the analytic function whose real part is each of the harmonic functions in Exercise 1.

C. Proofs

1 Suppose $f(z)$ is analytic inside and on a simple closed contour C, and that $|f(z)| \geq m > 0$ for $z \in C \cup I(C)$. Show that $|f(z)| > m$ for $z \in I(C)$ (the minimum modulus theorem).

2 Suppose $u(x, y)$ is harmonic inside and on a simple closed contour C, and that $u(x, y) \geq m$ for $z \in C \cup I(C)$. Show that $u(x, y) > m$ for $z \in I(C)$ (the minimum principle for harmonic functions).

3 Suppose $f(z)$ is entire, and that there exist finite positive constants K and M, and an integer k, so that $|f(z)|/|z|^k \leq M$ for all z such that $|z| > K$. Prove that $f(z)$ is a polynomial of degree k or less.

4 Prove d'Alembert's principle: If $f(z)$ is an analytic function and $|f(z)|$ assumes a minimum at a point z_0 inside a simple closed contour C, then $f(z_0) = 0$. (d'Alembert used this as a lemma in his proof of the fundamental theorem of algebra.)

5 Suppose that $f(z) = u(x, y) + iv(x, y)$ is analytic in z and that $z = x(\xi, \eta) + iy(\xi, \eta)$ is analytic in $\zeta = \xi + i\eta$. Then show that u and v, considered as functions of ξ and η, are harmonic functions of these variables. This important result shows that harmonicity is preserved in analytic transformations, which will be basic in our future studies of conformal mapping.

Chapter **3**

Taylor and Laurent Series and the Calculus of Residues

In this chapter we shall study and classify those functions which behave like polynomials (analytic functions) and those which behave like rational functions (meromorphic functions). A study of the properties of these functions will enable us to develop some highly useful applications: evaluation of difficult real and complex integrals and summation of certain series. These important results will be highly useful in our future applications of complex analysis to the study of fluid flows.

3.1. INTRODUCTION TO SEQUENCES AND SERIES*

A sequence, as in the real case, is simply a function whose domain is the positive integers—except that now the function is allowed to take on complex values. The customary notation for a sequence is $\{s_1, s_2, s_3, s_4, \ldots\}$ or simply $\{s_n\}$. An example of a sequence is

$$\{s_n\} = \{i^n\} = \{i, -1, -i, 1, i, -1, \ldots\}.$$

---

* This section can be skimmed over rather rapidly by those students already quite familiar with sequences, series, and convergence of sequences of functions, including uniform convergence.

TAYLOR AND LAURENT SERIES

As in the real case, we are primarily interested in sequences which converge, i.e., which become arbitrarily close to some value as n increases without limit. Convergence of a complex sequence can be defined either in terms of the convergence of two real sequences (the sequence of real and imaginary parts of $\{s_n\}$) or in terms of the absolute value of the difference between s_n and its limiting value, just as we did when we considered the limit of a function.

Thus, we have the following:

3.1.1. Definition The sequence $\{s_n\} = \{\sigma_n + i\tau_n\}$ is said to *converge to the limit* $s = \sigma + i\tau$ if the real sequence $\{|s_n - s|\}$ has the limit 0, or, equivalently, if the real sequence $\{\sigma_n\}$ has the limit σ and the real sequence $\{\tau_n\}$ has the limit τ. We then write

$$\lim_{n \to \infty} s_n = s.$$

That the two definitions are equivalent is a consequence of the sandwich principle.

EXAMPLE 1 The sequence given above, $\{i^n\}$, does not converge because neither the sequence of real parts $\{0, 1, 0, -1, 0, 1, \ldots\}$ nor the sequence of imaginary parts $\{1, 0, -1, 0, 1, 0, -1, \ldots\}$ converges. In the complex plane the points $\{i^n\}$ continually move around the unit circle and never get close to any one point in the complex plane. In fact, for any point ζ in the complex plane, there are arbitrarily large values of n for which $|\zeta - i^n| \geq 1$. To see this, let $\zeta = \alpha + i\beta$. Then $|\zeta - i^n|^2$ always has one of the values $(\alpha - 1)^2 + \beta^2$, $(\alpha + 1)^2 + \beta^2$, $\alpha^2 + (\beta - 1)^2$, or $\alpha^2 + (\beta + 1)^2$ no matter what n may be; and clearly for any value of α or β, n can be chosen so that at least one of these expressions is no smaller than 1.

EXAMPLE 2 On the other hand, the sequence

$$\left\{\left(\frac{i}{2}\right)^n\right\} = \left\{\frac{i}{2}, -\frac{1}{4}, -\frac{i}{8}, \frac{1}{16}, \frac{i}{32}, \ldots\right\}$$

converges, and its limit is zero.

This can be seen either be observing that

$$\left|\left(\frac{i}{2}\right)^n\right| = \frac{1}{2^n} \to 0$$

or by looking at the sequences of real and imaginary parts

$$\{0, -\tfrac{1}{4}, 0, \tfrac{1}{16}, 0, \ldots\}, \ \{\tfrac{1}{2}, 0, -\tfrac{1}{8}, 0, \tfrac{1}{32}, \ldots\}$$

and noting that both approach zero.

3.1. Introduction to Sequences and Series

A more complicated idea than a sequence of constants is a sequence of functions $\{f_n\}$. If each of the functions f_n is defined in a domain D, then for each z the sequence $\{f_n(z)\}$ is a sequence of constants which may or may not converge. In order for a sequence of functions to converge, we demand that it converge for each z in its domain:

3.1.2. Definition A sequence of functions $\{f_n\}$ in a domain D is said to *converge to a function f in D* if for each z in D, we have

$$\lim_{n \to \infty} f_n(z) = f(z).$$

EXAMPLE 3   One of the simplest sequences of functions is $\{z^n\}$. If $|z| = r$, then we see that

$$|z^n| = r^n,$$

which approaches zero if $r < 1$ and $+\infty$ if $r > 1$. If $r = 1$, then $z = e^{i\theta}$ and $z^n = e^{in\theta}$, so

$$z^n = \cos n\theta + i \sin n\theta.$$

By examining the real part (or the imaginary part) we see that no limit exists unless $\theta = 0$, i.e., $z = 1$. Hence we can draw the conclusion that the sequence $\{z^n\}$ has limit 0 if $|z| < 1$, limit 1 if $z = 1$, and has no limit otherwise.

EXAMPLE 4   Another interesting sequence of functions is the sequence $\{f_n\}$, where

$$f_n(z) = \begin{cases} nz & \text{if } |z| \leq 1/n \\ -nz + 2 & \text{if } 1/n < |z| < 2/n \\ 0 & \text{if } |z| > 2/n. \end{cases}$$

In this case, we see that

$$\lim_{n \to \infty} f_n(z) = 0$$

for *all* z: at the origin because $f_n(0) = 0$ for all n and at any $z \neq 0$ because $f_n(z) = 0$ for all n such that $2/n < |z|$, i.e., $n > 2/|z|$.

EXAMPLE 5   As a final example, we consider the sequence $\{f_n\}$ defined by

$$f_n(z) = \begin{cases} 0 & \text{if } |z| \leq 2 - 1/n \\ \tfrac{1}{2}n|z| - n + \tfrac{1}{2} & \text{if } 2 - 1/n < |z| < 2 + 1/n \\ 1 & \text{if } |z| \geq 2 + 1/n. \end{cases}$$

(See Figure 3.1.) One can easily see that

$$\lim_{n \to \infty} f_n(z) = \begin{cases} 0 & \text{if } |z| < 2 \\ \tfrac{1}{2} & \text{if } |z| = 2 \\ 1 & \text{if } |z| > 2. \end{cases}$$

114 TAYLOR AND LAURENT SERIES

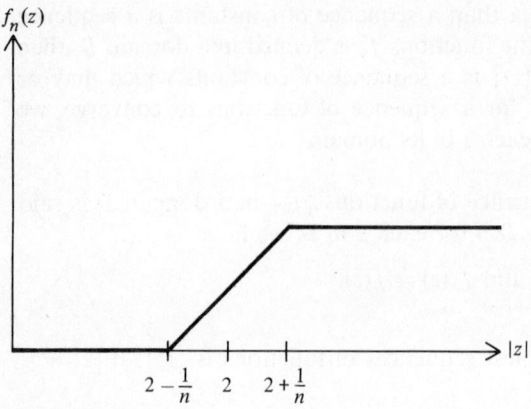

Figure 3.1

In Example 4 we have a sequence of noncontinuous functions whose limit is continuous. [To see that f_n is not continuous, note that as $z \to i/n$ from below, then $f_n(z) \to i$, while if $z \to i/n$ from above, then $f_n(z) \to 2 - i$.] On the other hand, each function in the sequence in Example 5 is continuous and yet the limit function is discontinuous. Thus it looks as if almost anything can happen when one takes the limit of a sequence of functions; and in fact this is indeed the case. Naturally the main cases in which we will develop an abiding interest will be those for which we can predict what will happen. It turns out that the crucial requirement will be that $|f_n(z) - f(z)|$ will become small simultaneously for all z in a set as n gets large, or, as we usually say, uniformly small.*

3.1.3. Definition We shall say that a sequence of functions $\{f_n\}$ on a set A *converges uniformly to a function* f if the least upper bound of the difference $|f_n(z) - f(z)|$ converges to zero, i.e., if the real sequence $\{\sigma_n\}$ defined by

$$\sigma_n = \operatorname*{lub}_{z \in A} |f_n(z) - f(z)|$$

converges to zero.

The reader should recall that the least upper bound of a real function g on a set B is defined to be the smallest number M for which $g(x) \leq M$ for all x in B. By a fundamental principle of real analysis, if a nonempty set of real numbers [in this case the values of $g(x)$] is bounded at all from above, then it does have a least upper bound.† The function g need not necessarily

---

* The reader who is already thoroughly familiar with the concept of uniform convergence in real analysis will find that the present discussion contains no new ideas.
† See Appendix A.

assume its least upper bound; for instance, 1 is the least upper bound of $1 - e^{-x}$ on $0 \leq x < \infty$, but $1 - e^{-x}$ never takes on the value 1. Thus, if the real function of $z = x + iy$ defined by $|f_n(z) - f(z)|$ is bounded for $z \in A$ (for each n), then the sequence σ_n in Definition 3.1.3 is well-defined.

Let us consider the functions in Examples 3, 4, and 5 to see if their convergence is uniform on those sets where they converge:

EXAMPLE 3 (Continued) We saw that $z^n \to 0$ whenever $|r| < 1$. Now

$$\underset{|z|<1}{\text{lub}} |z^n| = \underset{r<1}{\text{lub}} r^n = 1,$$

which does not converge to zero; hence z^n does not converge uniformly to 0 on $|z| < 1$. However, if $|z| \leq \rho < 1$, then

$$\underset{|z|\leq\rho}{\text{lub}} |z^n| = \underset{r\leq\rho}{\text{lub}} r^n = \rho^n,$$

and ρ^n is a sequence which approaches zero. Hence z^n converges uniformly to zero on any disk of radius less than 1.

EXAMPLE 4 (Continued) Here the sequence $f_n \to 0$ for all finite z, but it is immediately obvious that the convergence is not uniform since

$$\text{lub} |f_n(z)| \geq |f_n(\tfrac{1}{n})| = 1,$$

which again is a sequence not approaching zero. Convergence is uniform, however, for all z with $|z| \geq \rho > 0$ (why?).

EXAMPLE 5 (Continued) Once again the convergence is not uniform because this time

$$\text{lub} |f_n(z) - f(z)| = \tfrac{1}{2}$$

for all n, where $f(z)$ is the limit function. Convergence is uniform inside $|z| \leq \rho < 2$ or outside $|z| \geq R > 2$, or both, however (why?).

We have thus seen three examples of nonuniform convergence; note, however, that in each case the convergence *is* uniform in a subset of the plane. One may well ask why uniform convergence is so important; the answer is that nice things happen, as the next two theorems show.

3.1.4. Theorem Let $\{f_n\}$ be a sequence of continuous functions converging uniformly on a domain D to a limit function f. Then f is continuous on D.

116 TAYLOR AND LAURENT SERIES

Proof

$$|f(z) - f(z_0)| = |f(z) - f_n(z) + f_n(z) - f_n(z_0) + f_n(z_0) - f(z_0)|$$
$$\leq |f(z) - f_n(z)| + |f_n(z) - f_n(z_0)| + |f_n(z_0) - f(z_0)|$$
$$\leq |f_n(z) - f_n(z_0)| + 2 \operatorname*{lub}_{z \in D} |f_n(z) - f(z)|.$$

The last term has limit 0 as $n \to \infty$ so it can be made as small as desired; and because f_n is continuous, the first term can be made small by making $|z - z_0|$ sufficiently small. Hence the left-hand side can be made small, proving f continuous. QED

The next theorem is of even greater importance to us because our interest in analytic functions is of overriding importance.

3.1.5. Theorem Let $\{f_n\}$ be a sequence of analytic functions converging uniformly on a domain D to a limit function f. Then

1 f is analytic on D, and
2 f_n' converges to f' in D.

Proof 1. Let C be an arbitrary simple closed contour in D (of length L) with $I(C)$ also in D, and let z be an arbitrary point in $I(C)$. Then for each n we have

$$f_n(z) = \frac{1}{2\pi i} \oint_C \frac{f_n(\zeta)}{\zeta - z} d\zeta.$$

Let us consider the quantity

$$\frac{1}{2\pi i} \oint_C \frac{f(\zeta)}{\zeta - z} d\zeta.$$

The difference between this and the right-hand side of the preceding expression can be estimated as follows:

$$\left| \frac{1}{2\pi i} \oint_C \frac{f_n(\zeta)}{\zeta - z} d\zeta - \frac{1}{2\pi i} \oint_C \frac{f(\zeta)}{\zeta - z} d\zeta \right| = \frac{1}{2\pi i} \left| \oint_C \frac{f_n(\zeta) - f(\zeta)}{\zeta - z} d\zeta \right|$$
$$\leq \frac{L}{2\pi} \frac{\max_{\zeta \in C} |f_n(\zeta) - f(\zeta)|}{\min |\zeta - z|} \leq \frac{L}{2\pi \min |\zeta - z|} \operatorname*{lub}_{\zeta \in D} |f_n(\zeta) - f(\zeta)|,$$

which because of the uniform convergence clearly approaches zero. Hence

$$\lim_{n \to \infty} f_n(z) = \lim_{n \to \infty} \frac{1}{2\pi i} \oint_C \frac{f_n(\zeta)}{\zeta - z} d\zeta = \frac{1}{2\pi i} \oint_C \frac{f(\zeta)}{\zeta - z} d\zeta;$$

but since
$$\lim_{n \to \infty} f_n(z) = f(z),$$
we have
$$f(z) = \frac{1}{2\pi i} \oint_C \frac{f(\zeta)}{\zeta - z} d\zeta$$
for all simple closed contours C and all z in $I(C)$; therefore by Theorem 2.4.7, f is analytic.

2. Since f is analytic,
$$f'(z) = \frac{1}{2\pi i} \oint_C \frac{f(\zeta)}{(\zeta - z)^2} d\zeta;$$
and, of course,
$$f_n'(z) = \frac{1}{2\pi i} \oint_C \frac{f_n(\zeta)}{(\zeta - z)^2} d\zeta.$$
As in (1) it is a simple matter to show that the second sequence of integrals approaches the first, and hence that f' is the limit of f_n'.

The converses of these last two theorems are not true, since Example 4 showed us that the analytic function 0 is the limit of a sequence of noncontinuous functions. Neither can Theorem 3.1.4 be weakened, for Example 5 showed that the hypothesis of uniform convergence was necessary.

Before we leave the subject of sequences behind us, it behooves us to note that new sequences can be defined from existing ones by the simple process of adding the first n terms. Thus, if we have a given sequence of constants $\{a_n\}$, we can define a new sequence from it as follows:

$$s_1 = a_1$$
$$s_2 + a_1 + a_2$$
$$s_3 = a_1 + a_2 + a_3$$
$$\cdot$$
$$\cdot$$
$$\cdot$$
$$s_n = a_1 + a_2 + \cdots + a_n = \sum_{k=1}^{n} a_k.$$

If the new sequence $\{s_n\}$ has a limit s, then we write
$$s = \sum_{k=1}^{\infty} a_k;$$

the pair of sequences $\{a_n, s_n\}$, related as above, is called an infinite series, the a_k's are called the *terms* of the series, the s_n's are called the *partial sums*, and s is said to be the *sum* of the series. All this is quite analogous to the real case, and the terminology should be familiar to the student.

As in the case of sequences, we can reduce many of the questions in the complex case to similar questions involving real series. For example, if $a_k = \alpha_k + i\beta_k$, then the series $\sum a_k$ converges if and only if the two real series $\sum \alpha_k$ and $\sum \beta_k$ both converge. If it converges, then it converges to $\sum \alpha_k + i \sum \beta_k$. To prove this one need only consider the sequence of partial sums and recall that a sequence converges if and only if its real and imaginary parts do.

EXAMPLE 6 For example, we have

$$\sum_{n=1}^{\infty} \left(\frac{i}{2}\right)^n = \sum_{k=1}^{\infty} \frac{(-1)^k}{2^{2k}} + i \sum_{k=1}^{n} \frac{(-1)^{k-1}}{2^{2k-1}}$$

$$= \left(-\frac{1}{4}\right)\frac{1}{1+\frac{1}{4}} + i\left(\frac{1}{2}\right)\frac{1}{1+\frac{1}{4}} = -\frac{1}{5} + i\frac{2}{5}.$$

Thus, this complex series can be summed by breaking it up into its real and imaginary parts and summing each according to the rule for the real geometric series. This should prompt us to ask whether the rule

$$\sum_{k=0}^{\infty} ar^k = \frac{a}{1-r} \quad (|r| < 1)$$

carries over into the realm of the complex. In fact, it certainly does for this example, because

$$\frac{i}{2}\frac{1}{1-i/2} = \frac{i}{2-i} = \frac{i(2+i)}{5} = -\frac{1}{5} + \frac{2}{5}i,$$

which is the same result as above. Thus we are led to the following theorem.

3.1.6. Theorem If $|z| < 1$, then

$$\sum_{k=0}^{\infty} \alpha z^k = \frac{\alpha}{1-z}.$$

Proof Exactly as in the real case, we know that for $z \neq 1$

$$\sum_{k=0}^{n} z^k = 1 + z + z^2 + \cdots + z^n = \frac{1 - z^{n+1}}{1 - z},$$

so that when $|z| < 1$, we have

$$\lim_{n \to \infty} \frac{1 - z^{n+1}}{1 - z} = \frac{1}{1 - z} = \sum_{k=0}^{\infty} z^k;$$

both sides can be multiplied by a constant α to yield the result.

EXAMPLE 7

$$\sum_{k=0}^{\infty} \frac{(1 + i)^k}{3^k} = \frac{1}{1 - \frac{1}{3}(1 + i)} = \frac{3}{2 - i} = \frac{3(2 + i)}{5} = \frac{6}{5} + \frac{3}{5}i.$$

Naturally, in most cases it is not possible to find a concise expression for the limit of a sequence or for the sum of an infinite series. The actual numeric values, though, are not as important many times as just knowing whether or not the sequence or series converges; the following concept gives a criterion for convergence which (as we shall see) has many useful consequences.

3.1.7. Definition A sequence $\{s_n\}$ is said to satisfy the *Cauchy criterion* or to be a *Cauchy sequence* if for every $\varepsilon > 0$, there exists an integer $N = N(\varepsilon)$ such that $|s_n - s_m| < \varepsilon$ for all n and $m > N$. If $\{s_n\}$ is the sequence of partial sums of an infinite series, i.e.,

$$s_n = \sum_{k=1}^{n} a_k,$$

then the Cauchy criterion says that

$$|a_n + a_{n-1} + \cdots + a_{m+1}| < \varepsilon$$

for all $n > m > N$.

A rigorous formulation of the Cauchy criterion may not be easy to understand, but it can be explained in less precise terms as follows: $\{s_n\}$ satisfies the Cauchy criterion if, once we get far enough out in the sequence, any pair of terms beyond that point differs by an arbitrarily small amount. We may not be able to specify the limit s exactly, or in a finite number of steps; but the Cauchy criterion tells us that if we go far enough out in the sequence then we will know s to a certain number of decimal places. For example, it is not possible to write down a decimal expansion for π in a finite number of steps, but one can define a Cauchy sequence that converges to π by some algorithmic method, such as

$$s_1 = 3, s_2 = 3.1, s_3 = 3.14, s_4 = 3.141, s_5 = 3.1415, \ldots,$$

and, though we do not know π exactly, we do know it to any specified degree of accuracy if we proceed far enough out in the sequence. Thus we have

3.1.8. Theorem A sequence $\{s_n\}$ converges if and only if it is Cauchy.

The above argument indicates why a Cauchy sequence converges; because, computationally, one can calculate the limit to any specified degree of accuracy. We shall not attempt to make this part of the proof rigorous, however.

Conversely, if $\{s_n\}$ has a limit (say s), then

$$|s_n - s_m| = |s_n - s + s - s_m| \leq |s_n - s| + |s_m - s|;$$

the two latter terms can be made as small as desired, hence by the sandwich principle so can $|s_n - s_m|$. QED

The Cauchy criterion can be used to prove the well-known result (for real series that if $\sum a_n$ converges, then $a_n \to 0$. For,

$$|a_n| = |s_n - s_{n-1}| \to 0.$$

It is likewise useful for a few other important results which we shall get to when we have considered the following important concept:

3.1.9. Definition An infinite series $\sum a_n$ is said to *converge absolutely* if $\sum |a_n|$ converges.

Again the definition coincides with the familiar one in the real case. The handy thing about absolute convergence is that it reduces consideration of complex series to that of real series of positive terms, for which we already know many convergence tests. For this reason (as well as others) we shall rarely use any other kind of convergence. As in the real calculus, absolute convergence is a stronger requirement than convergence:

3.1.10. Theorem An absolutely convergent series converges.

Proof Let $\sum a_k$ be the series in question. We shall use the Cauchy criterion; by the triangle inequality,

$$|a_n + a_{n-1} + \cdots + a_{m+1}| \leq |a_n| + |a_{n-1}| + \cdots + |a_{m+1}|,$$

and since $\sum |a_k|$ converges, the Cauchy criterion implies that the right-hand side can be made arbitrarily small. Therefore so can the left, which implies that $\sum a_k$ converges. QED

3.1. INTRODUCTION TO SEQUENCES AND SERIES

It turns out that we can change the order of terms in absolutely convergent series without affecting the convergence or the value of the sum:

3.1.11. Theorem Let $\sum a_k$ be an absolutely convergent series and let $\sum a'_k$ be a series containing the same terms as the first except in a different order. Then $\sum a'_k$ converges absolutely and $\sum a'_k = \sum a_k$.

Proof By hypothesis, since $\sum a_k$ converges absolutely, there is an N so large that

$$\sum_{k=N+1}^{\infty} |a_k| < \varepsilon$$

for any given ε. (Why?) We then choose N_1 so large that all the numbers $a_1, a_2, \ldots, a_N$ are in the set $a'_1, a'_2, \ldots, a'_{N_1}$. Then for all $n \geq N$, the difference

$$\sum_{k=1}^{n} a_k - \sum_{k=1}^{N_1} a'_k$$

will contain only (at most) those terms a_k with subscripts $k > N$. For, by the construction of N_1, the terms a_1 up to a_N occur in both sums and hence cancel; possibly some other terms cancel as well. Hence

$$\left| \sum_{k=1}^{n} a_k - \sum_{k=1}^{N_1} a'_k \right| \leq \left| \sum_{k=N+1}^{n} a_k \right| \leq \sum_{k=N+1}^{n} |a_k| \leq \sum_{k=N+1}^{\infty} |a_k| < \varepsilon$$

Therefore in the limit as $n \to \infty$ we have

$$\left| \sum_{k=1}^{\infty} a_k - \sum_{k=1}^{N_1} a'_k \right| \leq \varepsilon,$$

which implies that $\sum a'_k$ converges to $\sum a_k$. QED

It is important that we be able to change the order of the terms of an absolutely convergent series, for otherwise it would be a nearly insuperable task to define uniquely the sum and product of two series.

3.1.12. Theorem Let $\sum a_k$ and $\sum b_k$ be two absolutely convergent series. Then

$$\sum_{k=1}^{\infty} a_k + \sum_{k=1}^{\infty} b_k = \sum_{k=1}^{\infty} (a_k + b_k)$$

and

$$\sum_{k=1}^{\infty} a_k \sum_{k=1}^{\infty} b_k = \sum_{k=1}^{\infty} \sum_{n=1}^{k} a_n b_{k-n+1}$$

$$= \sum_{k=1}^{\infty} (a_1 b_k + a_2 b_{k-1} + \cdots + a_k b_1).$$

Both the sum and the product series converge absolutely.

We could reasonably well define the sum and the product series as above even if they did not converge: The crucial content of the theorem is that they do converge absolutely whenever the original series do. The addition theorem is a sort of statement of commutativity for the terms of absolutely convergent series. The formula for the product is motivated by the fact that all possible products of terms of the two series occur, and it is somewhat easier to keep track of them if they are grouped in such a way that the sum of the subscripts of terms in a group is always the same: 2 in $a_1 b_1$, 3 in $a_1 b_2 + a_2 b_1$, 4 in $a_1 b_3 + a_2 b_2 + a_3 b_1$, and so on.

Proof The first part of the proof is to note that if $\{s_n\}$ and $\{t_n\}$ are the partial sums of $\sum a_k$ and $\sum b_k$, respectively, and if $s_n \to s$ and $t_n \to t$, then

$$s_n + t_n \to s + t, \qquad s_n t_n \to st.$$

The proofs of these facts are similar to the proofs for limits of functions in Section 2.1, so we shall not repeat them here.

Next, we note that the commutative law of addition implies that

$$s_n + t_n = \sum_{k=1}^{n} a_k + \sum_{k=1}^{n} b_k = \sum_{k=1}^{n} (a_k + b_k);$$

and then, taking the limit of both sides as $n \to \infty$, we have the first part of the theorem. The proof of the second part is quite similar; we group the terms of the product $s_n t_n$ so that the sum of the subscripts in each group is constant, then take the limit.

These facts on arithmetic manipulation of infinite series will be of the most use to us in the next section, when we consider series with variable terms. Nevertheless, it might be useful here to show a couple of examples:

EXAMPLE 8   We can sum certain series which are obvious sums or products by using the results of Theorem 3.1.12; for instance,

$$\sum_{k=0}^{\infty} \frac{3^k + 2^k}{6^k} = \sum_{k=0}^{\infty} \left(\frac{1}{2^k} + \frac{1}{3^k} \right) = \sum_{k=0}^{\infty} \frac{1}{2^k} + \sum_{k=0}^{\infty} \frac{1}{3^k} = \frac{1}{1 - \frac{1}{2}} + \frac{1}{1 - \frac{1}{3}}$$

$$= 2 + \frac{3}{2} = \frac{7}{2}.$$

As another example, consider

$$1 + \sum_{k=1}^{\infty} \frac{2^{k-1} + 2^{k-2} \cdot 3 + \cdots + 3^{k-1}}{2^k 3^k}$$

$$= 1 + \sum_{k=1}^{\infty} \left(\frac{1}{2 \cdot 3^k} + \frac{1}{2^2 3^{k-1}} + \cdots + \frac{1}{2^k 3} \right)$$

$$= \sum_{k=0}^{\infty} \frac{1}{2^k} \sum_{k=0}^{\infty} \frac{1}{3^k} = 2 \cdot \frac{3}{2} = 3.$$

Naturally this technique is not much help unless we can sum the simple series making up the more complicated one.

All the usual convergence tests for real series of positive terms can be carried over to the complex case in order to test complex series for absolute convergence. Here is a summary of some of these tests:

1 *The comparison test.* If $\sum c_n$ converges absolutely and if there exists an integer N such that $|a_n| \leq |c_n|$ for all $n \geq N$, then $\sum a_n$ converges absolutely.
2 *The ratio test.* If $\lim_{n \to \infty} |a_{n+1}/a_n| = r$, then $\sum a_n$ converges absolutely if $r < 1$ and diverges if $r > 1$.
3 *The root test.* If $\lim_{n \to \infty} \sqrt[n]{|a_n|} = r$, then $\sum a_n$ converges absolutely if $r < 1$, and diverges if $r > 1$.

Naturally there are many other convergence tests, some of them quite sophisticated.* The comparison test is the most important because out of it come most of the others. The usual series used for comparison are the p-series ($\sum n^{-p}$), and the geometric series. One uses the geometric series to derive the ratio and root tests. These latter two tests have the disadvantage that the limits may not exist; when this occurs, one must use the limit superior (or lim sup), which is simply the least upper bound of all the limit points. In the case of the ratio test, the quotient a_{n+1}/a_n may not even be defined for many n because of zero terms, in which case one must use the ratio of consecutive nonzero terms instead.

EXAMPLE 9   Let us investigate the convergence of

$$\sum_{n=0}^{\infty} \frac{1}{2^n + i}.$$

---

* See, for instance, K. Knopp, *Theorie und Anwendung der Unendlichen Reihen*, 5th ed., Springer-Verlag, Berlin, 1964. (In English translation, this is available as *Infinite Sequences and Series*, Dover, New York.)

Now

$$\left|\frac{1}{2^n+i}\right| = \frac{1}{\sqrt{2^{2n}+1}} < \frac{1}{2^n};$$

hence, since the terms of the series are less in absolute value than the terms of a convergent series, the series converges by the comparison test.

EXAMPLE 10 The series

$$\sum_{n=2}^{\infty}\left(\frac{i}{\ln n}\right)^n$$

converges since

$$\lim_{n\to\infty} \sqrt[n]{|a_n|} = \lim_{n\to\infty} \frac{1}{\ln n} = 0.$$

A. Exercises

1 Find the limits of the following sequences, or show that the limits do not exist (where square roots occur, assume that the values have arguments ≥ 0 and $< \pi$).

a $\{\sqrt{n+i} - \sqrt{n}\}$

b $\left\{\dfrac{n^2}{n^2+3} - \dfrac{2ni}{n+2i}\right\}$

c $\left\{\left(\dfrac{1+i}{\sqrt{2}}\right)^n\right\}$

d $\left\{\left(\dfrac{1+i}{2}\right)^n\right\}$

e $\left\{n\left(\dfrac{1+i}{2}\right)^n\right\}$

f $\{(\sqrt{n+2i} - \sqrt{n})\sqrt{n+1/2}\}$

g $\left\{\dfrac{i}{n^2} + \dfrac{2i}{n^2} + \dfrac{3i}{n^2} + \cdots + \dfrac{ni}{n^2}\right\}$

h $\left\{\dfrac{1+i}{n^2} + \dfrac{1+i}{(n+1)^2} + \cdots + \dfrac{1+i}{(2n)^2}\right\}$

2 Test the following series for convergence or divergence.

a $\sum_{n=1}^{\infty} (\sqrt{n+i} - \sqrt{n})$

b $\sum_{n=1}^{\infty} \left(\dfrac{1+i}{\sqrt{2}}\right)^n$

c $\sum_{n=1}^{\infty} \dfrac{n!}{(in)^n}$

d $\sum_{n=1}^{\infty} \dfrac{n}{2^n}$

e $\sum_{n=1}^{\infty} \dfrac{2n}{n^3 - i}$

f $\sum_{n=1}^{\infty} \dfrac{in - 2}{n \ln n}$

g $\sum_{n=1}^{\infty} \left\{\dfrac{i}{n^2} + \dfrac{2i}{n^2} + \dfrac{3i}{n^2} + \cdots + \dfrac{ni}{n^2}\right\}^n$

h $\sum_{n=1}^{\infty} e^{in} 3^{-n}$

3 Find the sums of the following series:

a $\sum_{n=4}^{\infty} \dfrac{3+i}{(2-i)^n}$

b $\sum_{n=0}^{\infty} \left(\dfrac{3}{1+3i}\right)^n$

c $\sum_{n=2}^{\infty} \dfrac{1}{n(n-1)}$

d $\sum_{n=1}^{\infty} \dfrac{2}{3^k(1+i)^{2k}}$

e $\sum_{n=0}^{\infty} \left(\dfrac{2-i}{5}\right)^n$

f $\sum_{n=3}^{\infty} \dfrac{1}{n(n-2)}$

B. Problems

1 Examine the following sequences of functions for convergence; state where they converge, where (if anywhere) the convergence is uniform, and what the limit function is.

a $\left\{\dfrac{2}{2-nz}\right\}$

b $\left\{\dfrac{2n+3z}{3n-z}\right\}$

c $\left\{\dfrac{3nz}{n^2-z^2}\right\}$

d $\{z(2-z) + z^2(2-z) + \cdots + z^n(2-z)\}$

e $\left\{\dfrac{z}{n}\right\}$

f $\left\{\dfrac{z^2+nz}{n}\right\}$

g $\left\{\dfrac{1}{n}\sin nz\right\}$

h $\{e^{-(z-n)^2}\}$

2 Give an example of a divergent series $\sum a_k$, where $\lim_{k\to\infty} a_k = 0$.

3 Prove that the geometric series $\sum z^k (|z| = r < 1)$ satisfies the Cauchy criterion; that is, given $\varepsilon > 0$, find an N such that $|s_n - s_m| < \varepsilon$ for all $n > m > N$, where $\{s_n\}$ is the sequence of partial sums of the series.

4 Suppose that $\lim_{n\to\infty} s_n = s$. Find

$$\lim_{n\to\infty} \dfrac{s_1 + s_2 + \cdots + s_n}{n}.$$

5 Find the sums of the following series:

a $\sum_{n=2}^{\infty} \dfrac{(1+i)^n + n^2 - 1}{(1+i)^n n(n-1)}$

b $\sum_{n=3}^{\infty} \left\{\dfrac{1}{3\cdot 1\cdot 3^{n-3}} + \dfrac{1}{4\cdot 2\cdot 3^{n-4}} + \cdots + \dfrac{1}{k(k-2)}\right\}.$

C. Proofs

1. Prove Theorem 3.1.5(1) in detail using rigorous epsilon–delta arguments.
2. Complete the proof of Theorem 3.1.5(2), again rigorously.
3. Let $\lim s_n = s$ and $\lim t_n = t$. Prove rigorously that
   a. $\lim(s_n + t_n) = s + t$
   b. $\lim(s_n t_n) = st$
   c. $\lim(s_n/t_n) = s/t$ (if $t \neq 0$).
4. Prove the ratio test, given that $\lim |a_{n+1}/a_n|$ exists.
5. Prove the root test, given that $\lim \sqrt[n]{|a_n|}$ exists.
6. Show that Theorem 3.1.5 can be extended under the same hypotheses to prove that (3) $f_n^{(k)}$ converges to $f^{(k)}$ for all k.

3.2. POWER SERIES AND TAYLOR'S THEOREM

An infinite series of the form

$$\sum_{k=0}^{\infty} a_k(z - z_0)^k,$$

where a_k ($k = 0, 1, 2, \ldots$) and z_0 are constants and z is a complex variable, is called a *power series*. The partial sums of a power series

$$f_n(z) = \sum_{k=0}^{n} a_k(z - z_0)^k$$

form a sequence of polynomials, which are entire functions. In this section we propose to examine the question of convergence of these sequences of polynomials and, hence, to evolve another characterization of analytic functions.

It is possible that a power series converges nowhere except at the point z_0 (where it obviously converges to a_0).

EXAMPLE 11 For example, the series

$$\sum_{n=0}^{\infty} n!(z - z_0)^n$$

converges only for $z = z_0$, since if $z \neq z_0$ the ratio test gives

$$\lim_{n \to \infty} \left| \frac{a_{n+1}(z - z_0)^{n+1}}{a_n(z - z_0)^n} \right| = \lim_{n \to \infty} (n+1)|z - z_0| = +\infty.$$

The fact that the limit is infinite for $z \neq z_0$ shows that the series diverges.

We are not really interested in power series which converge only at one point; hence, for the rest of this section we shall always assume that there is at least one $z_1 \neq z_0$ for which $\sum a_n(z_1 - z_0)^n$ converges. But then if this is so, there is a whole set of z for which the series converges:

3.2.1. Theorem Suppose $\sum a_n(z - z_0)^n$ converges for some $z_1 \neq z_0$. Then the series converges *absolutely* for all z in the open disk $|z - z_0| < |z_1 - z_0|$.

Proof Since $\sum a_n(z_1 - z_0)^n$ converges, its terms approach zero as $n \to \infty$. Hence, we can certainly find a positive number M, such that

$$|a_n(z_1 - z_0)^n| \leq M$$

for all values of n. Thus we have

$$|a_n| \leq \frac{M}{|z_1 - z_0|^n};$$

and, therefore,

$$\sum_{n=0}^{\infty} |a_n(z - z_0)^n| \leq \sum_{n=0}^{\infty} M \left| \frac{z - z_0}{z_1 - z_0} \right|^n.$$

Since $|z - z_0| < |z_1 - z_0|$, the latter is a convergent geometric series, whence the original series converges absolutely. QED

Thus, if a power series converges at some point other than at z_0, then it converges absolutely in an open disk around z_0; the radius of the largest disk for which this is true is called the radius of convergence. More precisely (or more correctly) is the following:

3.2.2. Definition Suppose that $\sum a_n(z - z_0)^n$ is a given power series. Let R be the least upper bound of the numbers r for which the series converges for at least one z_1 with $|z_1 - z_0| = r$. Then R is called *the radius of convergence of the given series*.*

We know, in fact, that the series converges *absolutely* inside the circle of radius R and diverges outside the circle. If $R = +\infty$, then, of course, the series converges absolutely for all finite values of z. Our next question should then be, how do we determine the radius of convergence of a given power series? Provided that the limits exist, the ratio test and the root test

---

* Note that R could be $+\infty$.

lead to two possibilities:

1 $R = \lim\limits_{n \to \infty} |a_n/a_{n+1}|$;

2 $R = \lim\limits_{n \to \infty} (1/\sqrt[n]{|a_n|})$.

To prove (1), we know by the ratio test that the series $\sum a_n(z - z_0)^n$ converges if

$$\lim_{n \to \infty} \left| \frac{a_{n+1}(z - z_0)^{n+1}}{a_n(z - z_0)^n} \right| < 1,$$

i.e., if

$$|z - z_0| < \lim_{n \to \infty} \left| \frac{a_n}{a_{n+1}} \right|,$$

and diverges otherwise; similarly for the root test. In many cases these two tests are not applicable, but the following *always* gives the radius of convergence:

$$R = \frac{1}{\limsup\limits_{n \to \infty} \sqrt[n]{|a_n|}}.$$

The result is called the Cauchy-Hadamard formula. We shall not prove it here, nor shall we work examples of such generality as to require this result.*

EXAMPLE 12 To find the radius of convergence of the series

$$\sum_{n=0}^{\infty} \frac{(z - 1)^n}{n 2^n},$$

the ratio test gives

$$R = \lim_{n \to \infty} \left| \frac{(n + 1) 2^{n+1}}{n 2^n} \right| = 2.$$

The root test (which, of course, *should* give the same answer!) gives

$$R = \lim_{n \to \infty} \sqrt[n]{n 2^n} = 2 \lim_{n \to \infty} n^{1/n} = 2.$$

(The limit of $n^{1/n}$ can be found by taking the logarithm, using L'Hôpital's rule, then exponentiating.) Hence, the given series converges absolutely for $|z - 1| < 2$.

---

* See, for example, Nevalinna and Paatero, *Introduction to Complex Analysis*, Addison-Wesley, Reading, 1969.

In fact, one can conclude even more than absolute convergence, for in a smaller closed disk around z_0, the convergence turns out to be uniform:

3.2.3. Theorem Let $\sum a_n(z - z_0)^n$ converge absolutely for $|z - z_0| < R$. Then for any $r < R$, the series converges *uniformly* for all z such that $|z - z_0| \leq r$.

Proof As before, let us write

$$f_n(z) = \sum_{k=0}^{n} a_k(z - z_0)^k, \qquad f(z) = \sum_{k=0}^{\infty} a_k(z - z_0)^k.$$

Then for all z in $|z - z_0| \leq r$, we have

$$|f(z) - f_n(z)| = \left| \sum_{k=n+1}^{\infty} a_k(z - z_0)^k \right| \leq \sum_{k=n+1}^{\infty} |a_k| r^k.$$

The last term represents the "tail" of an absolutely convergent series, which approaches zero as $n \to \infty$; it is independent of the particular z chosen, as long as it is in $|z - z_0| \leq r$; hence, in particular,

$$\operatorname{lub} |f(z) - f_n(z)| \leq \sum_{k=n+1}^{\infty} |a_k| r^k \to 0,$$

which proves the contention.

Because the f_n are polynomials, they are analytic functions, and, hence, their uniform limit f is analytic, by Theorem 3.1.5. Also, since $f_n(z_0) = a_0$ for all n and $f_n^{(k)}(z_0) = k! a_k$ for all $n \geq k \geq 1$, we have the following immediately:

3.2.4. Corollary $\sum a_k(z - z_0)^k$ represents an analytic function $f(z)$ inside its circle of convergence and, furthermore,

$$a_0 = f(z_0), \qquad a_k = \frac{f^{(k)}(z_0)}{k!} \quad (k = 1, 2, 3, \ldots).$$

If the radius of convergence is infinite, then the series converges to an entire function.

Thus a uniform limit of a sequence of polynomials is an analytic function. The apogee of this section is the fact that the converse of this theorem is also true:

130 TAYLOR AND LAURENT SERIES

3.2.5. Theorem *(Taylor's Theorem)* Let f be analytic in a domain D. Then if $z_0 \in D$,

$$f(z) = f(z_0) + \sum_{k=1}^{\infty} \frac{f^{(k)}(z_0)}{k!}(z - z_0)^k,$$

and the series converges absolutely for all z inside the largest circle around z_0 in which f is analytic or for all z if f is entire.

Proof Let Σ be the largest circle around z_0 inside which f is analytic, or the whole complex plane if f is entire, and let z be a point inside Σ. Let C be a circle in Σ around z_0 containing z in its interior; we suppose that C has radius ρ and that $|z - z_0| = r < \rho$ (Figure 3.2). Then by the Cauchy integral formula we have

$$f(z) = \frac{1}{2\pi i} \oint_C \frac{f(\zeta)}{\zeta - z} d\zeta = \frac{1}{2\pi i} \oint_C \frac{f(\zeta)}{\zeta - z_0 + z_0 - z} d\zeta$$

$$= \frac{1}{2\pi i} \oint_C \frac{f(\zeta)\, d\zeta}{(\zeta - z_0)[1 - (z - z_0)/(\zeta - z_0)]}.$$

For brevity, let us write

$$\alpha = \frac{z - z_0}{\zeta - z_0}$$

and note that for all ζ on C, we have

$$|\alpha| = \frac{r}{\rho} < 1.$$

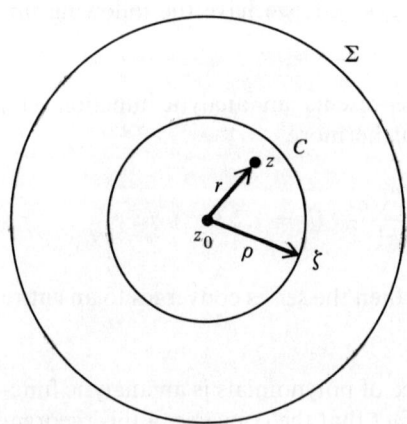

Figure 3.2

Furthermore,
$$\frac{1}{1-\alpha} = 1 + \alpha + \alpha^2 + \cdots + \alpha^n + \frac{\alpha^{n+1}}{1-\alpha}.$$

By putting this result into the expression for $f(z)$, we obtain

$$f(z) = \frac{1}{2\pi i} \oint_C \frac{f(\zeta)\,d\zeta}{\zeta - z_0}\left(1 + \alpha + \alpha^2 + \cdots + \alpha^n + \frac{\alpha^{n+1}}{1-\alpha}\right)$$

$$= \frac{1}{2\pi i} \oint_C \frac{f(\zeta)\,d\zeta}{\zeta - z_0} + \sum_{k=1}^n \frac{1}{2\pi i} \oint_C \frac{f(\zeta)\alpha^k\,d\zeta}{\zeta - z_0} + \frac{1}{2\pi i} \oint_C \frac{f(\zeta)\,d\zeta}{\zeta - z_0}\frac{\alpha^{n+1}}{1-\alpha}$$

$$= \frac{1}{2\pi i} \oint_C \frac{f(\zeta)\,d\zeta}{\zeta - z_0} + \sum_{k=1}^n \frac{(z - z_0)^k}{2\pi i} \oint_C \frac{f(\zeta)\,d\zeta}{(\zeta - z_0)^{k+1}} + R_n(z, z_0)$$

$$= f(z_0) + \sum_{k=1}^n \frac{f^{(k)}(z_0)}{k!}(z - z_0)^k + R_n(z, z_0).$$

We shall have completed the proof if we can show that $|R_n(z, z_0)|$ can be made small on the closed disk bounded by C for large n, that is, if

$$\lim_{k \to \infty}\left(\operatorname*{lub}_{z \in I(C)} |R_n(z, z_0)|\right) = 0.$$

Now

$$|R_n(z, z_0)| \leq \frac{1}{2\pi\rho} \cdot \frac{2\pi\rho M}{1 - r/\rho} \cdot \left(\frac{r}{\rho}\right)^{n+1},$$

where M is the maximum of $|f(\zeta)|$ on C; and because $r/\rho < 1$, the limit is zero.   QED

A special case of Taylor's theorem is the following:

3.2.6. Corollary (*Maclaurin's Theorem*)   If $f(z)$ is analytic in a domain D containing the origin, then

$$f(z) = \sum_0^\infty \frac{f^{(n)}(0)}{n!} z^n;$$

if $\sum$ is any circle around the origin with $I(\sum) \subset D$, then the series converges absolutely in $I(\sum)$. The series converges uniformly in any closed disk around the origin contained in D.

EXAMPLE 13   Find a Maclaurin series for $f(z) = e^z$. Since e^z is entire, it is certainly analytic in the domain consisting of the whole complex plane.

Furthermore, $d/dz(e^z) = e^z$ for every integer n, $d^n/dz^n(e^z) = e^z$. Thus $f^{(n)}(0) = e^0 = 1$; and, therefore,

$$e^z = \sum_{n=0}^{\infty} \frac{z^n}{n!}.$$

We can check quickly by the ratio test that $\sum z^n/n!$ converges absolutely for all infinite values of z, since it converges for

$$|z| < \lim_{n \to \infty} \left| \frac{a_n}{a_{n+1}} \right| = \lim_{n \to \infty} \left| \frac{1/n!}{1/(n+1)!} \right| = \lim_{n \to \infty} (n+1) = \infty.$$

Therefore, if $|z| < \infty$, then the power series for the exponential converges. Note that $e^z = \sum z^n/n!$ also gives the usual familiar power series for e^z if z is real. This fact, by now, should not be at all surprising.

EXAMPLE 14 Let us find the Maclaurin series for $f(z) = 1/(1-z)$. Note that $1/(1-z)$ is defined in the entire complex plane except at $z = 1$. Now $d/dz\,[1/(1-z)] = 1/(1-z)^2$; and it is not hard to see that for all integers n, $d^n/dz^n[1/(1-z)] = n!/(1-z)^{n+1}$. Therefore $f^{(n)}(0) = n!/(1-0)^{n+1} = n!$ and so

$$\frac{1}{(1-z)} = \sum_{0}^{\infty} \frac{n!}{n!} z^n = \sum_{0}^{\infty} z^n.$$

It can hardly be accidental that we have arrived at the same power series for $1/(1-z)$ by two different methods. In fact, the next theorem assures us that by whatever method we may arrive at a power series for an analytic function around a point, we will always obtain the same series.

3.2.7. Theorem Suppose that f is analytic at z_0 and that we have by some means arrived at a power series for $f(z)$:

$$f(z) = \sum_{n=0}^{\infty} a_n(z - z_0)^n.$$

Then this power series is, in fact, the Taylor series for f, that is, $a_n = f^{(n)}(z_0)/n!$ for all n.

Proof Since the two series both converge to f near z_0, we have

$$a_0 + a_1(z - z_0) + a_2(z - z_0)^2 + \cdots$$
$$= f(z_0) + f'(z_0)(z - z_0) + \frac{f''(z_0)}{2!}(z - z_0)^2 + \cdots$$

In the limit as $z \to z_0$, we find that

$$a_0 = f(z_0),$$

since a power series is analytic and, hence, continuous in its circle of convergence. Therefore,

$$a_1(z - z_0) + a_2(z - z_0)^2 + \cdots$$
$$= f'(z_0)(z - z_0) + \frac{f''(z_0)}{2!}(z - z_0)^2 + \cdots.$$

For $z \neq z_0$, we can cancel a common factor from both sides, obtaining

$$a_1 + a_2(z - z_0) + \cdots = f'(z_0) + \frac{f''(z_0)}{2!}(z - z_0) + \cdots.$$

In the limit as $z \to z_0$, we then have

$$a_1 = f'(z_0).$$

Therefore

$$a_2(z - z_0) + \cdots = \frac{f''(z_0)}{2!}(z - z_0) + \cdots.$$

Again we can cancel the common factor and proceed as above. A simple induction proof then leads to the conclusion that

$$a_n = \frac{f^{(n)}(z_0)}{n!}$$

for all n. QED

The importance of this theorem is that it greatly simplifies the procedures for finding the Taylor and Maclaurin series for many functions which are difficult to differentiate. It also leads to the following equally useful result:

3.2.8. Corollary If $f(z)$ is analytic in D, and $z_0 \in D$ with $f(z) = \sum_0^\infty a_n (z - z_0)^n$ for all z such that $|z - z_0| < r$, then the series for f can be differentiated term-by-term, so that

$$f'(z) = \sum_1^\infty n a_n (z - z_0)^{n-1},$$

and, more generally,

$$f^{(k)}(z) = \sum_k^\infty n(n-1) \cdots (n - k + 1) a_n (z - z_0)^{n-k}.$$

These series converge for all z with $|z - z_0| < r$.

Proof Since f is analytic, so is f', and, hence, both functions possess

134 TAYLOR AND LAURENT SERIES

Taylor series around $z = z_0$:

$$f(z) = \sum_{n=0}^{\infty} \frac{f^{(n)}(z_0)}{n!} (z - z_0)^n,$$

$$f'(z_0) = \sum_{n=0}^{\infty} \frac{(f')^{(n)}(z_0)}{n!} (z - z_0)^n$$

$$= \sum_{n=0}^{\infty} \frac{f^{(n+1)}(z_0)}{n!} (z - z_0)^n.$$

By the identity theorem for power series, $a_n = f^{(n)}(z_0)/n!$ and, therefore,

$$a_{n+1} = \frac{f^{(n+1)}(z_0)}{(n+1)!},$$

from which it follows that

$$(n+1) a_{n+1} = \frac{f^{(n+1)}(z_0)}{n!}.$$

Thus

$$f'(z_0) = \sum_{n=0}^{\infty} (n+1) a_{n+1} (z - z_0)^n = \sum_{n=1}^{\infty} n a_n (z - z_0)^{n-1},$$

the latter result having been obtained by changing the index of summation from n to $n - 1$; and this completes the proof.

Some of the techniques used to find series for functions without finding general formulas for the nth derivatives are the following:

1 Known series already derived, such as

$$e^z = 1 + z + \frac{z^2}{2} + \frac{z^3}{3!} + \cdots.$$

This series enables us to solve the following example, by the simple expedient of a substitution.

EXAMPLE 15 Find a Maclaurin series for $\exp z^2$. Any attempt to find a general formula for the nth derivative of $\exp z^2$ is going to prove rather difficult, as the first few calculations show:

$$\frac{d}{dz}(e^{z^2}) = 2z e^{z^2}; \qquad \frac{d^2}{dz^2}(e^{z^2}) = (4z^2 + 2) e^{z^2};$$

$$\frac{d^3}{dz^3}(e^{z^2}) = (8z^3 + 10z) e^{z^2}.$$

But we already know that $e^t = \sum t^n/n!$ for all complex t; in particular, if we put $t = z^2$, we obtain

$$e^{z^2} = \sum_{n=0}^{\infty} \frac{(z^2)^n}{n!} = \sum_{n=0}^{\infty} \frac{z^{2n}}{n!},$$

which must, therefore, be the Maclaurin series for $\exp z^2$.

2 The geometric series

$$\frac{1}{1-z} = 1 + z + z^2 + \cdots.$$

Again, substitutions are often useful, as the next example illustrates:

EXAMPLE 16 Find a Taylor series for $1/(1 - z)$ in powers of $(z - 2)$. Here we alter the denominator by the simple expedient of adding and subtracting 2:

$$\frac{1}{1-z} = \frac{1}{1-z+2-2} = \frac{1}{-1-(z-2)}$$

$$= -\frac{1}{1+(z-2)} = -\frac{1}{1-[-(z-2)]}.$$

The purpose of the last manipulation was to put the expression in the form $1/(1 - t)$, which has the power series $\sum t^n$. Namely we put $t = -(z - 2)$ and since

$$\frac{1}{1-[-(z-2)]} = \sum_{n=0}^{\infty} [-(z-2)]^n = \sum_{n=0}^{\infty} (-1)^n (z-2)^n,$$

we have

$$\frac{1}{1-z} = -\frac{1}{1-[-(z-2)]}$$

$$= -\sum_{n=0}^{\infty} (-1)^n (z-2)^n = \sum_{n=0}^{\infty} (-1)^{n+1} (z-2)^n.$$

3 The binomial theorem, which says that for any integer $n > 0$,

$$\frac{1}{(1+t)^n} = \sum_{k=0}^{\infty} \frac{(-1)^k (n+k-1)!}{(n-1)!\, k!} t^k \quad \text{for } |t| < 1.$$

This formula can be established by induction, or by the use of Corollary 3.2.8 by differentiating the geometric series for $1/(1 + t)$ term-by-term.

EXAMPLE 17 We can use the binomial theorem to find the Maclaurin series for $1/(1 + z^2)^2$ simply by substituting $t = z^2$ and $n = 2$ in the binomial formula:

$$\frac{1}{(1 + z^2)^2} = \sum_{k=0}^{\infty} \frac{(-1)^k (k + 1)!}{1! k!} z^{2k} = \sum_{k=0}^{\infty} (-1)^k (k + 1) z^{2k}.$$

The next group of theorems will give us yet other ways to find Taylor and Maclaurin series without differentiating the functions involved. Each theorem is followed by an example applying it to a particular problem.

3.2.9. Theorem Suppose $f(z) = \sum a_n(z - z_0)^n$ and $g(z) = \sum b_n (z - z_0)^n$ are two power series which converge absolutely in circles of radii R_1 and R_2, respectively, around z_0; then

$$f(z) + g(z) = \sum_{0}^{\infty} (a_n + b_n)(z - z_0)^n,$$

that is, we can add the two series term-by-term, and the series on the right, in general, converges absolutely in the smaller of the two circles.

Proof That the two series can be added term-by-term is a consequence of Theorem 3.1.12; that the series converges in the smaller of the two circles of convergence follows from the fact the sum of two analytic functions will be analytic in the intersection of their two domains.

EXAMPLE 18 To find a Taylor series for $1/(1 - z^2)$ around the point $z = 2$, we first break the expression up into partial fractions, then write it in terms of $(z - 2)$, as follows:

$$\frac{1}{1 - z^2} = \frac{1}{2}\left(\frac{1}{1 - z} + \frac{1}{1 + z}\right)$$

$$= -\frac{1}{2}\left(\frac{1}{1 + (z - 2)}\right) + \frac{1}{6}\frac{1}{[1 + (1/3)(z - 2)]}.$$

Then we apply the geometric series formula to each portion, as follows:

$$\frac{1}{1 + (z - 2)} = \sum_{0}^{\infty} (-1)^n (z - 2)^n,$$

which converges for $|z - 2| < 1$, and

$$\frac{1}{1 + 1/3 (z - 2)} = \sum_{0}^{\infty} \frac{(-1)^n}{3^n} (z - 2)^n,$$

which converges for $|z - 2| < 3$. Therefore,

$$\frac{1}{1-z^2} = -\frac{1}{2}\sum_0^\infty (-1)^n (z-2)^n + \frac{1}{6}\sum_0^\infty \frac{(-1)^n}{3^n}(z-2)^n$$

$$= \frac{1}{2}\sum_0^\infty (-1)^n \left(\frac{1}{3^{n+1}} - 1\right)(z-2)^n$$

is the Taylor series for our given function, which must by Theorem 3.2.9 converge in the smaller of the two circles, namely $|z - 2| < 1$. Note that it would have been impossible (in a practical sense) to obtain this Taylor series by repeatedly differentiating $1/(1 - z^2)$, since the expressions quickly become quite complicated.

3.2.10. Theorem Two power series $f(z) = \sum a_n(z - z_0)^n$ and $g(z) = \sum b_n(z - z_0)^n$, as in Theorem 3.2.9, can be multiplied term-by-term, yielding

$$f(z)g(z) = \sum_{k=0}^\infty \sum_{l=0}^\infty a_k b_l (z - z_0)^{k+l} = \sum_{n=0}^\infty \sum_{k=0}^\infty a_k b_{n-k}(z - z_0)^n,$$

which converges absolutely for $|z - z_0| < \min(R_1, R_2)$.

Proof Analogous to Theorem 3.2.9.

EXAMPLE 19 Find the first few terms of the Maclaurin series for $\sin z/(1 - z)$. Since we know that

$$\sin z = z - \frac{z^3}{3!} + \frac{z^5}{5!} + \cdots, \qquad \frac{1}{1-z} = 1 + z + z^2 + z^3 + \cdots,$$

we have

$$\frac{\sin z}{1-z} = \left(z - \frac{z^3}{3!} + \frac{z^5}{5!} + \cdots\right)(1 + z + z^2 + z^3 + \cdots)$$

$$= z + z^2 + \left(1 - \frac{1}{3!}\right)z^3 + \left(1 - \frac{1}{3!}\right)z^4 + \cdots.$$

EXAMPLE 20 Find a Maclaurin series for $1/(1 + z^2)^2$. We have already done this once, using the binomial theorem (Example 17). We can equally well do this by noting first that

$$\frac{1}{1+z^2} = \sum_{n=0}^\infty (-1)^n z^{2n}$$

by the geometric series formula; then, because of Corollary 3.2.8, we can differentiate both sides, differentiating the series term-by-term, to obtain

$$-\frac{2z}{(1+z^2)^2} = \sum_{n=0}^{\infty} 2n(-1)^n z^{2n-1} = \sum_{n=1}^{\infty} 2n(-1)^n z^{2n-1}.$$

Dividing both sides by $-2z$ and changing the index of summation from n to $n+1$, we obtain

$$\frac{1}{(1+z^2)^2} = \sum_{n=1}^{\infty} n(-1)^{n-1} z^{2n-2} = \sum_{n=0}^{\infty} (n+1)(-1)^n z^{2n},$$

which is the same result we obtained before.

Up until now we have constrained ourselves to discussing functions which are analytic at a point z_0 and inside some circle around that point. Most of the functions which we have considered are entire; or, at the very worst, like $1/(1-z)$ and $1/(1-z^2)$, are analytic except at one or two points. These points where a function fails to be defined are also important, and we shall therefore give them a name:

3.2.11. Definition An *isolated singular point* (or *isolated singularity*) of a single-valued analytic function f is a point z_0, such that f is defined and analytic in some open disk $\sum = \{z : |z - z_0| < \delta\}$ around z_0 except at the point z_0 itself.

EXAMPLE 21   The familiar function

$$f(z) = \frac{1}{1-z}$$

is analytic everywhere except at $z = 1$. This point is certainly an *isolated singular point* since f is analytic at every point in *any* open disk around $z = 1$ except at $z = 1$ itself.

EXAMPLE 22 $z = +1$ and $z = -1$ are isolated singular points of the function $1/(z^2 - 1)$ because they are its only singular points, and each can be enclosed by a circle which excludes the other; any circle of radius less than 2 would do (Figure 3.3).

EXAMPLE 23   A more difficult example is the function $1/\sin(1/z)$. The denominator is zero for $z = \pm 1/\pi, \pm 1/2\pi, \pm 1/3\pi, \ldots$, so each of these is a singular point; then, in addition, the denominator fails to be defined at all for $z = 0$; hence this also is a singular point. The singularities at $+1/n\pi$ and $-1/n\pi$ ($n = 1, 2, 3, \ldots$) are isolated, since a circle of radius less than

$$\frac{1}{n\pi} - \frac{1}{(n+1)\pi} = \frac{1}{n(n+1)\pi}$$

3.2. POWER SERIES AND TAYLOR'S THEOREM

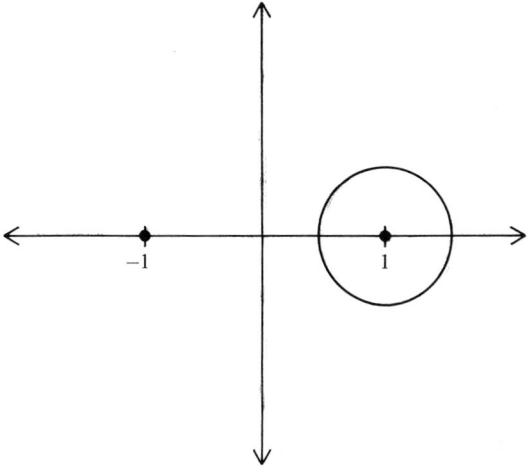

Figure 3.3

around each will exclude all the other singularities (Figure 3.4). However, $z = 0$ is *not* an isolated singularity, since any circle of whatever radius centered at $z = 0$ will contain an infinite number of the points $\pm 1/n\pi$.

Suppose, now, that f is analytic everywhere except at a finite number of isolated singular points and, in particular, that it is analytic at $z = z_0$. Then we have the following:

3.2.12. Theorem Under the above conditions, the Taylor series for f around z_0,

$$f(z) = \sum_{n=0}^{\infty} \frac{f^{(n)}(z_0)}{n!}(z - z_0)^n$$

converges absolutely in a circle of radius r around z_0, where r is the distance between z_0 and the *nearest* singular point.

Proof We have already shown that the Taylor series converges absolutely inside any circle around z_0 in which f is analytic. The largest possible such circle must be one which contains one or more singular points on its bound-

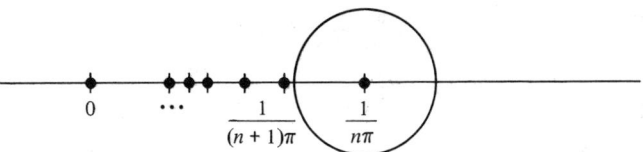

Figure 3.4

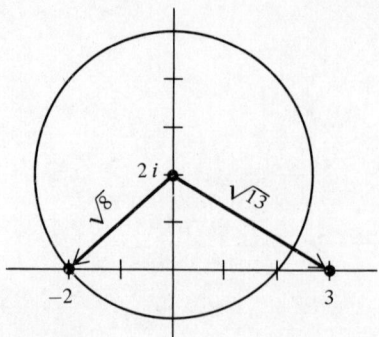

Figure 3.5

ary. If there are none (i.e., f is entire), then we know that the series is absolutely convergent for all z.

Thus, for example, it is no accident that the Maclaurin series for $1/(1-z)$, the geometric series $\sum z^n$, converges only for $|z|<1$. The unit circle around the origin contains the singular point $z=1$ on its boundary. $1/(1-z)$ would not be analytic at every point inside a larger circle. Likewise, for instance, we can predict with utmost confidence that the Taylor series for

$$\frac{1}{(z+2)(z-3)}$$

around the point $z=2i$ will converge inside a circle around $2i$ with radius equal to the minimum of $|2i+2|$ and $|2i-3|$, or $\sqrt{8}$ and $\sqrt{13}$, namely $\sqrt{8}$. A circle of this radius contains the nearest singularity, $z=-2$, on its boundary (Figure 3.5).

A. Exercises

1. Determine the radius of the circle inside which the following series converge:

   a $\sum_{0}^{\infty} \dfrac{z^n}{2^n + i}$

   b $\sum_{0}^{\infty} \dfrac{(z-1)^n}{n^\alpha}$ $(Re\,\alpha \geq 1)$

   c $\sum_{0}^{\infty} nz^n$

   d $\sum_{0}^{\infty} \dfrac{z^{2n}}{\sqrt{2n+1}}$

   e $\sum_{0}^{\infty} \dfrac{(z+i)^n}{(n+1)(n+2)}$

   f $\sum_{0}^{\infty} \dfrac{(z-1)^n}{1\cdot 3\cdot 5\cdot 7\cdots(2n+1)}$

   g $\sum_{0}^{\infty} \dfrac{n(-1)^n(z+2)^n}{4^n(n^2+1)^{5/2}}$

   h $1 + \dfrac{z}{2} + z^2 + \dfrac{z^3}{8} + z^4 + \dfrac{z^5}{32} + \cdots$

i $\displaystyle\sum_{0}^{\infty} \frac{2^n n! z^n}{(2n+1)!}$

j $\displaystyle\sum_{0}^{\infty} \frac{n! z^n}{n^n}$

k $\displaystyle\sum_{n=0}^{\infty}\sum_{k=0}^{n}\left(\frac{1}{2^k(n-k)!}\right)z^n$

l $\displaystyle\sum_{n=0}^{\infty}\left(\frac{1}{n}+\frac{1}{n!}\right)z^n$

2 Expand in a Taylor series about the indicated points and determine the radius of convergence:

a $e^{-z}, z = 0, z = 1$.
b $z^4 + 3z^2 - 2z + 5, z = -3$
c $(z+2)e^{3z}, z = 0$
d $\dfrac{1}{(z^2 - 5z + 4)}, z = 2$
e $\sinh z, z = 0$
f $\log(1+z), z = 0$ (use the branch for which $\log 1 = 0$)
g $\cos z, z = \frac{1}{2}\pi$
h $\sqrt{1+z}, z = 0$ (use the branch for which $\sqrt{1} = +1$)
i $\dfrac{1}{(2+3z^2)^4}; z = 0$
j $\dfrac{(2z+1)}{(z^2+2z+1)}; z = 0$. HINT: $2z+1 = 2(z+1) - 1$
k $\dfrac{(2z+1)}{(z^2+5z+4)}; z = 0$
l $\dfrac{z}{(5-6z+3z^2)^4}; z = 1$

3 Determine the radius of convergence of the Taylor expansions of the following functions *without* finding the actual series:

a $e^{-z^2} \sinh(z+1), z = 0$
b $\dfrac{\sin z}{z(z+1)}, z = i$
c $\dfrac{z+3}{z^2 - 5z + 4}, z = 2$
d $\dfrac{z+1}{\sin z}, z = -1$
e $e^{1/z}, z = 3i$
f $\dfrac{z^2 + 1}{\sin z}, z = \frac{1}{2}$

B. Problems

Find the first three terms of the Taylor series of each of the functions in Exercise 3, about the indicated point.

C. Proofs

Prove that the binomial theorem given prior to Example 16 is valid. You may use any result from this chapter in your proof.

3.3. LAURENT SERIES EXPANSIONS FOR ANALYTIC FUNCTIONS WITH ISOLATED SINGULARITIES

In this section we shall continue our study of functions which are analytic except for isolated singular points. We saw in the previous section that such a function can be expanded in a Taylor series around any point which is *not* a singular point of f. An attempt to expand a function f in a domain which contains singular points of f results in series with negative powers:

3.3.1. Theorem *(Laurent's Theorem)* Let f be analytic in the interior of an arbitrary annulus A centered at z_0, that is, $A = \{z: R_1 < |z - z_0| < R_2\}$. Then f can be expanded in a series of the form

$$f(z) = \sum_{n=0}^{\infty} a_n(z - z_0)^n + \sum_{n=1}^{\infty} b_n(z - z_0)^{-n}, \qquad (3.1)$$

and the series converges absolutely in A and uniformly in any closed concentric annulus $A' = \{z: R_1 < r_1 \leq |z - z_0| \leq r_2 < R_2\}$ contained in A. If $\sum_1$ represents the circle of radius r_1 and $\sum_2$ represents the circle of radius r_2, then the coefficients are given by

$$a_n = \frac{1}{2\pi i} \oint_{\Sigma_2} \frac{f(\zeta)\,d\zeta}{(\zeta - z_0)^{n+1}}, \qquad b_n = \frac{1}{2\pi i} \oint_{\Sigma_2} \frac{f(\zeta)\,d\zeta}{(\zeta - z_0)^{-n+1}}.\quad *$$

The series (3.1) is called *the Laurent series for f in the annulus A*.

COMMENTS This very general theorem contains Taylor's theorem as a special case ($R_1 = 0$, z_0 is not a singular point, and all b_n's vanish). In the case when $R_1 = 0$ and z_0 *is* a singularity, then the series converges in a "punctured disk" $\{z: 0 < |z - z_0| < R_2\}$ centered at z_0, where R_2 is the distance to the nearest singularity from z_0 (if any). In the most general case, the circle of radius R_1 contains a singularity on its circumference, R_2 contains another, and there are none in between. The point z_0 may or may not be a singular point (Figure 3.6 illustrates many of these cases).

Sketch of a Proof The proof is quite similar to Taylor's theorem except for the fact that A has two boundary curves. We let z be any point in A with $|z - z_0| = \rho$, then let $\sum_1$ be a circle around z_0 with radius $r_1 < \rho$,

---

* The value of the coefficients is independent of the choice of r_1 and r_2, inasmuch as any r_1 with $R_1 < r_1 < R_2$, or any r_2 with $R_1 < r_2 < R_2$, will do.

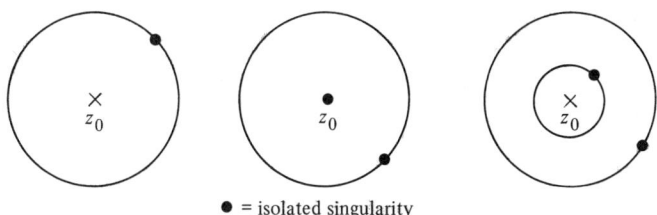

• = isolated singularity

Figure 3.6

and let Σ_2 be a circle around z_0 with radius $r_2 > \rho$. By connecting Σ_1 and Σ_2 by two radial lines and letting these two lines merge as in the proof of Cauchy's integral formula (Figure 3.7), we find that

$$f(z) = \frac{1}{2\pi i} \oint_{\Sigma_2} \frac{f(\zeta)\,d\zeta}{\zeta - z} - \frac{1}{2\pi i} \oint_{\Sigma_1} \frac{f(\zeta)\,d\zeta}{\zeta - z}.$$

As in the proof of Taylor's theorem, we let

$$\alpha = \frac{z - z_0}{\zeta - z_0},$$

so that in the first integral we can write

$$\frac{1}{\zeta - z} = \frac{1}{\zeta - z_0 + z_0 - z} = \frac{1}{(\zeta - z_0)(1 - \alpha)}$$
$$= \frac{1}{\zeta - z_0}\left(1 + \alpha + \cdots + \alpha^n + \frac{\alpha^{n+1}}{1 - \alpha}\right).$$

The derivation of the formula for the a_n's and the proof that the remainder term goes to zero now proceeds exactly as in the proof of Taylor's theorem, since $|\alpha| < 1$ here as there. To derive the formula for the b_n's, we let

$$\beta = \frac{\zeta - z_0}{z - z_0},$$

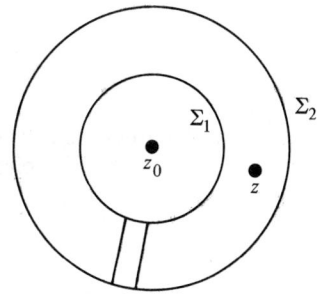

Figure 3.7

144 TAYLOR AND LAURENT SERIES

and write in the second integral

$$\frac{1}{\zeta - z} = \frac{1}{\zeta - z_0 + z_0 - z} = -\frac{1}{(z - z_0)(z - \beta)}$$

$$= -\frac{1}{z - z_0}\left(1 + \beta + \cdots + \beta^n + \frac{\beta^{n+1}}{1 - \beta}\right).$$

Since $|\beta| < 1$, we can derive the formula for the b_n's and show that the remainder term goes to zero much as in Taylor's theorem.

In general, the formulas for computing the coefficients of the Laurent expansion in Theorem 3.3.1 are not particularly useful. In many special cases it is relatively simple to calculate these coefficients by other methods; one of the most important of these is the case in which only a finite number of the b_n's are nonzero.

3.3.2. Definition Let z_0 be an isolated singularity of a single-valued analytic function f and suppose that there is a positive integer k for which

$$C_0 = \lim_{z \to z_0} (z - z_0)^k f(z)$$

exists, and is finite and nonzero. Then z_0 is said to be a *pole of order k of f*.

When z_0 is a pole of order k, then it is a straightforward matter to obtain the Laurent series for f in the punctured disk around z_0, for $(z - z_0)^k f(z)$ is analytic with a *removable singularity* at z_0. This means that if we define a new function $g(z)$ by

$$g(z_0) = C_0, \qquad g(z) = (z - z_0)^k f(z) \quad \text{for } z \neq z_0,$$

then g is analytic at z_0. To see this, we note that g is analytic for $z \neq z_0$ in every neighborhood of z_0; we need only prove that it is analytic at z_0 as well. By Theorem 2.4.7, it will suffice to prove that

$$\frac{1}{2\pi i} \oint_C \frac{g(z)\, dz}{z - z_0} = g(z_0) = C_0$$

for every simple closed contour C surrounding z_0 but no other singularity of g. Without loss of generality (Lemma 2.4.4), we can take C to be a circle $\{z : |z - z_0| = \delta\}$. Then, since

$$C_0 = \frac{1}{2\pi i} \oint_C \frac{C_0\, dz}{z - z_0},$$

we have

$$\left| \frac{1}{2\pi i} \oint_C \frac{g(z)\,dz}{z-z_0} - C_0 \right| = \left| \frac{1}{2\pi} \oint_C \frac{[g(z)-C_0]\,dz}{z-z_0} \right|$$
$$\leq \max_{z \in C} |g(z) - C_0|.$$

This last difference can be made as small as desired by taking δ sufficiently small, since by definition

$$C_0 = \lim_{z \to z_0} g(z).$$

Because

$$\frac{1}{2\pi i} \oint_C \frac{g(z)\,dz}{z-z_0}$$

has the same value for any such circle C however small and since this value can be made arbitrarily close to C_0, we conclude that

$$C_0 = \frac{1}{2\pi i} \oint_C \frac{g(z)\,dz}{z-z_0}.$$

Therefore g is analytic at z_0, as claimed. As a result, g has a Taylor series near z_0 of the form

$$g(z) = (z-z_0)^k f(z) = C_0 + C_1(z-z_0) + \cdots + C_k(z-z_0)^k + \cdots;$$

thus, for $z \neq z_0$ we can write

$$f(z) = \frac{C_0}{(z-z_0)^k} + \frac{C_1}{(z-z_0)^{k-1}} + \cdots + C_k + C_{k+1}(z-z_0) + \cdots.$$

Now if we rename the coefficients

$$b_n = C_{k-n} \qquad (n = 1, 2, \ldots, k)$$

and

$$a_n = C_{k+n} \qquad (n = 0, 1, 2, \ldots),$$

then we see that

$$f(z) = \sum_{n=1}^{k} b_n(z-z_0)^{-n} + \sum_{n=0}^{\infty} a_n(z-z_0)^n.$$

The expression that we have thus obtained for the Laurent expansion of f near z_0 is unique, because the identity theorem for power series says that the Taylor series for $g(z) = (z-z_0)^k f(z)$ is unique; furthermore, the coefficients a_n and b_n are the same as given by Laurent's theorem, since

$$C_m = \lim_{z \to z_0} \frac{1}{m!} \frac{d^m}{dz^m} [(z - z_0)^k f(z)]$$

$$= \frac{1}{2\pi i} \oint_C \frac{(\zeta - z_0)^k f(\zeta) \, d\zeta}{(\zeta - z_0)^{m+1}} = \frac{1}{2\pi i} \oint_C \frac{f(\zeta) \, d\zeta}{(\zeta - z_0)^{m-k+1}},$$

where the integration is extended around any simple closed contour around z_0 in the circle of convergence of the series for g. Thus, in particular,

$$b_n = C_{k-n} = \lim_{z \to z_0} \frac{1}{(k-n)!} \frac{d^{k-n}}{dz^{k-n}} [(z - z_0)^k f(z)] \tag{3.2}$$

$$= \frac{1}{2\pi i} \oint \frac{f(\zeta) \, d\zeta}{(\zeta - z_0)^{-n+1}}$$

and

$$a_n = C_{k+n} = \lim_{z \to z_0} \frac{1}{(k+n)!} \frac{d^{k+n}}{dz^{k+n}} [(z - z_0)^k f(z)] \tag{3.3}$$

$$= \frac{1}{2\pi i} \oint \frac{f(\zeta) \, d\zeta}{(\zeta - z_0)^{n+1}},$$

precisely the formulas given by Laurent's theorem. We therefore have

3.3.3. Corollary If $f(z)$ has a pole of order k at z_0, then the Laurent expansion of f near z_0 contains no negative powers beyond $(z - z_0)^{-k}$, and its coefficients are given by formulas (3.2) and (3.3).

In addition to the foregoing result, there is a version of the identity theorem, for Laurent series, which says that any two expansions of the same function in the same annulus and about the same point must be identical, however they were arrived at; hence, any such series *is* the Laurent series. Likewise there are analogous results concerning the addition, multiplication, and differentiation of Laurent series term-by-term. We trust that the reader was sufficiently convinced by our proofs for Taylor series in Section 3.2, that we can forego the details of the proofs. We use these results much as with Taylor series, as the following examples indicate.

EXAMPLE 24   Let us find a Laurent series for $e^{1/z}$ about the point $z = 0$. To do this, we simply note that $e^t = \sum_0^\infty t^n/n!$ for all finite complex values of t, so that

$$e^{1/z} = \sum_0^\infty \frac{1}{z^n n!} = 1 + \frac{1}{z} + \frac{1}{2! z^2} + \cdots$$

and this Laurent series converges for all values of z with $|z| > 0$.

EXAMPLE 25 Find a Laurent series about $z = 1$ for $1/[(z - 1)(z - 2)]$. To do this, we first use partial fractions to obtain

$$\frac{1}{(z - 1)(z - 2)} = -\frac{1}{z - 1} + \frac{1}{z - 2}.$$

The first fraction is already a Laurent series about $z = 1$ (of one term). $1/(z - 2)$ has no singularity at $z = 1$, so we can expand it in a Taylor series about this point, for instance by using the geometric series:

$$\frac{1}{(z - 2)} = -\frac{1}{(2 - z)} = -\frac{1}{1 + (1 - z)}$$

$$= -\sum_0^\infty (-1)^n (1 - z)^n = -\sum_0^\infty (z - 1)^n.$$

Putting these two results together,

$$\frac{1}{(z - 1)(z - 2)} = -\frac{1}{z - 1} - \sum_0^\infty (z - 1)^n = -\sum_{n=-1}^\infty (z - 1)^n,$$

and the combined result converges for $0 < |z - 1| < 1$.

Before we leave this example, we note that we should be able to find a Laurent expansion for $1/[(z - 1)(z - 2)]$ which converges for $|z - 1| > 1$, since there are no singular points of the function in this region (Figure 3.8). Again, since

$$\frac{1}{(z - 1)(z - 2)} = -\frac{1}{z - 1} + \frac{1}{z - 2},$$

the problem reduces to that of determining a Laurent expansion for $1/(z - 2)$

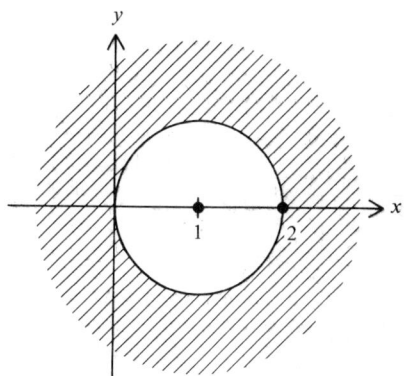

Figure 3.8

in powers of $z - 1$ which converges for $|z - 1| > 1$. To do this, we see that

$$\frac{1}{z-2} = -\frac{1}{1-(z-1)} = \frac{1}{z-1}\left(\frac{1}{1-[1/(z-1)]}\right).$$

Now since $|z - 1| > 1$, we have

$$\left|\frac{1}{z-1}\right| < 1$$

and, hence, the quantity in square brackets can be expanded in a convergent geometric series

$$\frac{1}{1-[1/(z-1)]} = \sum_{n=0}^{\infty}\left(\frac{1}{z-1}\right)^n.$$

Therefore

$$\frac{1}{z-2} = \frac{1}{z-1}\sum_{n=0}^{\infty}\left(\frac{1}{z-1}\right)^n = \sum_{n=1}^{\infty}\left(\frac{1}{z-1}\right)^n.$$

Putting these results together,

$$\frac{1}{(z-1)(z-2)} = -\frac{1}{z-1} + \sum_{n=1}^{\infty}\left(\frac{1}{z-1}\right)^n = \sum_{n=2}^{\infty}\left(\frac{1}{z-1}\right)^n.$$

Note that these two series which we have obtained in the different domains manifest quite different behavior. The first contains only one negative power; the second contains no positive powers whatever.

EXAMPLE 26 Find the first three nonzero terms of the Laurent expansion of $1/\sin z$ about the origin. We do this as follows: Note that $z/\sin z$ is analytic (with a removable singularity) near $z = 0$, and

$$\frac{z}{\sin z} \cdot \sin z = z.$$

By writing

$$\frac{z}{\sin z} = \sum_{n=0}^{\infty} a_n z^n,$$

we have

$$(a_0 + a_1 z + a_2 z^2 + a_3 z^3 + \cdots)\left(0 + z + 0 - \frac{z^3}{3!} + 0 + \frac{z^5}{5!} + \cdots\right)$$

$$= 0 + z + 0 + 0 \cdots$$

Equating coefficients of like powers of z on both sides, we obtain

$$a_0 \cdot 0 = 0$$
$$a_0 \cdot 1 + a_1 \cdot 0 = 1$$
$$a_0 \cdot 0 + a_1 \cdot 1 + a_2 \cdot 0 = 0$$
$$a_0 \left(-\frac{1}{3!}\right) + a_1 \cdot 0 + a_2 \cdot 1 + a_3 \cdot 0 = 0$$
$$a_0 \cdot 0 + a_1 \left(-\frac{1}{3!}\right) + a_2 \cdot 0 + a_3 \cdot 1 + a_4 \cdot 0 = 0$$
$$a_0 \cdot \frac{1}{5!} + a_1 \cdot 0 + a_2 \left(-\frac{1}{3!}\right) + a_3 \cdot 0 + a_4 \cdot 1 + a_5 \cdot 0 = 0,$$

and solving simultaneously, we obtain $a_0 = 1$, $a_1 = 0$, $a_2 = 1/3!$, $a_3 = 0$, $a_4 = 7/360$, and hence

$$\frac{1}{\sin z} = \frac{1}{z} + \frac{1}{6}z + \frac{7}{360}z^3 + \cdots.$$

This Laurent expansion will converge for $0 < |z| < \pi$, since $z = \pm \pi$ are the nearest singular points to the origin.

EXAMPLE 27 Expand

$$\frac{z}{(z-1)(z-2)}$$

in a Laurent series valid for $1 < |z| < 2$. Note that the annulus in which the series is to be valid is centered at $z = 0$, so our series must be in powers of z. Also, neither of the singularities is at the center of the annulus; in fact, one is on the inner boundary and one is on the outer (Figure 3.9). Now

$$\frac{z}{(z-1)(z-2)} = -\frac{1}{z-1} + \frac{2}{z-2},$$

so let us look at each term individually. First,

$$-\frac{1}{z-1} = -\frac{1}{z}\left(\frac{1}{1-1/z}\right),$$

and since $|z| > 1$ we have $|1/z| < 1$, so the quantity in square brackets can be expanded in a geometric series

$$-\frac{1}{z-1} = -\frac{1}{z}\sum_0^\infty \frac{1}{z^n} = -\sum_0^\infty \frac{1}{z^{n+1}} = -\sum_{n=1}^\infty \frac{1}{z^n}.$$

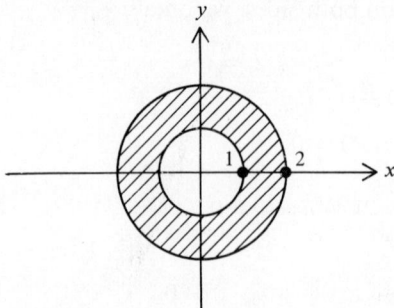

Figure 3.9

For the second term,

$$\frac{2}{z-2} = -1/2\left(\frac{2}{1-(z/2)}\right) = -\frac{1}{1-(z/2)},$$

and since $|z| < 2$, we have $|z/2| < 1$, so the quantity in large parentheses can be expanded in a geometric series, giving

$$\frac{2}{z-2} = -\sum_{n=0}^{\infty} \frac{z^n}{2^n}.$$

By putting these two results together, we obtain

$$\frac{z}{(z-1)(z-2)} = -\sum_{0}^{\infty} \frac{z^n}{2^n} + \sum_{1}^{\infty} \frac{1}{z^n},$$

a Laurent series valid for $1 < |z| < 2$.

If a function $f(z)$ possesses a convergent Laurent series in a neighborhood of an isolated singular point z_0, then the series of *negative* powers of $(z - z_0)$ in this expansion is called the *principal part* of the Laurent series for f around z_0. [If the series has no principal part, then it is a Taylor series and $f(z)$ is, of course, analytic at z_0.] In Example 24, the Laurent series we obtained had an infinite number of negative powers; in the next two examples, the Laurent series had only a finite number of negative powers. We emphasize strongly that we are speaking here of that Laurent expansion which is valid in a *deleted neighborhood* of a *singular point* z_0.* For, we saw that the function

$$\frac{1}{(z-1)(z-2)}$$

has a Laurent expansion for $|z - 1| > 1$ with an infinite number of negative powers; but, of course, that particular series does not converge for values

---

* This means for all z sufficiently close but not equal to z_0.

of z close to 1. (Refer again to Figure 3.8.) Depending on whether the principal part does or does not contain an infinite number of terms, we have special names for these situations:

3.3.4. Definition Suppose $f(z)$ has a Laurent expansion

$$\sum_0^\infty a_n(z-z_0)^n + \sum_0^\infty b_n(z-z_0)^{-n}$$

in a deleted neighborhood of $z = z_0$. If the principal part

$$\sum_1^\infty b_n(z-z_0)^{-n}$$

contains an infinite number of terms, i.e., $b_n \neq 0$ for infinitely many n, then z_0 is said to be an *essential singularity of $f(z)$*. If the principal part terminates after a finite (nonzero) number of terms, and if b_k is the last nonzero coefficient, then z_0 is said to be a pole of order k of $f(z)$. (See Definition 3.2.3 and Corollary 3.3.3.) In either case, the coefficient

$$b_1 = \frac{1}{2\pi i} \oint_\Sigma f(\zeta)\, d\zeta,$$

the coefficient of $(z-z_0)^{-1}$, is called the *residue* of $f(z)$ at z_0.

Thus, $e^{1/z}$ has an essential singularity at the origin, and its residue is 1. $1/[(z-1)(z-2)]$ has a pole of order 1 at $z = 1$, and the residue there is -1. We shall have considerably more to say about poles and residues and their applications, in the next two sections of this chapter.

Another elegant little application of the theory of this section is the following:

3.3.5. Theorem Suppose that

$$R(z) = \frac{p(z)}{q(z)}$$

is a rational function, p having degree $n \geq 0$ and q having degree $m > 0$. Suppose the zeros of q are $z_1, z_2, \ldots, z_l$ with orders $k_1, k_2, \ldots, k_l$, respectively. Then $R(z)$ can be decomposed into partial fractions, that is, it can be written as a sum of the form

$$R(z) = h(z) + \sum_{i=1}^{l} \left(\frac{c_{i1}}{z-z_i} + \frac{c_{i2}}{(z-z_i)^2} + \cdots + \frac{c_{ik_i}}{(z-z_i)^{k_i}} \right),$$

where $h(z)$ is a polynomial of degree $n - m$ if $n - m \geq 0$, identically zero otherwise. The constants in the expression are given by

$$c_{ij} = \lim_{z \to z_i} \frac{1}{(k_i - j)!} \frac{d^{k_i - j}}{dz^{k_i - j}} \{(z - z_i)^{k_i} [R(z) - h(z)]\}.$$

(There is a similar formula for the coefficients of h. See Proof 1 at the end of Section 3.4.)

Proof It is clear from the discussion following Definition 3.3.2 that in a neighborhood of z_i, $R(z)$ has a Laurent expansion of the form

$$R(z) = \frac{c_{11}}{(z - z_1)} + \frac{c_{12}}{(z - z_1)^2} + \cdots + \frac{c_{1k_1}}{(z - z_1)^{k_1}} + S(z),$$

where $S(z)$ is a power series in $(z - z_1)$. In fact, the function defined by

$$\tilde{S}(z) = R(z) - \left(\frac{c_{11}}{(z - z_1)} + \cdots + \frac{c_{1k_1}}{(z - z_1)^{k_1}} \right)$$

is analytic in a neighborhood of z_1, but has singularities at the other singularities of $R(z)$, namely $z_2, \ldots, z_l$. Thus $\tilde{S}(z)$, near z_2, has a Laurent series expansion near z_2 similar to the above:

$$\tilde{S}(z) = \left(\frac{c_{21}}{(z - z_2)} + \cdots + \frac{c_{2k_2}}{(z - z_2)^{k_2}} \right) + T(z),$$

where $T(z)$ is analytic in a neighborhood of z_2 but has singularities at $z_3, z_4, \ldots, z_l$. We continue this process until we have eliminated all the singularities of $R(z)$; then the function

$$h(z) = R(z) - \sum_{i=1}^{l} \left(\frac{c_{i1}}{(z - z_1)} + \cdots + \frac{c_{ik_i}}{(z - z_i)^{k_i}} \right)$$

is *entire*. If we divide both sides by z^{n-m} (or 1 if $n - m < 0$) and let $|z| \to \infty$, then it is clear that the right-hand side is bounded in absolute value. Hence, by Liouville's theorem (as extended in Proof 3, Section 2.5), $h(z)$ is at worst a polynomial of degree $n - m$. In the case $n - m < 0$, $|h(z)| \to 0$ as $|z| \to \infty$ and, hence, h is identically zero.

A. Exercises

1 Find Laurent expansions for the following functions which are valid in a neighborhood of the point indicated, and state the region of convergence.

  a $\dfrac{1}{z(z-2)}; z = 0$

  b $\dfrac{e^{z^2}}{z^3}; z = 0$

c $\sin\left(\dfrac{\pi}{z-1}\right); z = 1$ **d** $\dfrac{z}{z^2+1}; z = i$

e $\dfrac{\sin z}{z^2}; z = 0$ **f** $\dfrac{1}{z^3 - 3z - 2}; z = -1$

g $\dfrac{z^2 + 2}{z^2(z-1)}; z = 0$ **h** $\dfrac{1}{z(z-1)^2}; z = 0$

i $\dfrac{1}{z(4-z)}; z = 0$ **j** $\dfrac{1}{z(4-z)}; z = 4$

2 Find Laurent expansions for the following functions valid in the regions indicated:

a $\dfrac{1}{(z-1)}: |z| > 1; |z+1| > 2$

b $\dfrac{z}{z^2+1}: |z-3| > \sqrt{10}$

c $\dfrac{1}{z(z-1)}; 0 < |z| < 1; |z| > 1; |z-2| > 2$

d $\dfrac{1}{z(4-z)}: |z| > 4$

e $\dfrac{3z-3}{(2z-1)(z-2)} : \dfrac{1}{2} < |z-1| < 1$

f $\dfrac{z}{z^2+1}: |z-i| > 2$

B. Problems

1 Expand $e^{z/(z-1)}$ in a Laurent series about $z = 1$.

2 Describe the singular points of all the functions in Part A; that is, are they essential singularities or poles, and if poles, of what order?

3 Find by any method the first few nonzero germs (for both positive and negative powers) in the Laurent expansions of the following near the point or in the region indicated, and determine whether the singularity is a pole or an essential singularity:

a $\dfrac{1}{\sin z}, z = 0$ **b** $\tan z, z = \pi/2$

c $\dfrac{1}{\sin z \cos z}, z = 0$ **d** $\sin\left(z^2 + \dfrac{1}{z^2}\right), z = 0$

e $\dfrac{1}{(e^z - 1)}, z = 0$ **f** $\dfrac{1}{(z-1)\sin \pi z}, z = 1$

g $\dfrac{\sin z}{z^2(1-z)}; z = 0$ **h** $\dfrac{\cos \pi z}{z - 1}: |z - 1| > 1$

i $\dfrac{\arctan z}{z^4}: z = 0$ (use the branch for which arctan 0 = 0.)

C. Proof

1. Prove that the singularities of a rational function are all poles of finite order.
2. Find a counterexample that illustrates that a function which is analytic in the plane except for poles need not be a rational function.
3. Fill in the details in the proof of Theorem 3.3.1.

3.4. THE CALCULUS OF RESIDUES

Functions which are analytic except for *poles*, together with their residues at these poles, play such an important part in the study of complex variables from this point forward that we shall give them a special name:

3.4.1. Definition A function $f(z)$, which is analytic everywhere in a domain D except for *poles*, is said to be *meromorphic* in D.

The basis for this study of residues which we are about to undertake is the formula

$$\oint_C \frac{\alpha \, d\zeta}{(\zeta - z_0)} = 2\pi i \alpha, \qquad z_0 \in I(C),$$

which the reader should recall from Lemma 2.4.4. We intend to extend and generalize this formula as much as possible. First, we shall need the following

3.4.2. Lemma If $f(z)$ is a meromorphic function with a pole at $z = z_0$, if

$$f(z) = \sum_{n=0}^{\infty} a_n(z - z_0)^n + \sum_{n=1}^{k} b_n(z - z_0)^{-n}$$

is the Laurent expansion for $0 < |z - z_0| < r$, and if C is an arc in $\{z: 0 < R_1 \leq |z - z_0| \leq R_2 < r\}$, then we have

$$\int_C f(z) \, dz = \sum_0^{\infty} a_n \int_C (z - z_0)^n \, dz + \sum_1^{k} b_n \int_C (z - z_0)^{-n} \, dz,$$

that is, a Laurent series can be integrated term-by-term in its region of convergence.

3.4. THE CALCULUS OF RESIDUES

Proof Let $\varepsilon > 0$ be given, and assume that the length of C is L. Then since the Laurent series for f converges uniformly on C, it is possible to find an N so large that

$$f(z) = \sum_0^N a_n(z - z_0)^n + \sum_1^k b_n(z - z_0)^{-n} + R_N(z),$$

where $|R_N(z)| < \varepsilon/L$ for all z on C. Then

$$\int_C f(z)\, dz = \sum_0^N a_n \int_C (z - z_0)^n\, dz + \sum_1^k b_n \int_C (z - z_0)^{-n} + \int_C R_N(z)\, dz,$$

since the interchanging of integration and summation is permitted for finite sums. But

$$\left| \int_C R_N(z)\, dz \right| \leq \int_C |R_N(z)||dz| < \frac{\varepsilon}{L} \cdot L = \varepsilon.$$

Therefore, the sum of the integrals converges to the integral of the sum, and this completes the proof.

We are now ready to prove the first important proposition of the calculus of residues:

3.4.3. Lemma Suppose that $f(z)$ is meromorphic in a domain D, and that z_0 is a pole of f in D. If C is a simple closed contour in D with $z_0 \in I(C)$, and with no other singularities of f contained in $C \cap I(C)$, then

$$\oint_C f(z)\, dz = 2\pi i b_1,$$

where b_1 is the residue of f at $z = z_0$.

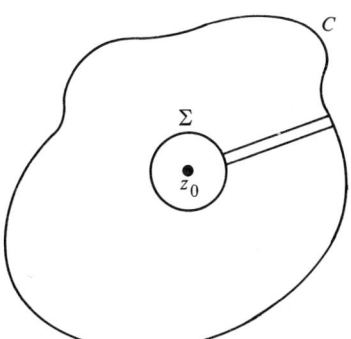

Figure 3.10

156 TAYLOR AND LAURENT SERIES

Proof Let $\Sigma = \{z: |z - z_0| = r\}$, where r is chosen so small that $\Sigma \subset I(C)$ (Figure 3.10). For $z \in I(\Sigma)$, $f(z)$ possesses a convergent Laurent expansion

$$f(z) = \sum_0^\infty a_n (z - z_0)^n + \sum_1^k b_n (z - z_0)^{-n}.$$

By Lemma 3.4.2 we can integrate the series term-by-term around Σ, obtaining

$$\oint_\Sigma f(z)\,dz = \sum_0^\infty a_n \oint_\Sigma (z - z_0)^n\,dz + \sum_1^k b_n \oint_\Sigma (z - z_0)^{-n}\,dz.$$

We showed in Chapter 2 that

$$\oint_\Sigma (z - z_0)^n\,dz = 0 \qquad (n = 0, 1, 2, \ldots),$$

$$\oint_\Sigma (z - z_0)^{-n}\,dz = 0 \qquad (n = 2, 3, \ldots),$$

and we know that

$$\oint_\Sigma (z - z_0)^{-1}\,dz = 2\pi i.$$

Therefore,

$$\oint_\Sigma f(z)\,dz = 2\pi i b_1,$$

since only the term in $(z - z_0)^{-1}$ contributes to a nonzero result.* To complete the proof, we perform our usual trick of connecting Σ and C with a pair of lines (Figure 3.10) and noting that around the resulting closed curve Γ, $\oint_\Gamma f(z)\,dz = 0$, since $f(z)$ is analytic in $I(\Gamma)$. But in the limit as the lines merge, the integrals back and forth along K cancel each other, from which

$$\int_C f(z)\,dz - \int_\Sigma f(z)\,dz = 0,$$

which implies the result.

The theorem which we are aiming for now follows readily from Lemma 3.4.3:

3.4.4. Theorem *(The Residue Theorem)* Suppose $f(z)$ is meromorphic in a domain D and that C is a simple closed contour in D on which f is analytic. If $z_1, z_2, \ldots, z_k$ are the only poles of f lying in $I(C)$, with residues

---

* This is why b_1 is called the "residue."

3.4. THE CALCULUS OF RESIDUES

$b_1^{(1)}, b_1^{(2)}, \ldots, b_1^{(k)}$, respectively, then

$$\oint_C f(z)\, dz = 2\pi i (b_1^{(1)} + b_1^{(2)} + \cdots + b_1^{(k)}),$$

i.e., the integral of a meromorphic function around a closed curve is equal to $2\pi i$ times the sum of the residues at the poles of f in $I(C)$.*

Proof (See Figure 3.11 for $k = 3$.) We draw around each z_i a circle $\sum_i = \{z : |z - z_i| = r_i\}$, where r_i is so small that each $\sum_i$ is contained in $I(C)$ and is also small enough so that no $\sum_i$ intersects or contains any of the other $\sum_j$. Then we connect $\sum_1$ to C by means of a line K_1, $\sum_2$ to $\sum_3$ by means of a line $K_2, \ldots, \sum_{k-1}$ to $\sum_k$ by means of a line K_k, and examine the resulting closed curve Γ (Figure 3.12, $k = 3$). $f(z)$ is analytic in $I(\Gamma)$, so

$$\oint_\Gamma f(z)\, dz = 0.$$

But in the integration around Γ, we traverse each K_i once in each direction, so that the integrals back and forth along the K_i's cancel and we are left with

$$\oint_\Gamma f(z)\, dz = \oint_C f(z)\, dz - \oint_{\Sigma_1} f(z)\, dz - \oint_{\Sigma_2} f(z)\, dz - \cdots - \oint_{\Sigma_k} f(z)\, dz = 0$$

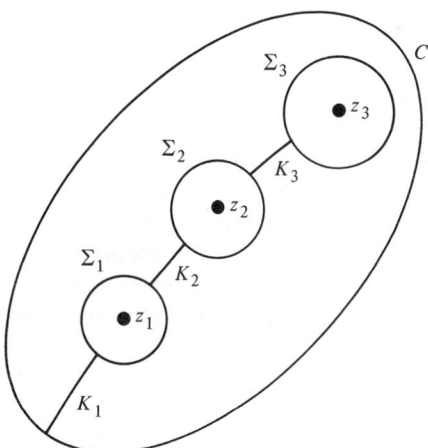

Figure 3.11

---

* There can only be a finite number of poles of f inside C; if there were an infinite number, then they would have to have a limit point, which would be a nonisolated singularity, which is not allowed. (See Appendix A.)

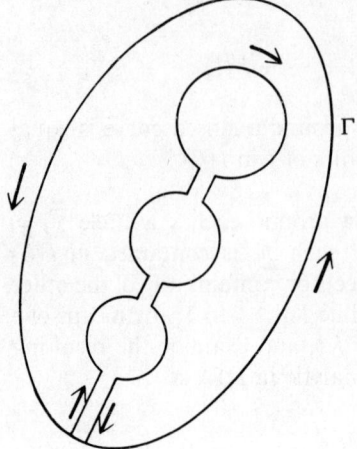

Figure 3.12

From Lemma 3.4.3

$$\oint_{\Sigma_i} f(z)\,dz = 2\pi i b_1^{(i)},$$

and from this we have

$$\oint_C f(z)\,dz - 2\pi i(b_1^{(1)} + b_1^{(2)} + \cdots + b_1^{(k)}) = 0,$$

from which the theorem follows immediately.

Cauchy's theorem and the residue theorem provide the means to evaluate the integrals of many complex functions around closed curves, provided that we can find the residues at the poles inside the curves. Let us now seek to develop techniques by which we can find these residues. Naturally, if we can obtain the Laurent expansion of a function near a point, then the residue is simply the coefficient of $(z - z_0)^{-1}$. Thus, for example, since

$$e^{1/z} = 1 + \frac{1}{z} + \frac{1}{(2z)^2} + \cdots,$$

we immediately read off that the residue of $e^{1/z}$ at $z = 0$ is 1, the coefficient of $1/z$. In fact, if a function has an essential singularity at a point as $e^{1/z}$ does at $z = 0$, just about the only way to obtain the residue at that point is to find the Laurent expansion and read off the coefficient b_1 (see, however, Proof 1, Section 3.4).

3.4. THE CALCULUS OF RESIDUES

However, most of the functions which we shall encounter in practice possess poles of finite order. Indeed, if f has a pole of order k at $z = z_0$, then (3.2), with $n = 1$, gives us an immediate formula for the residue

$$b_1 = \frac{1}{(k-1)!} \frac{d^{k-1}}{dz^{k-1}} [(z-z_0)^k f(z)] \bigg|_{z=z_0},$$

where, in general, we must take the *limit* as $z \to z_0$ in order to find the residue.

EXAMPLE 28 Find the residue of

$$f(z) = \frac{z}{z^2 - 2z - 3} \quad \text{at} \quad z = -1.$$

Note first that since the denominator factors into $(z-3)(z+1)$, we have

$$(z+1)f(z) = (z+1)\frac{z}{(z+1)(z-3)} = \frac{z}{z-3} \quad (z \neq -1),$$

so that

$$\lim_{z \to -1} (z+1)f(z) = \lim_{z \to -1} \frac{z}{z-3} = \frac{1}{4}.$$

EXAMPLE 29 Find the order of the pole of $f(z) = 1/(z \sin z)$ at $z = 0$, and find the residue at the pole. To find the order of the pole, we must find the smallest integer N so that $\lim z^N f(z)$ is finite and nonzero. (Why??) Now

$$\lim_{z \to 0} z^N f(z) = \lim_{z \to 0} \frac{z^{N-1}}{\sin z} = \lim_{z \to 0} \frac{(N-1)z^{N-2}}{\cos z},$$

which has a finite limit for $N = 2$ and for no smaller N. Thus $f(z)$ has a pole of order 2 at $z = 0$. Next we define $F(z) = z^2 f(z) = z/\sin z$, and to find the residue at $z = 0$ we need to differentiate this $N - 1$ times, i.e., once, and take the limit as $z \to 0$. Since

$$F'(z) = \frac{\sin z - z \cos z}{\sin^2 z},$$

there are several ways we can proceed. First, we could find the limit by using L'Hôpital's rule:

$$\lim_{z \to 0} F'(z) = \lim_{z \to 0} \frac{\cos z - \cos z + z \sin z}{2 \sin z \cos z} = \lim_{z \to 0} \frac{z}{2 \cos z} = 0;$$

and hence the residue is zero. Second, we could write out the first few terms of the Maclaurin series in numerator and denominator:

$$F'(z) = \frac{[z - (z^3/3!) + \cdots] - z[1 - (z^2/2!) + \cdots]}{[z - (z^3/3!) + \cdots]^2}$$

$$= \frac{[(1/2!) - (1/3!)]z^3 + \cdots}{z^2 + \cdots},$$

for which the limit is obviously zero. Third, and simplest, we need only realize that $f(z)$ is an even function and, hence, its residue must be zero along with all the coefficients of odd powers of z.

EXAMPLE 30 Find the order of the pole and the residue at the origin of

$$f(z) = \frac{1}{z(\cos^2 z - 1)}.$$

If we follow the method of the previous example, using the auxiliary function F, we find that the calculation quickly becomes quite involved. An easier method is to work with the first few terms of the Maclaurin expansion for $\cos z$:

$$\cos z = 1 - \frac{z^2}{2} + \frac{z^4}{24} - \cdots,$$

so that

$$\cos^2 z = 1 - z^2 + \frac{z^4}{4} + \frac{z^4}{12} + \cdots = 1 - z^2 + \frac{z^4}{3} + \cdots$$

and, thus,

$$f(z) = \frac{1}{z(-z^2 + \tfrac{1}{3}z^4 + \cdots)} = -\frac{1}{z^3}\frac{1}{1 - \tfrac{1}{3}z^2 + \cdots},$$

from which we immediately read off that the order of the pole is 3. Next, for z sufficiently close to zero, we can apply the geometric series

$$\frac{1}{1 - \tfrac{1}{3}z^2 + \cdots} = 1 + \left(\frac{z^2}{3} + \cdots\right) + \left(\frac{z^2}{3} + \cdots\right)^2 + \cdots$$

and, hence,

$$f(z) = -\frac{1}{z^3}\left(1 + \frac{z^2}{3} + \cdots\right) = -\frac{1}{z^3} - \left(\frac{1}{3z}\right) + \cdots$$

and we see that the residue is $-\tfrac{1}{3}$.

An alternative method of working this example is to note the trigonometric identities

$$\cos^2 z - 1 = -\sin^2 z, \qquad -\sin^2 z = \tfrac{1}{2}\cos 2z - \tfrac{1}{2},$$

3.4. THE CALCULUS OF RESIDUES 161

then use the Maclaurin series for $\cos 2z$,

$$\frac{1}{2}\cos 2z - \frac{1}{2} = \left[\frac{1}{2}\left(1 - \frac{(2z)^2}{2!} + \frac{(2z)^4}{4!} - \cdots\right) - \frac{1}{2}\right] = \left(-z^2 + \frac{z^4}{3} + \cdots\right),$$

which is the same as the expansion obtained above. This example serves to illustrate how very many methods there are of arriving at the same result. The student should not be reluctant to use any mathematically valid method that comes to mind.

It is clear that it is easiest to find the residues at the poles of rational functions, or at least of functions which contain only polynomials in their denominators. As we saw in Theorem 3.3.5, rational functions can always be broken up by partial fractions into finite Laurent series from which it is easy to read off the residues; we do this by noting that if a polynomial $q(z)$ has a zero of order 1 at z, then $q(z) = (z - z_0)q_1(z)$, where $q_1(z_0) \neq 0$. From this we have $q'(z) = q_1(z) + (z - z_0)q'_1(z)$, whence $q'(z_0) = q_1(z_0) \neq 0$. Thus, if

$$f(z) = \frac{p(z)}{q(z)}$$

has a pole of first order at $z = z_0$, this means that

$$f(z) = \frac{p(z)}{(z - z_0)q_1(z)}$$

and

$$b_1 = \lim_{z \to z_0}(z - z_0)f(z) = \lim_{z \to z_0}\frac{p(z)}{q_1(z)},$$

and, therefore, the residue at a first order pole $z = z_0$ of a rational function $f(z)$ is

$$b_1 = \lim_{z \to z_0}(z - z_0)f(z) = \frac{p(z_0)}{q'(z_0)}.$$

This very simple formula for the residue at a first order pole holds even if the numerator $p(z)$ is not a polynomial, as long as $p(z)$ is analytic at z_0.

What if $q(z)$ has a zero of order 2? Then, by arguments similar to the above, $q(z) = (z - z_0)^2 q_2(z)$, where $q_2(z_0) \neq 0$, and it follows in like manner that $q''(z_0) \neq 0$, while $q'(z_0) = 0$. Thus, near z

$$q(z) = \frac{(z - z_0)^2}{2!}q''(z_0) + \frac{(z - z_0)^3}{3!}q'''(z_0) + \cdots.$$

Hence

$$b_1 = \lim_{z \to z_0} \frac{d}{dz}[(z-z_0)^2 f(z)] = \lim_{z \to z_0} \frac{d}{dz}\left(\frac{p(z)}{\frac{1}{2}q''(z_0) + \frac{1}{6}(z-z_0)q'''(z_0) + \cdots}\right)$$

Using the quotient rule for differentiation, we obtain

$$b_1 = \lim_{z \to z_0} \frac{[\frac{1}{2}q''(z_0) + \cdots]p'(z) - p(z)[\frac{1}{6}q'''(z_0) + \cdots]}{[\frac{1}{2}q''(z_0) + \cdots]^2}$$

$$= \frac{\frac{1}{2}q''(z_0)p'(z_0) - \frac{1}{6}p(z_0)q'''(z_0)}{\frac{1}{4}[q''(z_0)]^2}$$

giving finally that

$$b_1 = \frac{6q''(z_0)p'(z_0) - 2p(z_0)q'''(z_0)}{3[q''(z_0)]^2}.$$

Formulas may be computed similarly for higher-order poles, but they are considerably more complicated, so that further pursuit of such detail becomes tedious and self-defeating.

EXAMPLE 31   Find the residue of

$$f(z) = \frac{e^z}{3z - 1}$$

at the pole $z = \frac{1}{3}$. To do this we use the formula for first-order poles: Here $p(z) = e^z$ and $q(z) = 3z - 1$, so that $q'(z) = 3$, and, therefore,

$$b_1 = \tfrac{1}{3}e^{1/3}.$$

EXAMPLE 32   Find the residue of

$$f(z) = \frac{z - 1}{z^3 - z^2 - 5z - 3}$$

at $z = -1$. Since $z^3 - z^2 - 5z - 3 = (z + 1)^2(z - 3)$, the pole is of order 2. And $p(z) = z - 1$, $p'(z) = 1$, $q'(z) = 3z^2 - 2z - 5$, $q''(z) = 6z - 2$, and $q'''(z) = 6$. Hence, using the formula for poles of order 2,

$$b_1 = \frac{6(-8)1 - 2(-2)(6)}{3(-8)^2} = \frac{-48 + 24}{192} = -\frac{24}{192} = -\frac{1}{8}.$$

Of course even in this example it is much simpler to calculate

$$\lim_{z \to -1} \frac{d}{dz}[(z + 1)^2 f(z)],$$

which is

$$\lim_{z \to -1} \frac{d}{dz}\left(\frac{z-1}{z-3}\right) = \lim_{z \to -1} \frac{d}{dz}\left(1 + \frac{2}{z-3}\right) = -\frac{2}{(z-3)^2}\bigg|_{-1} = -\frac{1}{8}.$$

We are now reasonably well-prepared to find the residues of meromorphic functions at poles; and using these techniques, we can evaluate many integrals of meromorphic functions around closed curves. For, according to the residue theorem, all we need to do to evaluate such an integral is (1) identify the poles which lie inside the curve; (2) find the residues at these poles; (3) add these numbers together and multiply by $2\pi i$.

EXAMPLE 33 Evaluate

$$\oint_C \frac{e^z \, dz}{z^3 - 6z^2 + 11z - 6},$$

where C is the locus of those z for which $|z - 1| + |z - 2| = \frac{3}{2}$. This is an ellipse with foci at $z = 1$ and $z = 2$, and as $q(z) = z^3 - 6z^2 + 11z - 6 = (z - 1)(z - 2)(z - 3)$, the two poles $z = 1$ and $z = 2$ are contained in $I(C)$, while the third pole is not. (The reader should verify this.) Since $q'(z) = 3z^2 - 12z + 11$, we have $q'(1) = 3 - 12 + 11 = 2$ and $q'(2) = 12 - 24 + 11 = -1$. Hence the residue at $z = 1$ is $e/2$ and at $z = 2$ is $e^2/(-1) = -e^2$. Therefore

$$\oint_C \frac{e^z \, dz}{z^3 - 6z^2 + 11z - 6} = 2\pi i \left(\frac{e}{2} - e^2\right).$$

EXAMPLE 34 Evaluate

$$\oint_{|z|=1} \frac{\cosh z}{z^3} \, dz.$$

Since

$$\cosh z = 1 + \frac{z^2}{2!} + \cdots,$$

the residue at $z = 0$ is $\frac{1}{2}$. There are no other poles, because $\cosh z$ is entire; hence the value of the integral is πi.

A. Exercises

Find the poles and the residues at the poles of the following functions:

1 $\dfrac{\sin z}{z^4}$

2 $\dfrac{e^z}{z^n}$ (n positive integer)

3 $\dfrac{z^2 + 3z + 1}{(z^2 - 1)(z - 1)}$

4 $\dfrac{e^z}{\cosh z}$

5 $\left(\dfrac{z - 1}{z + 1}\right)^3$

6 $\dfrac{z^2 + 2z + 2}{(z - 2)^2 (z^2 - 1)}$

7 $\dfrac{z + 2}{z^2 - 3z}$

8 $\dfrac{z^2 + 2z + 1}{(z - 2)^2 (z^2 - 1)}$

9 $\dfrac{e^z}{z^2 + 1}$

10 $\dfrac{z^3 + 1}{(z - 2)(z^2 + 4)}$

B. Problems

Evaluate the following integrals using the residue theorem.

1 $\displaystyle\oint_{|z|=1} \dfrac{\cos z}{z^3}\, dz$

2 $\displaystyle\oint_C \dfrac{z^3 - z + 1}{z^2 (z^2 - 1)}\, dz$, where C is (a) $\{z : |z| = 2\}$; (b) $\{z : |z| = 1/2\}$.

3 $\displaystyle\oint_{|z|=2} \dfrac{e^z (z^2 + 1)\, dz}{z^3 + z^2 + z + 1}$ HINT: $(z - 1)(z^3 + z^2 + z + 1) = z^4 - 1$.

4 $\displaystyle\oint_{|z|=1} \dfrac{\sin \pi z}{z^2}\, dz$

5 $\displaystyle\oint_C \dfrac{(3z^2 + 1)\, dz}{z^2 (z^2 + 4)(z - 2)^2}$, where C is (a) the unit circle;
(b) $\{z : |z| = 3\}$;
(c) $\{z : |z - 2| = 1\}$.

C. Proofs

1 Suppose $f(z)$ is analytic except for an isolated essential singularity at $z = z_0$ and that

$$f(z) = \sum_0^k a_n (z - z_0)^n + \sum_0^\infty b_n (z - z_0)^{-n},$$

that is, the power series portion of the Laurent series for $f(z)$ terminates with the term for $n = k$. Prove that

$$b_1 = \dfrac{1}{(k + 1)!} \lim_{z \to z_0} \dfrac{d^{k+1}}{dz^{k+1}} (z - z_0)^k f\!\left(z_0 + \dfrac{1}{z - z_0}\right).$$

Thus, if either the principal part or the power series part terminates after a finite number of terms, we have a formula for b_1.

2 Let

$$f(z) = \frac{a_{n-1}z^{n-1} + z_{n-2}z^{n-2} + \cdots + a_0}{b_n z^n + b_{n-1}z^{n-1} + \cdots + b_0}.$$

By considering $\int_{|z|=R} f(z)\, dz$ and letting $R \to \infty$, show that the sum of all the residues of $f(z)$ is a_{n-1}/b_n. (Naturally we assume $a_{n-1}b_n \neq 0$).

3.5. APPLICATIONS OF THE CALCULUS OF RESIDUES

Now we are ready to consider three applications of the calculus of residues which will enable us to solve many problems, all concerning *real* variables, which were previously beyond our reach. These are: (1) the evaluation of certain real improper integrals; (2) the evaluation of certain real integrals involving multiple-valued functions; and (3) the summation of certain real infinite series. We shall approach each of these problems by means of examples, after which we shall summarize the techniques used in a more formal theory.

EXAMPLE 35 Evaluate

$$\int_0^\infty \frac{dx}{x^4 + 1}.$$

We do this by a method which we shall call *contour integration*. That is, consider

$$\oint_C \frac{dz}{z^4 + 1},$$

where C is a finite closed curve including a portion of the x axis. We evaluate this complex integral using the calculus of residues, then let the contour

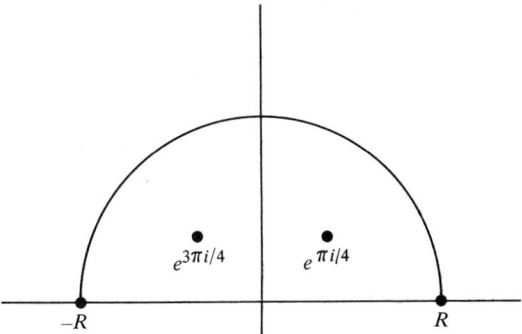

Figure 3.13

increase in size and hope that everything in sight approaches a finite limit. In this case, we shall let C consist of the portion of the real axis from $-R$ to R and of the semicircle $z = Re^{i\theta}$ ($0 \leq \theta \leq \pi$) (Figure 3.13). We take R so large that the two poles of $1/(z^4 + 1)$ in the upper half plane ($e^{\pi i/4}, e^{3\pi i/4}$) are contained in $I(C)$. Then we note that the residues at these poles are

$$\frac{1}{4[\exp(\pi i/4)]^3} = \frac{1}{4}e^{-3\pi i/4},$$

$$\frac{1}{4[\exp(3\pi i/4)]^3} = \frac{1}{4}e^{-9\pi i/4} = \frac{1}{4}e^{-\pi i/4}.$$

Therefore, since

$$e^{-3\pi i/4} = \cos\frac{3\pi}{4} - i\sin\frac{3\pi}{4} = -\frac{\sqrt{2}}{2} - i\frac{\sqrt{2}}{2},$$

$$e^{-\pi i/4} = \cos\frac{\pi}{4} - i\sin\frac{\pi}{4} = \frac{\sqrt{2}}{2} - i\frac{\sqrt{2}}{2},$$

we have

$$\oint_C \frac{dz}{z^4 + 1} = \frac{2\pi i}{4}\left(-\frac{\sqrt{2}}{2} - i\frac{\sqrt{2}}{2} + \frac{\sqrt{2}}{2} - i\frac{\sqrt{2}}{2}\right) = \frac{\pi i(-i\sqrt{2})}{2} = \frac{\sqrt{2}\pi}{2}.$$

On the other hand,

$$\oint_C \frac{dz}{z^4 + 1} = \int_{-R}^{R} \frac{dx}{x^4 + 1} + \int_0^{\pi} \frac{iRe^{i\theta}\,d\theta}{R^4 e^{4i\theta} + 1} = \frac{\sqrt{2}\pi}{2}.$$

In the limit as $R \to \infty$, the integral after the first equality approaches

$$\int_{-\infty}^{\infty} \frac{dx}{x^4 + 1} = 2\int_0^{\infty} \frac{dx}{x^4 + 1},$$

an absolutely convergent real integral; and the second integral is estimated as follows:

$$\left|\int_0^{\pi} \frac{iRe^{i\theta}\,d\theta}{R^4 e^{4i\theta} + 1}\right| \leq \int_0^{\pi} \frac{R\,d\theta}{|R^4 e^{i\theta} + 1|} \leq \frac{R\pi}{R^4 - 1},$$

and since

$$\lim_{R \to \infty} \frac{R\pi}{R^4 - 1} = 0,$$

we obtain

$$\oint_C \frac{dz}{z^4 + 1} = \lim_{R \to \infty} \int_{-R}^{R} \frac{dx}{x^4 + 1} + \int_0^{\pi} \frac{iRe^{i\theta}\,d\theta}{R^4 e^{4i\theta} + 1} = 2\int_0^{\infty} \frac{dx}{x^4 + 1} = \frac{\sqrt{2}\pi}{2}.$$

and as a result

$$\int_0^\infty \frac{dx}{x^4+1} = \frac{\sqrt{2}\pi}{4}.$$

In fact, any real improper integral along the real axis from $-\infty$ to ∞, in which the integrand is a rational function whose denominator has *no real zeros* and whose numerator has degree at least two less than the denominator, can be evaluated simply by adding the residues at the poles in the *upper half plane* and multiplying by $2\pi i$. This conclusion is embodied in the following theorem:

3.5.1. Theorem Let $p(z) = a_n z^n + a_{n-1} z^{n-1} + \cdots + a_0$ and $q(z) = b_m z^m + \cdots + b_0$ be two polynomials with real coefficients and with $n \leq m - 2$, and suppose that $q(z)$ has no real zeros. Then if $c_1, c_2, \ldots, c_k$ are the residues at the poles of $p(z)/q(z)$ in the upper half plane, we have

$$\int_{-\infty}^{\infty} \frac{p(x)\,dx}{q(x)} = 2\pi i(c_1 + c_2 + \cdots + c_k).$$

Proof Proceeding as in example 15.1, we obtain

$$\oint_C \frac{p(z)\,dz}{q(z)} = \int_{-R}^{R} \frac{p(x)\,dx}{q(x)} + \int_0^\pi \frac{p(Re^{i\theta})\,iRe^{i\theta}\,d\theta}{q(Re^{i\theta})} = 2\pi i(c_1 + \cdots + c_k).$$

Now

$$\lim_{R \to \infty} \int_{-R}^{R} \frac{p(x)\,dx}{q(x)} = \int_{-\infty}^{\infty} \frac{p(x)\,dx}{q(x)},$$

which exists as an improper integral because the integrand behaves for large values of x like $1/x^2$; and because the denominator does not vanish, the integrand is finite for every finite x. Furthermore,

$$\left| \int_0^\pi \frac{p(Re^{i\theta})\,iRe^{i\theta}\,d\theta}{q(Re^{i\theta})} \right| \leq \frac{\pi R |a_n R^n + \cdots + a_0|}{|b_m R^m + \cdots + b_0|}$$

$$\leq \frac{\pi R(|a_n| R^n + \cdots + |a_0|)}{|b_m| R^m - |b_{m-1}| R^{m-1} - \cdots - |b_0|},$$

and since $n + 1 < m$, this expression approaches zero as $R \to \infty$. This completes the proof. QED.

Note that $2\pi i(c_1 + c_2 + \cdots + c_k)$ must of necessity turn out to be a real number, since the original integral is real, always assuming, of course, that the coefficients of p and q are real.

168 TAYLOR AND LAURENT SERIES

EXAMPLE 36 Evaluate
$$\int_{-\infty}^{\infty} \frac{dx}{(x^2+1)(x^2+2x+2)}.$$

The integrand satisfies all the conditions of Theorem 3.5.1, so all we need to do is to find the residues of
$$\frac{1}{(z^2+1)(z^2+2z+2)}$$
at the poles $z = i$ and $z = -1 + i$ in the upper half plane. These are
$$\lim_{z \to i} \frac{1}{(z+i)(z^2+2z+2)} = \frac{1}{2i(-1+2i+2)} = \frac{1}{2(-2+i)} = \frac{-2-i}{10},$$
and
$$\lim_{z \to -1+i} \frac{1}{(z+i)(z-i)(z+i+1)} = \frac{1}{(-1+2i)(-1)(2i)} = \frac{1}{2(2+i)} = \frac{2-i}{10}.$$
Hence
$$\int_{-\infty}^{\infty} \frac{dx}{(x^2+1)(x^2+2x+2)} = 2\pi i \left(\frac{-2-i}{10} + \frac{2-i}{10} \right) = 2\pi i \left(\frac{-i}{5} \right) = \frac{2}{5}\pi.$$

We can also evaluate the real integrals involving certain rational functions of the trigonometric functions from 0 to 2π. Let us first illustrate this by means of an example.

EXAMPLE 37 Evaluate $\int_0^{2\pi} d\theta/(2 + \sin \theta)$. To do this, we recall that
$$\sin \theta = \frac{e^{i\theta} - e^{-i\theta}}{2i}.$$

On the unit circle, of course, $z = e^{i\theta}$ and $z^{-1} = e^{-i\theta}$ and $dz = ie^{i\theta} d\theta = iz\, d\theta$. Thus we can consider the given integral to be an integral of a function of z around the unit circle, as follows:
$$\int_0^{2\pi} \frac{d\theta}{2 + \sin \theta} = \oint_{|z|=1} \frac{dz/iz}{2 + (z - z^{-1})/(2i)} = 2 \oint_{|z|=1} \frac{dz}{z^2 + 4iz - 1}.$$

The zeros of the denominator are
$$z = \frac{-4i \pm \sqrt{-16+4}}{2} = -2i \pm \sqrt{3}i.$$

The root $(-2 + \sqrt{3})i$ is contained inside the unit circle, and the residue

3.5. APPLICATIONS OF THE CALCULUS OF RESIDUES

at this pole is

$$\lim_{z \to (-2+\sqrt{3})i} \left(\frac{1}{z + (2 + \sqrt{3})i} \right) = \frac{1}{2\sqrt{3}i} = -\frac{i}{2\sqrt{3}}.$$

The other root is outside the unit circle and makes no contribution. Hence,

$$\int_0^{2\pi} \frac{d\theta}{2 + \sin \theta} = 2\pi i \left(-\frac{i}{2\sqrt{3}} \right) = \frac{\pi}{\sqrt{3}}.$$

Based on the technique used in this last example, we can now formulate a set of rules for evaluating integrals of the form $\int_0^{2\pi} R(\sin \theta, \cos \theta) \, d\theta$, where R is a rational function *whose denominator does not vanish on the path of integration*:

1. Make the substitution $\cos \theta = (z + z^{-1})/2$, $\sin \theta = (z - z^{-1})/2i$, $d\theta = dz/iz$, that is, consider

$$\oint_{|z|=1} R\left(\frac{(z - z^{-1})}{2i}, \frac{(z + z^{-1})}{2} \right) \frac{dz}{iz}.$$

2. After simplifying the resulting expression, calculate the residues $c_1, c_2, \ldots, c_k$ at the poles inside the unit circle.

3. The integral will then equal $2\pi i (c_1 + c_2 + \cdots + c_k)$.

EXAMPLE 38 Evaluate

$$\int_0^{2\pi} \frac{\cos 2\theta \, d\theta}{3 - 2 \cos 2\theta}.$$

We obtain

$$\int_0^{2\pi} \frac{\cos 2\theta \, d\theta}{3 - 2 \cos 2\theta} = \oint_{|z|=1} \frac{\frac{1}{2}(z^2 + z^{-2})(dz/iz)}{3 - (z^2 + z^{-2})} = \frac{-1}{(2i)} \oint_{|z|=1} \frac{(z^4 + 1) \, dz}{z^5 - 3z^3 + z}$$

$$= \frac{-1}{(2i)} \oint_{|z|=1} \frac{(z^4 + 1) \, dz}{z(z^4 - 3z^2 + 1)}.$$

The integrand has a simple pole at $z = 0$ with residue 1, and the factor $z^4 - 3z^2 + 1$ yields

$$z^2 = \frac{3 \pm \sqrt{9 - 4}}{2} = \frac{3 \pm \sqrt{5}}{2}$$

of which only $\frac{1}{2}(3 - \sqrt{5})$ is contained in the unit circle, so that $z = \pm \sqrt{(3 - \sqrt{5})/2}$ are first-order poles inside the unit circle. Their residues are

$$\lim_{z \to \alpha} \frac{z^4 + 1}{z(z^2 - \beta^2)(z + \alpha)}, \quad \lim_{z \to -\alpha} \frac{z^4 + 1}{z(z^2 - \beta^2)(z - \alpha)},$$

where $\alpha = \sqrt{(3 - \sqrt{5})/2}$ and $\beta = -\sqrt{(3 - \sqrt{5})/2}$; upon carrying out the details, both values turn out to be $-3\sqrt{5}/10$. Hence,

$$\int_0^{2\pi} \frac{\cos 2\theta \, d\theta}{3 - 2\cos 2\theta} = -\frac{2\pi i}{2i}\left(1 - \frac{3\sqrt{5}}{5}\right) = \frac{\pi(5 - 3\sqrt{5})}{5}.$$

A third type of real integral which can be evaluated by similar methods is one of the form $\int_{-\infty}^{\infty} R(x)\cos \beta x \, dx$ or $\int_{-\infty}^{\infty} R(x)\sin \beta x \, dx$, where $\beta > 0$ where $R(x)$ is a rational function with real coefficients having no real poles and such that its denominator exceeds its numerator in degree by at least one.* We evaluate integrals of this sort by means of contour integration involving the function $R(z)e^{i\beta z}$. When we obtain the result, we then take either the real or the imaginary part, according to the cosine or the sine appearing in the original integral. The following theorem assures us that the integral over the large semicircle of Figure 3.13 goes to zero:

3.5.2. Theorem   Let

$$R(z) = \frac{a_n z^n + \cdots + a_0}{b_m z^m + \cdots + b_0}$$

with $m \geq n + 1$. If $\beta > 0$, then

$$\lim_{R \to \infty} \int_C R(z)e^{i\beta z} \, dz = 0,$$

where C is $\{z : z = Re^{i\theta}, 0 \leq \theta \leq \pi\}$,

Proof On C,

$$|R(z)| = \left|\frac{a_n R^n e^{in\theta} + \cdots + a_0}{b_m R^m e^{im\theta} + \cdots + b_0}\right| \leq \frac{|a_n|R^n + \cdots + |a_0|}{|b_m|R^m - \cdots - |b_0|}$$

and

$$\exp[i\beta z] = \exp[i\beta Re^{i\theta}] = \exp[i\beta R(\cos\theta + i\sin\theta)]$$
$$= \exp[i\beta R\cos\theta]\exp[-\beta R\sin\theta],$$

so

$$|e^{i\beta z}| = e^{-\beta R \sin\theta}$$

---

* The restriction $\beta > 0$ is actually no handicap at all, since $\cos(\beta x) = \cos(-\beta x)$ and $\sin(\beta x) = -\sin(-\beta x)$.

3.5. APPLICATIONS OF THE CALCULUS OF RESIDUES

Now $\sin \theta \geq 2\theta/\pi$ for $0 \leq \theta \leq \pi/2$, since the sine curve is concave downwards and lies above the line connecting $(0, 0)$ and $(\pi/2, 1)$ (Figure 3.14). So

$$|e^{i\beta z}| = e^{-\beta R \sin \theta} \leq e^{-2\beta R \theta/\pi}$$

and

$$\left| \int_C R(z) \exp[i\beta z] \, dz \right| \leq \frac{|a_n| R^n + \cdots + |a_0|}{|b_m| R^m - \cdots - |b_0|} \left| \int_C \exp[i\beta R e^{i\theta}] \, iR \exp[i\theta] \, d\theta \right|$$

$$\leq \frac{|a_n| R^n + \cdots + |a_0|}{|b_m| R^m - \cdots - |b_0|} R \int_0^\pi \exp[-\beta R \sin \theta] \, d\theta.$$

By using this, we have

$$\int_0^\pi e^{-\beta R \sin \theta} \, d\theta = \int_0^{\pi/2} e^{-\beta R \sin \theta} \, d\theta + \int_{\pi/2}^\pi e^{-\beta R \sin \theta} \, d\theta$$

$$= \int_0^{\pi/2} e^{-\beta R \sin \theta} \, d\theta + \int_0^{\pi/2} e^{-\beta R \sin(\theta + \pi/2)} \, d\theta$$

$$= 2 \int_0^{\pi/2} e^{-\beta R \sin \theta} \, d\theta \leq 2 \int_0^{\pi/2} e^{-2\beta R \theta/\pi} \, d\theta$$

$$= \pi/(\beta R) [1 - e^{-\beta R}].$$

So finally,

$$\left| \int_C R(z) e^{i\beta z} \, dz \right| \leq \frac{\pi}{\beta} (1 - e^{-\beta R}) \frac{|a_n| R^{n+1} + \cdots + |a_0| R}{|b_m| R^{m+1} - \cdots - |b_0| R},$$

and since by hypothesis $m > n$, the limit of this quantity as $R \to \infty$ is 0.

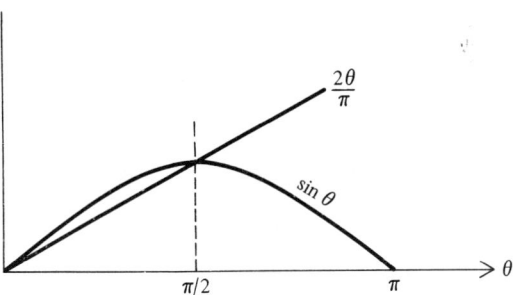

Figure 3.14

172 TAYLOR AND LAURENT SERIES

EXAMPLE 39 Evaluate $\int_{-\infty}^{\infty} \cos 2x\, dx/(1 + x^2)$. By the methods of contour integration, using the contour of Figure 3.13, we have

$$\oint_C \frac{e^{2iz}\, dz}{1 + z^2} = 2\pi i\, \frac{e^{-2}}{2i} = \pi e^{-2},$$

since the residue of

$$\frac{e^{2iz}}{1 + z^2}$$

at the pole $z = i$ is $e^{-2}/2i$. But

$$\oint_C \frac{e^{2iz}\, dz}{1 + z^2} = \int_{-R}^{R} \frac{e^{2ix}\, dx}{1 + x^2} + \int_0^\pi \frac{e^{2iRe^{i\theta}} iRe^{i\theta}\, d\theta}{R^2 e^{2i\theta} + 1}.$$

By Theorem 3.5.2, the last integral goes to zero as $R \to \infty$. Hence

$$\int_{-\infty}^{\infty} \frac{\cos 2x\, dx}{x^2 + 1} = \operatorname{Re} \int_{-\infty}^{\infty} \frac{e^{2ix}\, dx}{x^2 + 2} = \pi e^{-2}.$$

Sometimes similar integrals can be evaluated by slightly altering the contour of Figure 3.13; an important example of this is the following:

EXAMPLE 40 Evaluate $\int_0^\infty (\sin x/x)\, dx$. We note that since $\sin x/x$ is even, $\int_0^\infty (\sin x/x)\, dx = \frac{1}{2} \int_{-\infty}^{\infty} (\sin x/x)\, dx$. However, when we look at $\oint_C (e^{iz}/z)\, dz$ around the contour of Figure 3.13, we notice that the integrand has a pole at the point $z = 0$ on the path of integration. To avoid this difficulty we modify the contour by adding a small semicircle of radius δ around the origin (Figure 3.15). Thus C consists of the real axis from $-R$ to $-\delta$ and from δ to R, the semicircle $z = \delta e^{i\theta}$ (θ running from π to 0), and the semicircle $z = Re^{i\theta}$ ($0 \leq \theta \leq \pi$). We shall use this contour and then let $\delta \to 0$ and $R \to \infty$. We have

$$\oint_C \frac{e^{iz}\, dz}{z} = \left(\int_{-R}^{-\delta} + \int_{\delta}^{R} \right) \frac{e^{ix}\, dx}{x} + \int_\pi^0 \frac{e^{i\delta e^{i\theta}} i\delta e^{i\theta}\, d\theta}{\delta e^{i\theta}}$$

$$+ \int_0^\pi \frac{e^{iRe^{i\theta}} iRe^{i\theta}\, d\theta}{Re^{i\theta}} = 0,$$

since the integrand has no poles in $I(C)$. The integral over the large semicircle goes to zero as $R \to \infty$, by Theorem 3.5.2. As for the one over the small semicircle,

$$\int_\pi^0 \frac{e^{i\delta e^{i\theta}} i\delta e^{i\theta}\, d\theta}{\delta e^{i\theta}} = i \int_\pi^0 e^{i\delta e^{i\theta}}\, d\theta.$$

Since, as $\delta \to 0$, the integrand of this last integral approaches 1, we might

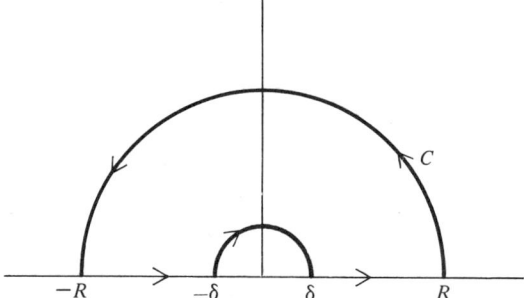

Figure 3.15

suspect that

$$\lim_{\delta \to 0} i \int_\pi^0 e^{i\delta e^{i\theta}} \, d\theta = -\pi i.$$

In fact, this is true, since

$$i \int_\pi^0 e^{i\delta e^{i\theta}} \, d\theta + i\pi = i \left(\int_\pi^0 e^{i\delta e^{i\theta}} \, d\theta - \int_\pi^0 d\theta \right) = i \left(\int_\pi^0 (e^{i\delta e^{i\theta}} - 1) \, d\theta \right).$$

Now

$$e^{i\delta e^{i\theta}} = e^{i\delta(\cos\theta + i\sin\theta)} = e^{i\delta\cos\theta} e^{-\delta\sin\theta} = e^{-\delta\sin\theta}[\cos(\delta\cos\theta) + i\sin(\delta\cos\theta)]$$

and, hence,

$$e^{i\delta e^{i\theta}} - 1 = [e^{-\delta\sin\theta}\cos(\delta\cos\theta) - 1] + ie^{-\delta\sin\theta}\sin(\delta\cos\theta).$$

The modulus of this quantity goes to zero uniformly as $\delta \to 0$, since

$$\left|e^{i\delta e^{i\theta}} - 1\right| = e^{-\delta\sin\theta} \sqrt{[\cos(\delta\cos\theta) - 1]^2 + \sin^2(\delta\cos\theta)}$$

$$= e^{-\delta\sin\theta} \sqrt{2 - 2\cos(\delta\cos\theta)}$$

$$= e^{-\delta\sin\theta} 2 \left|\sin[(\delta\cos\theta)/2]\right|.$$

This can easily be estimated now, since $\sin\theta \geq -1$ and, hence,

$$e^{-\delta\sin\theta} \leq e^{\delta};$$

furthermore, since $|\sin x| \leq x$ for real x, we have

$$\left|\sin[(\delta\cos\theta)/2]\right| \leq \tfrac{1}{2}\delta|\cos\theta| \leq \tfrac{1}{2}\delta.$$

Therefore,

$$\left|e^{i\delta e^{i\theta}} - 1\right| \leq \tfrac{1}{2}\delta e^{\delta},$$

an estimate which does not depend on θ and which approaches zero as

$\delta \to 0$. Hence

$$\lim_{\delta \to 0} i \int_\pi^0 e^{i\delta e^{i\theta}} \, d\theta = -\pi i$$

and, therefore, by taking imaginary parts this time, we have

$$\int_0^\infty \frac{\sin x}{x} \, dx = \frac{\pi}{2}.*$$

The next item on our program is to consider a somewhat more difficult task: the evaluation of real integrals involving functions whose complex counterparts turn out to be multiple-valued.

EXAMPLE 41 Let us attempt to evaluate the integral

$$\int_0^\infty \frac{x^{-\alpha} \, dx}{1 + x}, \qquad 0 < \alpha < 1.$$

(As we shall see shortly, the existence of this integral is a routine matter.) We encounter a number of obstacles in attempting to evaluate this integral. The integrand is not an even function, so a contour like that in Figure 3.15 will not work; besides, the zero of the denominator at $x = -1$ is on the path of integration. Furthermore, the complex function $z^{-\alpha}/(1 + z)$ is multiple-valued, so whatever contour we use, it must not completely encircle the origin. Therefore let us try the contour C illustrated in Figure 3.16.

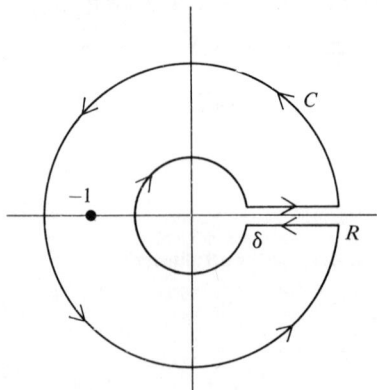

Figure 3.16

---

* This integral is not, however, absolutely convergent. See R. C. Buck, *Advanced Calculus*, 2nd Ed., McGraw-Hill, New York, 1965 (pp. 141–142).

3.5. Applications of the Calculus of Residues

We integrate along a segment above the real axis from $\delta + i\varepsilon$ ($\varepsilon > 0$, small) to $R + i\varepsilon$, then around the circle to radius R to $R - i\varepsilon$, then from $R - i\varepsilon$ to $\delta - i\varepsilon$, then counterclockwise around the circle of radius δ to $\delta + i\varepsilon$, our starting point. We choose $\delta < 1 < R$ so that the pole of the integrand at -1 is inside the contour. Inside this contour, it is possible to define a single-valued branch of $z^{-\alpha}$ and we do this by setting $\arg(-1) = \pi$ so that at $z = -1$,

$$z^{-\alpha} = e^{-\alpha \log z} = e^{-\alpha \pi i}.$$

The residue of the integrand at the pole $z = -1$ is then simply $e^{-\alpha \pi i}$, so that

$$\oint_C \frac{z^{-\alpha} \, dz}{1 + z} = 2\pi i e^{-\pi i \alpha}.$$

We then do as we often have before and let $\varepsilon \to 0$. The two straight line segments approach the segment of the real axis between δ and R, and we have*

$$\lim_{\varepsilon \to 0} \int_{\delta + i\varepsilon}^{R + i\varepsilon} \frac{z^{-\alpha}}{1+z} \, dz = \int_\delta^R \frac{x^{-\alpha}}{1+x} \, dx$$

and

$$\lim_{\varepsilon \to 0} \int_{R - i\varepsilon}^{\delta - i\varepsilon} \frac{z^{-\alpha}}{1+z} \, dz = \int_R^\delta \frac{x^{-\alpha} e^{-2\pi i \alpha}}{1+x} \, dx,$$

where $x^{-\alpha}$ represents the unique positive real $-\alpha$ power of the positive real number x. Note that the two integrals will not cancel each other, because the argument of z approaches zero from above the axis and 2π from below. Thus

$$\lim_{\varepsilon \to 0} \oint_C \frac{z^{-\alpha} \, dx}{1+z} = \int_\delta^R \frac{x^{-\alpha} \, dx}{1+x} + \int_R^\delta \frac{x^{-\alpha} e^{-2\pi i \alpha} \, dx}{1+x} + \int_0^{2\pi} \frac{R^{-\alpha} e^{-i\alpha\theta} i R e^{i\theta} \, d\theta}{1 + Re^{i\theta}}$$

$$+ \int_{2\pi}^0 \frac{\delta^{-\alpha} e^{-i\alpha\theta} i \delta e^{i\theta} \, d\theta}{\delta e^{i\theta} + 1} = 2\pi i e^{-\pi i \alpha}.$$

Since, on the two circles, $z = Re^{i\theta}$ and $z = \delta e^{i\theta}$, respectively. The first two terms to the right of the first equal sign have the limit

$$\lim_{\substack{\delta \to 0 \\ R \to 0}} (1 - e^{-2\pi i \alpha}) \int_\delta^R \frac{x^{-\alpha} \, dx}{1+x} = (1 - e^{-2\pi i \alpha}) \int_0^\infty \frac{x^{-\alpha} \, dx}{1+x},$$

which is a convergent improper integral, since at the lower limit the integrand behaves like $x^{-\alpha}$ and at the upper limit like $x^{-1-\alpha}$, both of which

---

* We have not proved this (see Proof 2, Section 3.5) but it seems plausible!

integrate to functions having finite limits. For the third integral,

$$\left| \int_0^{2\pi} \frac{R^{-\alpha} e^{-i\alpha\theta} iRe^{i\theta} \, d\theta}{Re^{i\theta} + 1} \right| \leq R^{-\alpha} \frac{R}{R-1} 2\pi,$$

which goes to zero as $R \to \infty$. Similarly

$$\left| \int_{2\pi}^0 \frac{\delta^{-\alpha} e^{-i\alpha\theta} i\delta e^{i\theta} \, d\theta}{\delta e^{i\theta} + 1} \right| \leq \frac{\delta^{1-\alpha}}{1-\delta} 2\pi,$$

which goes to zero as $\delta \to 0$. Hence

$$(1 - e^{-2\pi i\alpha}) \int_0^\infty \frac{x^{-\alpha} \, dx}{1 + x} = 2\pi i (e^{-\pi i\alpha}),$$

so that

$$\int_0^\infty \frac{x^{-\alpha} \, dx}{1 + x} = \frac{2\pi i (e^{-\pi i\alpha})}{1 - e^{-2\pi i\alpha}};$$

and, since

$$\frac{2\pi i e^{-\pi i\alpha}}{1 - e^{-2\pi i\alpha}} = \pi \bigg/ \left(\frac{e^{\pi i\alpha} - e^{-\pi i\alpha}}{2i} \right) = \frac{\pi}{\sin \pi \alpha},$$

we finally have

$$\int_0^\infty \frac{x^{-\alpha} \, dx}{1 + x} = \frac{\pi}{\sin \pi \alpha} \qquad (0 < \alpha < 1).$$

The procedure with an integral of this sort, then, is to encircle the branchpoint by two circles, one of large radius R and one of small radius δ, and to proceed one way along the "top" of the relevant portion of the axis and the opposite way along the "bottom," then take the limit as $\delta \to 0$ and $R \to \infty$. The integrals along the axis do not cancel because of the multiple-valued behavior of the function. It may also be necessary to indent the line along the axis if the axis contains a singularity of the function (see the problems at the end of this section).

Finally, let us consider applications of the calculus of residues to the summation of certain infinite series. Suppose, for instance, that we wish to find the sum of the series

$$\sum_{n=-\infty}^\infty f(n),$$

which we assume, of course, is a convergent real series. At first glance it does not appear that this problem has anything to do with functions of a complex variable. However, suppose that $f(z)$, if considered as a function of a *complex* variable, is a meromorphic function, defined at each integer

3.5. APPLICATIONS OF THE CALCULUS OF RESIDUES

n, and decreasing rapidly enough at the integers so that $\sum f(n)$ is a convergent series of real terms.

If we could find a meromorphic function $g(z)$ with first-order poles at the integers n and with residue 1 at each pole, then by the calculus of residues we would have

$$\frac{1}{2\pi i}\oint_{C_N} g(z) f(z) dz = \sum_{-N}^{N} f(n),$$

where C_N is any simple closed contour containing the integers from $-N$ to $+N$ in its interior. In general, it turns out, we cannot do quite this well, because, since f is meromorphic, the product $g(z) f(z)$ may have poles in C_N at points other than $0, \pm 1, \ldots, \pm N$. Because of this, we shall instead obtain an expression of the form

$$\frac{1}{2\pi i}\oint_{C_N} g(z) f(z) dz = \sum_{-N}^{N} f(n) + d_1 + d_2 + \cdots + d_k,$$

where $d_1, \ldots, d_k$ are the residues of gf at the *other* poles inside C_N.

Therefore in our attempts to evaluate the sum $\sum f(n)$, in addition to the fact that $f(n)$ approaches zero as $n \to \infty$ rapidly enough for the series to converge, we shall also assume that f has at most a *finite* number of poles. Under this assumption, C_N could be chosen large enough to contain all the poles. Then, if we can choose C_N in such a manner that

$$\lim_{N\to\infty}\oint_{C_N} g(z) f(z) dz = 0,$$

we shall have summed the series $\sum f(n)$, since

$$\sum_{-\infty}^{\infty} f(n) = -d_1 - d_2 - \cdots - d_k.$$

As a first step in solving this problem, we note that $\pi \cos \pi z = \pi \cos \pi z / \sin \pi z$ has simple poles at the integers, and that the residue at each of these poles is

$$\pi \cos \pi n \lim_{z\to n} \frac{(z-n)}{\sin \pi z} = \frac{\pi \cos \pi n}{\cos \pi n} = 1.$$

Hence $\pi \cot \pi z$ is a good candidate for our function $g(z)$.

Next, for our contour C_N, we choose a square with sides parallel to the axes and passing through the points $\pm(N + \frac{1}{2}), \pm(N + \frac{1}{2})i$ (Figure 3.17). Now, since we want $\oint_{C_N} g(z) f(z) dz$ to go to zero as $N \to \infty$, let us examine the behavior of $\pi \cot \pi z$, our choice for $g(z)$, on this contour.

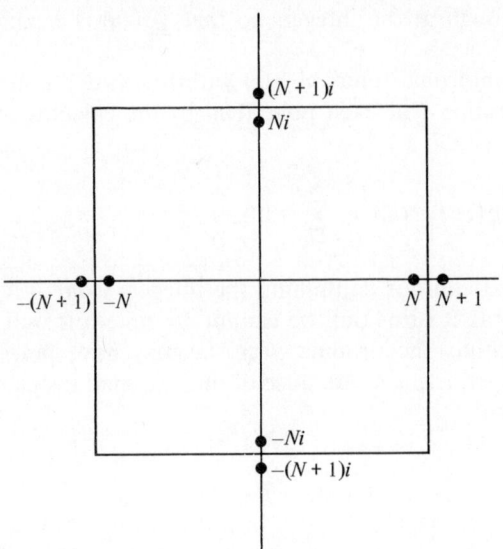

Figure 3.17

First, on the upper portion where $y > \frac{1}{2}$, we have

$$|\cot \pi z| = \left|\frac{e^{\pi i z} + e^{-\pi i z}}{e^{\pi i z} - e^{-\pi i z}}\right| \leq \frac{|e^{\pi i z}| + |e^{-\pi i z}|}{||e^{\pi i z}| - |e^{-\pi i z}||}$$

$$= \frac{e^{-\pi y} + e^{\pi y}}{e^{\pi y} - e^{-\pi y}} = \frac{1 + e^{-2\pi y}}{1 - e^{-2\pi y}} \leq \frac{1 + e^{-\pi}}{1 - e^{-\pi}},$$

so that $\pi \cot \pi z$ is bounded by a constant independent of N on this part of the square. The same bound holds for $y < -\frac{1}{2}$, as can be shown by a similar calculation. For $-\frac{1}{2} \leq y \leq \frac{1}{2}$, there are two possibilities: $z = N + \frac{1}{2} + iy$ and $z = -N - \frac{1}{2} + iy$. Consider the first one:

$$|\cot \pi z| = |\cot \pi(N + \tfrac{1}{2} + iy)| = \left|\cot\left(\frac{\pi}{2} + \pi i y\right)\right| = |\tanh \pi y| \leq \tanh\frac{\pi}{2},$$

a constant again independent of N, thanks to the periodicity of $\cot \pi z$. The same bound holds for $z = -N - \frac{1}{2} + iy$, so we finally have that on C_N there exists a constant M independent of N so that

$$|\pi \cot \pi z| \leq M$$

for all $z \in C_N$. Since the length of C_N is $8N + 4$, we have

$$\left|\oint_{C_N} \pi \cot \pi z f(z)\, dz\right| \leq M(8N + 4) \max_{z \in C_N} |f(z)|.$$

3.5. APPLICATIONS OF THE CALCULUS OF RESIDUES

For this quantity to approach zero as $N \to \infty$, $|f(z)|$ must go to zero fast enough to overtake the growth of the factor $(8N + 4)$. In particular, if there is a constant K, such that

$$\max_{z \in C_N} |f(z)| \leq \frac{K}{N^{\alpha+1}}$$

for some $\alpha > 0$, then the integral does go to zero. It is not unreasonable to demand this behavior of f, since after all $\sum f(n)$ must converge. We sum this up as the following:

3.5.3. Theorem If $f(z)$ is meromorphic in the plane with a finite number of poles (none of them at an integer), if

$$\sum_{-\infty}^{\infty} f(n)$$

is a convergent series, and if there exist positive constants K and α so that

$$\max_{z \in C_N} |f(z)| \leq \frac{K}{N^{\alpha+1}},$$

then

$$\sum_{-\infty}^{\infty} f(n) = -(d_1 + d_2 + \cdots + d_k),$$

where the d_i are the residues of $\pi \cot \pi z f(z)$ at the poles of $f(z)$.

Let us work an example illustrating this.

EXAMPLE 42 Find the value of

$$\sum_{1}^{\infty} \frac{1}{n^2}.$$

This sum is

$$\frac{1}{2} \sum_{-\infty}^{\infty}{}' \frac{1}{n^2},$$

where the prime on the summation sign means that the term for $n = 0$ is missing. Since $f(z) = 1/z^2$ is the simplest choice, we have

$$|f(z)| \leq \frac{\sqrt{2}}{(N + 1/2)^2} \leq \frac{\sqrt{2}}{N^2},$$

on C_N, a bound of the required form. Now $\pi \cot \pi z f(z) = (\pi \cot \pi z)/z^2$ has a pole at the origin, for which we must find the residue. We see that

$$\frac{\pi \cot \pi z}{z^2} = \frac{\pi \cos \pi z}{z^2 \sin \pi z} = \frac{\pi}{z^2} \frac{[1 - (\pi^2 z^2/2!) + \cdots]}{[\pi z - (\pi^3 z^3/3!) + \cdots]} = \frac{1}{z^3}\left(1 - \frac{\pi^2 z^2}{3!} + \cdots\right)$$

from which the residue at the origin is $-\pi^2/3$. Hence

$$\sum_{-\infty}^{\infty}{}' \frac{1}{n^2} = \frac{\pi^2}{3}$$

(where the prime means that the $n = 0$ term is omitted) and, therefore,

$$\sum_{1}^{\infty} \frac{1}{n^2} = \frac{\pi^2}{6}.$$

A. Exercises

1 Evaluate the following integrals:

a $\quad \displaystyle\int_{-\infty}^{\infty} \frac{x^2 \, dx}{(x^2 + 1)(x^2 + x + 1)}$

b $\quad \displaystyle\int_{-\infty}^{\infty} \frac{dx}{(x^2 + a^2)(x^2 + b^2)}, a > b > 0$

c $\quad \displaystyle\int_{-\infty}^{\infty} \frac{(x + 1) \, dx}{x^4 + 1}$

d $\quad \displaystyle\int_{0}^{\infty} \frac{dx}{x^4 + x^2 + 1}$

e $\quad \displaystyle\int_{-\infty}^{\infty} \frac{dx}{(x^2 + x + 2)^2}$

f $\quad \displaystyle\int_{0}^{\infty} \frac{dx}{(x^2 + 1)^3}$

2 Evaluate the following:

a $\quad \displaystyle\int_{0}^{2\pi} \frac{d\theta}{a + b \sin \theta}, \quad a > |b|.$

b $\quad \displaystyle\int_{0}^{\infty} \frac{\sin^2 x}{x^2} \, dx$

c $\quad \displaystyle\int_{0}^{\infty} \frac{\cos x \, dx}{x^2 + 1}$

d $\quad \displaystyle\int_{0}^{2\pi} \frac{d\theta}{a + b \sin \theta + c \cos \theta}$

$\qquad\qquad\qquad\qquad\qquad\qquad (a^2 > b^2 + c^2)$

e $\quad \displaystyle\int_{0}^{\infty} \frac{\sin \alpha x \, dx}{x} \, (\alpha > 0)$

f $\quad \displaystyle\int_{-\infty}^{\infty} \frac{\sin x \, dx}{x^2 + 2x + 2}$

B. Problems

1 Sum the following:

a $\quad \displaystyle\sum_{1}^{\infty} \frac{1}{(n^2 + 1)^2}$

b $\quad \displaystyle\sum_{1}^{\infty} \frac{1}{n^4}$

c $\sum_{1}^{\infty} \dfrac{1}{n^6}$

d $\sum_{1}^{\infty} \dfrac{(-1)^n}{n^2}$ HINT: Instead of $\pi \cot \pi z$ use $\pi \csc \pi z$.

e $\sum_{1}^{\infty} \dfrac{1}{n^2 + a^2}$

f $\sum_{1}^{\infty} \dfrac{\sin^2 n\theta}{n^2 + a^2}$ $(0 < \theta < \pi)$

2 Evaluate the following:

a $\displaystyle\int_0^{\infty} \dfrac{\log x\, dx}{x^2 + a^2}$

b $\displaystyle\int_0^{\infty} \dfrac{\log(1+x)\, dx}{1 + x^2}$

c $\displaystyle\int_0^{\infty} \dfrac{x^{-\alpha}\, dx}{x^2 + 2x \cos \beta + 1}$, $0 < |\alpha| < 1$, $0 < |\beta| < \pi$

C. Proofs

1 Show that problems involving alternating series, as Problem 1d above, can be solved by using $\pi \csc \pi z$ by following through the same arguments as in the text.

2 Prove that
$$\lim_{\varepsilon \to 0} \int_{\delta + i\varepsilon}^{R + i\varepsilon} \dfrac{(x + i\varepsilon)^{-\alpha}}{1 + x + i\varepsilon}\, dx = \int_{\delta}^{R} \dfrac{x^{-\alpha}}{1 + x}\, dx.$$
HINT: Integrate $z^{-\alpha}/(1 + z)$ around the rectangle with vertices δ, R, $R + i\varepsilon$, $\delta + i\varepsilon$.

3 Suppose f is analytic in the sector $0 \leq r < \infty$, $0 \leq \theta \leq \alpha$, and that there exist constants $K > 0$, $k > 0$, $\delta > 1$, $\beta > -1$, such that $|f(z)| \leq Mz^{-\delta}$ for $|z| > K$ and $|f(z)| \leq Mz^{\beta}$ for $|z| < k$, where M is positive and real. Prove that $\int_0^{\infty} f(z)\, dz$ (along any ray from 0 to ∞ in the sector) is well-defined, i.e., is independent of the particular ray along which the integration proceeds.

Chapter **4**

The Elements of Conformal Mapping

The study called "conformal mapping" is as close as we can get to a graphical study of the geometry of complex functions. In order to graph complex functions, one would need four dimensions for the coordinates (x, y, u, v), which is in no way possible. So instead, we study the w plane in order to ascertain what sort of distortion the function f works in taking the z plane into the w plane. And, just as analytic geometry has its applications in the solution of many problems of mathematical physics, so also does conformal mapping. We shall consider one elegant application, to two-dimensional fluid flow, in Chapter 5.

4.1. ANALYTIC AND UNIVALENT MAPPINGS

A function f from the z plane to the w plane is frequently called a *mapping*. This term is used because the range of f is a "map" of the domain of f in much the same way as a page in an atlas is a map of a portion of the earth's surface; the analogy makes even more sense if f is one-to-one, so that no part of the range is a map of more than one part of the domain. It turns out that the distribution of zeros of an analytic function (and its derivative) is crucial in this study.

184 THE ELEMENTS OF CONFORMAL MAPPING

4.1.1. Definition If f is defined in a domain D and if $z_0 \in D$ is such that $f(z_0) = 0$, then z_0 is said to be an *isolated zero* of f if there exists a real number $\delta < 0$ so that the entire interior of the circle $\Sigma = \{z : |z - z_0| = \delta\}$ is contained in D, and so that $f(z) \neq 0$ for $z \in I(\Sigma)$, $z \neq z_0$.

The concept of an isolated zero is thus quite analogous to the concept of an isolated singularity—there is some small circle around the isolated zero which contains no other zero of f.

4.1.2. Lemma If f is analytic in a domain D, then either f is identically zero or else all its zeros in D are isolated.

Proof Let us assume, contrary to the lemma, that f is not identically zero, but that it possesses a nonisolated zero $z_0 \in D$. If we consider a circle $\Sigma_1 : \{z : |z - z_0| = \delta\}$ with δ small enough so that $\Sigma_1 \cup I(\Sigma_1) \subset D$, then there must be another zero $z_1 \neq z_0$ of f in Σ_1. In $\Sigma_2 = \{z : |z - z_0| = \delta/2\}$, there is a zero $z_2 \neq z_0$ of f. In general, $\Sigma_n = \{z : |z - z_0| < \delta/2^{n-1}\}$ contains a zero $z_n \neq z_0$ of $f(z)$. Furthermore, by the definition of a convergent sequence, the sequence $\{z_1, z_2, z_3, \ldots\}$ converges to z_0. Now since $f(z)$ is analytic, it possesses a Taylor series expansion $f(z) = \sum_{n=0}^{\infty} a_n (z - z_0)^n$ around z_0 which converges uniformly in $\Sigma_1 \cup I(\Sigma_1)$. Since $f(z)$ is not identically zero, only at most a finite number of the first coefficients, say $a_0, a_1, \ldots, a_n$, can vanish, so that

$$f(z) = a_{n+1}(z - z_0)^{n+1} + a_{n+2}(z - z_0)^{n+2} + \cdots$$
$$= (z - z_0)^{n+1}[a_{n+1} + a_{n+2}(z - z_0) + \ldots],$$

where $a_{n+1} \neq 0$. Let us denote by $\phi(z)$ the function in the square brackets. Since this series converges uniformly in $\Sigma_1 \cup I(\Sigma_1)$, and represents an analytic and continuous function there, we can find a real $\varepsilon > 0$ so small that

$$|\phi(t) - a_{n+1}| < \tfrac{1}{2}|a_{n+1}|$$

for all z inside $\Gamma = \{z : |z - z_0| = \varepsilon\}$. This simply says that the tail end of the series for $\phi(z)$ gets arbitrarily close to zero as z gets close to z_0. Now, since $f(z_k) = (z_k - z_0)^{n+1} \phi(z_k)$ and $z_k - z_0 \neq 0$, we must have $\phi(z_k) = 0$ for all k. But for sufficiently large k, we can find a z_k inside Γ; simply take k so large that $\delta/2^{k-1} < \varepsilon$. We then have

$$|\phi(z_k) - a_{n+1}| = |0 - a_{n+1}| = |a_{n+1}| < \tfrac{1}{2}|a_{n+1}|.$$

This is a manifest contradiction, since $a_{n+1} \neq 0$. Hence, one must conclude that since $f(z)$ is not identically zero, it is impossible for $f(z)$ to have a non-isolated zero at z_0. QED.

This lemma will be quite important when we begin to study what an analytic function does to a curve C in the z plane, i.e., what the image under

f of this curve becomes in the w plane. But first we need to specify exactly what sort of analytic functions we are talking about.

4.1.3. Definition Let f be an analytic function in a domain D, which has the property that if $f(z_1) = f(z_2)$, then $z_1 = z_2$. Then $f(z)$ is said to be a *univalent (or schlicht) function in D*.

In other words, a univalent function assumes each value in its range precisely once. In the calculus of real variables, f would be called *one-to-one* or *injective*. Univalent functions, or at least analytic functions which are univalent in certain domains, will be quite important in the study of mappings produced by analytic functions.

EXAMPLE 1 Clearly the functions $f_1(z) = z$ and $f_2(z) = 1/z$ are univalent in the entire complex plane and in the plane exclusive of the origin, respectively. On the other hand, no other integer power of z is univalent. For example, z^2 is clearly *not* univalent, since $(-a)^2 = a^2$ for any nonzero value of a. Similarly, z^k is "k-valent" because the statement that $z_1^k = z_2^k$ implies only that $z_1 = z_1 e^{2n\pi i/k}$, where n can be 0, 1, 2, …, $k-1$. In short, there are k different points in the plane where z^k takes on each nonzero value.

EXAMPLE 2 $f(z) = e^z$ is not univalent in the whole plane, since it is periodic with period $2\pi i$, that is, $e^z = e^{z+2\pi i}$. But in the infinite strip in which $0 \leq \operatorname{Im} z < 2\pi$, for instance, e^z is univalent, since $e^{z_1} = e^{z_2}$ implies $e^{x_1}(\cos y_1 + i \sin y_1) = e^{x_2}(\cos y_1 + i \sin y_2)$, and since the two moduli must be equal, we have $e^{x_1} = e^{x_2}$ and, thus, $x_1 = x_2$; and $\cos y_1 = \cos y_2$ and $\sin y_1 = \sin y_2$ implies that $y_1 = y_2 \pm 2n\pi$, but $n = 0$ is the only possible choice if $0 \leq y < 2\pi$. Therefore $z_1 = z_2$.

EXAMPLE 3 If $f(z) = (az + b)/(cz + d)$ with $ad - bc \neq 0$ (the so-called bilinear or Möbius function considered in Example 8, Section 1.4), then $f(z) = az/d + b/d$ if $c = 0$ and $f(z) = a/c + (bc - ad)/(c^2 z + cd)$ if $c \neq 0$. In either of these two cases, since only the first power of z is involved, $f(z)$ is clearly univalent.

Univalent functions are particularly important because of their action on arcs, namely, they map arcs into arcs:

4.1.4. Lemma If $C = \{z: z = z(t) = x(t) + iy(t), a \leq t \leq b\}$ is an arc in the domain D of a univalent function f, then $f(C) = \{w: w(t) = f(z(t))\}$ is an arc in the w plane.

Proof Since C is an arc, it does not cross itself, so that $z(t_1) = z(t_2)$ implies that $t_1 = t_2$, for all t_1 and t_2 between a and b. Thus if $w(t_1) = w(t_2)$, we have $f(z(t_1)) = f(z(t_2))$, so, since f is univalent, $z(t_1) = z(t_2)$, whence $t_1 = t_2$.

Thus, if $f(C)$ is, in fact, an arc, then it does not cross itself. So we must now show that $f(C)$ is an arc. To do this, we must demonstrate that $w(t)$ is differentiable and has a nonzero derivative for $a \leq t \leq b$ except possibly at a finite number of points (recall Definition 2.3.1). First, we note that

$$w(t) = f(z(t))$$

is a composite function; f is analytic and, hence, differentiable for all z on C, and z is differentiable except possibly at a finite number of points. Hence, by the chain rule,

$$w'(t) = f'(z) z'(t).$$

$z'(t)$ is defined and nonzero except at a finite number of points; $f'(z)$ cannot be zero on C more than a finite number of times since the zeros of analytic functions are isolated. Therefore $f(C)$ is an arc.   QED.

The following corollary is immediate:

4.1.5. Corollary Under the hypotheses of Lemma 4.1.4, if C is a simple closed contour, then $f(C)$ is a simple closed contour.

Thus, univalent functions preserve simple curves, that is, a univalent map of a simple curve is a simple curve; and furthermore, closed curves are also preserved. As we know from the Jordan curve theorem, a simple closed curve divides the rest of the complex plane into two disjoint sets, which we call the interior and the exterior of the curve. It turns out that univalent functions essentially preserve this property also.

4.1.6. Lemma Let f be a univalent function in a domain containing a simple closed contour C and its interior. Then f takes the interior of C into the interior of $f(C)$ and the exterior of C into the exterior of $f(C)$. If f has a pole in the interior of C then it takes the interior of C into the exterior of $f(C)$ and vice versa.

The proof of this lemma is quite difficult and profound, like that of the Jordan curve theorem itself: In fact, to use a typical excuse, it is beyond the scope of this book. However, most of our work will be with examples where one can reasonably easily verify the truth of this lemma. No doubt its truth seems intuitively evident to the reader; however, we have mentioned it precisely because the most evident results are often those most in need of proof. Finally, the fact that this theorem is true will make it somewhat easier to study the effects of mappings upon certain closed curves; we need only study the image of one point in $I(C)$, for instance, in order to know which region in the w plane is the image of all of $I(C)$.

There is one more eventuality which is worth mentioning. It is possible that f is meromorphic and has a pole *on* the simple closed curve C. If this is the case, then $f(C)$ will be a curve of infinite extent which divides the w

4.1. ANALYTIC AND UNIVALENT MAPPINGS

plane into two disjoint sets, one of which is the image of $I(C)$, and the other of which is the image of $E(C)$ (Figure 4.1).

We are now prepared to study conformal mappings, those mappings of the z plane into the w plane produced (as it will turn out) by analytic functions with nonvanishing derivatives. Such mappings are called conformal ("conforming") precisely because of their action on the angles between curves: just as a mercator map of the earth's surface (for example) preserves the angles between lines of longitude and parallels of latitude (although not the overall shape), so conformal mappings preserve angles, although *globally* they may produce considerable distortion.

Let us first recall (Figure 4.2) that if two smooth curves intersect, then the angle α between the curves at the point of intersection is defined to be the angle between the tangents to the two curves at this point; and, given the equations of the two curves, it is a straightforward calculation to find α.

EXAMPLE 4 For example, consider the parabola $y = 20x^2/9$ and the unit circle $x^2 + y^2 = 1$ (Figure 4.3). They intersect at two points, one of which is $(\frac{3}{5}, \frac{4}{5})$. At this point the slope m_2 of the circle is $m_2 = -\frac{3}{4}$, and the slope m_1 of the parabola is $m_1 = \frac{8}{3}$. Thus, the tangent of the angle α between them is

$$\tan \alpha = \frac{m_2 - m_1}{1 + m_1 m_2} = \frac{41}{12}.$$

From this one could use a book of tables to determine α as desired. If the equation of a curve is given in parametric form $x = x(t)$, $y = y(t)$, then the reader should recall that the slope of the curve at a given point is

$$\frac{dy}{dx} = \frac{dy/dt}{dx/dt}.$$

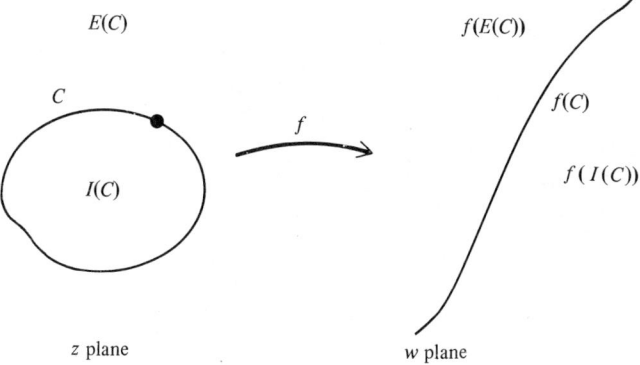

Figure 4.1

188 THE ELEMENTS OF CONFORMAL MAPPING

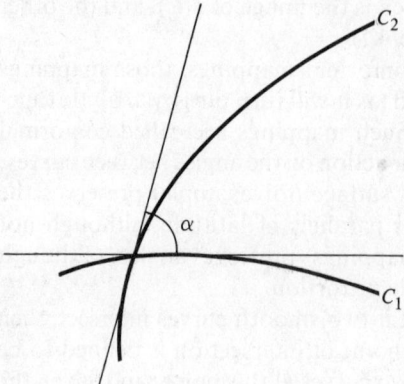

Figure 4.2

4.1.7. Definition If f is a continuous complex function with the property that it preserves curves near a point z_0* and that the angle between any two curves intersecting at z_0 is equal to the angle between the image curves at $f(z_0)$ (measured in the same direction), then f is said to be a *conformal mapping* at z_0.

The next theorem tells us of the relationship between conformal mappings and analytic functions.

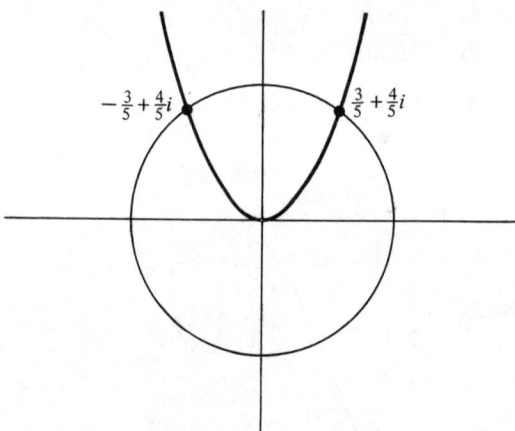

Figure 4.3

---

* That is, if C is an arc of a curve near z_0, then $f(C)$ is an arc of a curve near $f(z_0)$.

4.1. ANALYTIC AND UNIVALENT MAPPINGS

4.1.8. Theorem If f is univalent in a domain D then f' never vanishes in D, and f is conformal at each point z_0 in D.

Proof We shall not attempt a rigorous proof of the first half of the theorem. Heuristically, the reason that f' cannot vanish is that if it does vanish at some point z_0, then the power series for f near z_0 must look like

$$f(z) = a_0 + a_k(z - z_0)^k + \cdots, \quad k \geq 2.$$

For z near z_0, this mapping looks like a power map $(z - z_0)^k$ with $k \geq 2$, which is not univalent (see Example 1).

To prove the second half, let C_1 and C_2 be any two arcs intersecting at z_0, and let $f(C_1)$ and $f(C_2)$ be the image arcs intersecting at $f(z_0)$. If α_1 and α_2 are the angles that the tangents to C_1 and C_2 at z_0 make with the x axis, and if β_1 and β_2 are the angles that the tangents to $f(C_1)$ and $f(C_2)$ at $f(z_0)$ make with the u axis, then our task is to show that $\beta_2 - \beta_1 = \alpha_2 - \alpha_1$. It is crucial in our proof that $f'(z_0) \neq 0$, so that $f(z_0)$ is not a critical point of $f(C_1)$ and $f(C_2)$, where the tangent is not defined.

By appealing to the definition of the tangent to a curve at a point as the limiting position of a chord through it and an adjacent point (Figure 4.4), we can write

$$\alpha_1 = \lim_{z_1 \to z_0} \arg(z_1 - z_0), \qquad \alpha_2 = \lim_{z_2 \to z_0} \arg(z_2 - z_0),$$

where z_1 approaches z_0 along C_1 and z_2 approaches z_0 along C_2. Similarly, in the w plane,

$$\beta_1 = \lim_{z_1 \to z_0} \arg[f(z_1) - f(z_0)], \qquad \beta_2 = \lim_{z_2 \to z_0} \arg[f(z_2) - f(z_0)].$$

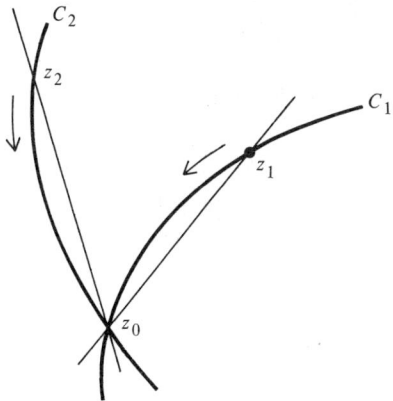

Figure 4.4

190 THE ELEMENTS OF CONFORMAL MAPPING

Then

$$\beta_2 - \beta_1 - (\alpha_2 - \alpha_1) = \lim_{z_2 \to z_0} \arg[f(z_2) - f(z_0)]$$
$$- \lim_{z_1 \to z_0} \arg[f(z_1) - f(z_0)] - [\lim_{z_2 \to z_0} \arg(z_2 - z_0) - \lim_{z_1 \to z_0} \arg(z_1 - z_0)]$$
$$= \lim_{z_2 \to z_0} \left(\arg \frac{f(z_2) - f(z_0)}{z_2 - z_0} \right) - \lim_{z_1 \to z_0} \left(\arg \frac{f(z_1) - f(z_0)}{z_1 - z_0} \right).$$

But $f(z)$ was given to be analytic at $z = z_0$, and we have also clearly specified that $f'(z_0) \neq 0$; hence $f'(z_0)$ has an argument which is uniquely defined regardless of the direction in which the limit is taken, provided, of course, that we restrict ourselves to values of the argument such that $-\pi \leq \arg f'(z_0) < \pi$ (for instance). The proof would fail here if $f'(z_0) = 0$ precisely because the argument of zero is undefined. Hence the proof is complete. QED.

EXAMPLE 5 An example of a function which produces a conformal mapping everywhere except at $z = 0$ is the function $w = 1/z$. In particular, let us recall the result of Example 4, where the two curves $y = 20x^2/9$ and $x^2 + y^2 = 1$ intersected at the point $(\frac{3}{5}, \frac{4}{5})$ at an angle whose tangent was $\frac{41}{12}$. In the w plane, since

$$w = u + iv = \frac{x - iy}{x^2 + y^2}$$

the parabola goes into the curve with parametric equations

$$u = \frac{81}{81x + 400x^3}, \quad v = -\frac{180}{81 + 400x^2},$$

which are obtained simply by substituting $20x^2/9$ for y and simplifying. Likewise, the upper half of the unit circle in the z plane goes into the lower half of the unit circle in the w plane, which is parametrized by

$$u = x, \quad v = -\sqrt{1 - x^2}.$$

See Figure 4.5. Clearly the point of intersection in the w plane is $(\frac{3}{5}, -\frac{4}{5})$. Since for the image of the parabola,

$$\frac{du}{dx} = -\frac{81(81 + 1200x^2)}{(81x + 400x^3)^2}, \quad \frac{dv}{dx} = \frac{180(800x)}{(81 + 400x^2)^2},$$

its slope is given by

$$\frac{dv}{du} = \frac{dv/dx}{du/dx} = -\frac{16000x^3}{729 + 10800x^2}$$

which at the point of intersection is

$$m_1 = -\tfrac{128}{171}.$$

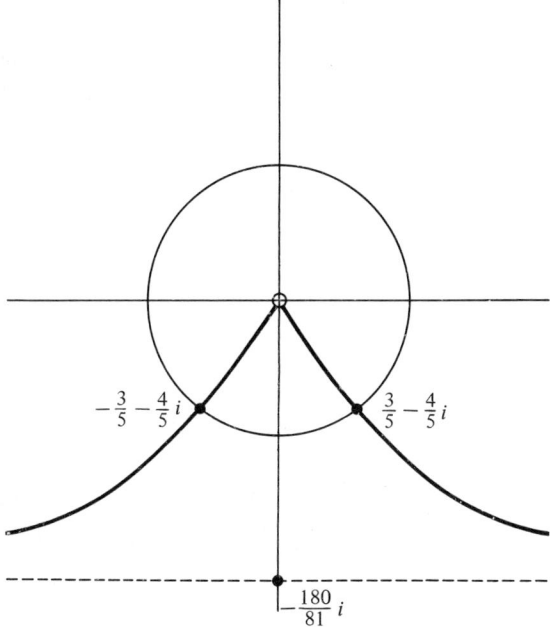

Figure 4.5

The slope of the image circle at the point of intersection is
$$m_2 = \tfrac{3}{4},$$
so the angle β between the image curves has tangent
$$\tan \beta = \frac{m_2 - m_1}{1 + m_1 m_2} = \frac{41}{12},$$
just as was the case with the originals.

It turns out that the converse of Theorem 4.1.8 is also true; that is, a continuous complex function which produces a conformal mapping at a point is analytic and has a nonvanishing derivative at that point; but as we shall not need this result, we shall not go into further details on it. Suffice it to say that conformality is a criterion for analyticity and univalency, at least locally.

A reasonable question to which we shall need an answer, however, is the following: What happens if f is analytic at z_0, but $f'(z_0) = 0$? As it turns out, the mapping does fail to be conformal at z_0, but it does very predictable things to angles at z_0:

4.1.9. Theorem If f is analytic in a domain D and if at $z_0 \in D$ we have $f'(z_0) = f''(z_0) = \cdots = f^{(k)}(z_0) = 0$, with $f^{(k+1)}(z_0) \neq 0$, and if α is the angle between two smooth simple curves C_1 and C_2 intersecting at z_0, then $f(C_1)$ and $f(C_2)$ intersect at an angle equal to $(k+1)\alpha$.

Proof We choose points z_1 and z_2 as in the previous proof, and look at the Taylor series for f around z_0 evaluated at these two points:

$$f(z_2) - f(z_0) = a_{k+1}(z_2 - z_0)^{k+1}(1 + \cdots),$$
$$f(z_1) - f(z_0) = a_{k+1}(z_1 - z_0)^{k+1}(1 + \cdots),$$

so that

$$\arg[f(z_2) - f(z_0)] = \arg a_{k+1} + \arg(z_2 - z_0)^{k+1} + \eta_1,$$
$$\arg[f(z_1) - f(z_0)] = \arg a_{k+1} + \arg(z_1 - z_0)^{k+1} + \eta_2,$$

where η_1 and η_2 are numbers which go to zero as z_1 and z_2 approach z_0 (since the argument of 1 is zero, and 1 is the limit of the remainder of the Taylor series in the above formulas). These two equations and those following are actually modulo 2π, which will, of course, not affect the validity of our conclusions. Thus

$$\arg[f(z_2) - f(z_0)] - \arg[f(z_1) - f(z_0)] + \eta_2 - \eta_1 =$$
$$(k+1)[\arg(z_2 - z_0) - \arg(z_1 - z_0)],$$

and in the limit as z_1 and z_2 approach z_0, we, therefore, have

$$\beta_2 - \beta_1 = (k+1)(\alpha_2 - \alpha_1),$$

which is the desired result. Notice that if $k = 0$ then this is precisely the result of Theorem 4.1.8; there *is* justice in the universe!

An example of a function which behaves like this is, of course, $f(z) = z^n = r^n e^{in\theta}$ near $z = 0$, where n is an integer. We immediately note that if $\arg z = \theta$, then $\arg f(z) = n\theta$. If $\theta > 2\pi/n$, then, of course, $\arg f(z)$ exceeds 2π. Thus we see that this function, in fact, has the effect of wrapping the complex plane around the origin n times. Notice that Corollary 4.1.5 fails, for instance, if we consider the effect of the mapping $f(z) = z^2$ on the curve $\{z : z(t) = \cos t + i \sin t, 0 \leq t \leq 2\pi\}$, that is, the unit circle. The effect of squaring z is to double the argument; thus, the image of the unit circle in the $w = f(z)$ plane is $\{w : w(t) = \cos 2t + i \sin 2t, 0 \leq t \leq 2\pi\}$. The image curve is no longer a *simple* closed curve, because as t runs between 0 and 2π, w traverses the unit circle *twice*. This is an example of what we mean when we say that z^n wraps the complex plane around the origin n times.

To carry this example a bit further, note that z^2 is not univalent (we might say that it is 2-valent) in any domain which contains the origin. Furthermore, it is not univalent in any domain *not* containing the origin, if said domain contains values of z of the same modulus whose arguments differ by π (Figure 4.6). The value of z^2 is the same at these two points and univalency fails.

Likewise, for general n, z^n is not univalent in any domain containing the origin or containing two points with the same modulus whose arguments differ by an integer multiple of $2\pi/n$ (Figure 4.7). Again the value of z^n will be the same for all such points.

These relatively simple examples serve to illustrate that univalency and conformality at a point are closely related, that a function whose derivative vanishes at a point of its domain is not univalent in that domain, and that a mapping fails to be conformal near a point where the derivative vanishes. Of course, the vanishing of the derivative is not the only reason that a function might fail to be univalent, for we have seen a periodic function like e^z is not univalent, though its derivative e^z vanishes nowhere.

It is important to remember that conformality is a local property, that is, one which holds at or near an individual point, and for which we have only to examine some small neighborhood of the point in question. Thus, if f is analytic in a domain D, it produces a conformal mapping at every point of D except those *isolated* points which are zeros of the *analytic* function f'. And at these isolated points, it multiplies angles by a positive integer which is equal to the order of the lowest nonvanishing derivative at that point. We shall need this later in the chapter when we consider the Schwarz–Christoffel transformation, which is designed to map the upper half plane into polygons, that is, to take straight angles along the real axis and bend them in such a way as to cause them to coincide with the angles at the vertices of a polygon.

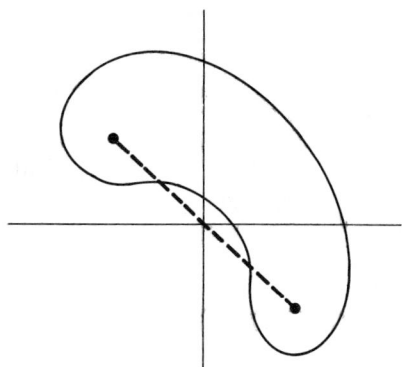

Figure 4.6

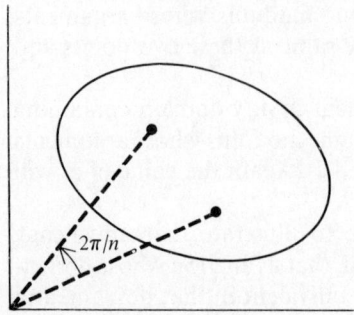

Figure 4.7

A. Exercises

Find domains in which the functions (a) $\sin z$, (b) $\cos z$, (c) $\sinh z$, (d) $\cosh z$ are univalent.

B. Problems

1. Into what curve does the function $1/(1 - z)$ map the unit circle? Into what domain does it map the interior of the circle? Where is this mapping conformal?

2. Into what curve does the function $1/(1 - z)$ map the straight line $y = x$? Show that the angle between this line and the unit circle at their point of intersection $(\sqrt{2}/2, \sqrt{2}/2)$ is preserved.

C. Proofs

1. If $w = (az + b)/(cz + d)$ with $ad - bc \neq 0$ and if $z = (\alpha\zeta + \beta)/(\gamma\zeta + \delta)$ with $\alpha\delta - \beta\gamma \neq 0$, then show that w is a bilinear function of ζ. Thus, the composition of two bilinear functions is bilinear.

2. Show that if $z = f(\zeta)$ is univalent in a domain D, and if $w = g(z)$ is univalent in $f(D)$, then w is a univalent function of ζ in D.

3. Show graphically the images of the straight lines $y = 2x + 1$, $x + y = 1$ in the w plane under the mapping $w = z^2$. Is the angle between these lines preserved? Explain.

4.2. SOME EXAMPLES OF CONFORMAL MAPPINGS

Our next order of business is to examine the conformal mappings induced by the elementary analytic functions with which we are familiar. In this way we can build a mental "library" of familiar mappings, which many times will enable us to solve conformal mapping problems almost at a glance.*

Four simple operations on the z plane will obviously preserve angles between curves; we call these operations simple because they are well known from elementary analytic geometry. These are: rotation of the axes about the origin; translation of the origin to another point; a uniform contraction or expansion of coordinates (change of scale); and reflection through a line. Reflection through a line, however, does not preserve the *direction* or *sense* of the angle between two curves, which we demand for conformality. For example, the mapping $f(z) = \bar{z}$ takes the lines $y = x$ and $x = 0$ into $y = -x$ and $x = 0$, in that order, so that the angle between them goes from $\pi/4$ to $-\pi/4$. Hence we shall have to eliminate this particular map from consideration. However, any number of the first three operations performed in succession will produce a conformal mapping, i.e., will preserve angles (and their senses) between curves everywhere in the plane. Let us examine each in turn.

1. *Rotation* (about the origin). Rotation of the complex plane about the origin takes a complex number $z = re^{i\theta}$ into a number the same distance r from the origin, but with its argument diminished by an amount equal to the angle of rotation ϕ (measured in the positive sense), that is, $z = re^{i\theta}$ goes into $w = re^{i(\theta - \phi)} = e^{-i\phi}re^{i\theta}$. Thus any rotation of axes about the origin corresponds simply to multiplication by a complex number of absolute value one.

2. *Translation.* Translation of the complex plane adds a constant α to $x = \text{Re}(z)$ and another constant β to $y = \text{Im}(z)$, so that $z = x + iy$ goes into $w = (x + \alpha) + i(y + \beta)$. If we write $a = \alpha + i\beta$, then we see that every translation can be written in the form $w = z + a$.

3. *Scale change.* An expansion or contraction of the coordinate system in the complex plane multiplies $x = \text{Re}(z)$ and $y = \text{Im}(z)$ by the same positive constant γ, so that z goes into $w = \gamma x + i\gamma y = \gamma z$.

By inspection, we see that each of the above mappings, $w = e^{i\phi}z$, $w = z + a$, and $w + \gamma z$, is conformal everywhere, since they are univalent and the derivatives do not vanish. In fact, if a and b are any two complex constants with $a \neq 0$, then the linear function $w = az + b$ is a conformal mapping for every finite z, since it is univalent in the entire plane and $w'(z) = a \neq 0$. This linear function $w = az + b$ actually consists of (1) an

---
* An atlas of conformal maps is available in G. A. Korn and T. M. Korn, *Mathematical Handbook for Scientists and Engineers*, McGraw-Hill, New York, 1961, pp. 205–211.

expansion of coordinates by the amount $|a|$; (2) a rotation of coordinates through the angle $-\arg a$; and (3) a translation of coordinates by the amount b. Hence every linear mapping is conformal and can be expressed as a combination of the three basic operations; and conversely, any combination of any number of the three basic operations is a linear function and is conformal in the whole plane. Indeed, geometric figures are transformed into *similar* figures by a linear mapping, so, of course, the angles between curves are preserved.

Now let us turn to the nonlinear function $w = 1/z$. Since $w' = -1/z^2$ is defined and nonzero everywhere except at the origin, the mapping induced by $w = 1/z$ must be conformal everywhere except at the origin. Let us examine this mapping a bit more carefully: if $z = re^{i\theta}$, then $w = 1/z = (1/r)e^{-i\theta}$. Thus, the effect of this mapping is to take z into the point which has argument equal to $-\arg z$ and modulus equal to $1/|z|$ (Figure 4.8). In particular, if $|z| = r = 1$, then $w = e^{-i\theta}$, so that points on the *unit circle* go into their conjugates, which are also on the unit circle. If $r < 1$, then $|w| = 1/r > 1$, so points inside the unit circle in the z plane go into points outside the unit circle in the w plane, and conversely. Thus the mapping $w = 1/z$, which is called an inversion, has the effect of turning the plane inside out and upside down with respect to the unit circle.

It is not hard to see that the Möbius transformation $w = (az + b)/(cz + d)$ with $ad - bc \neq 0$ is a combination of the four operations: inversion, rotation, translation, and expansion. To see this we simply recall that,

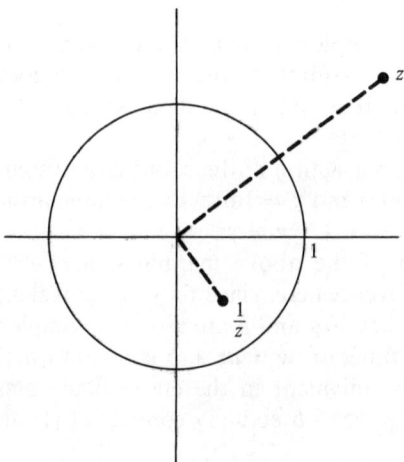

Figure 4.8

4.2. Some Examples of Conformal Mappings

according as $c = 0$ or $c \neq 0$, we have

$$w = \frac{a}{d}z + \frac{b}{d} \quad \text{or} \quad w = \frac{a}{c} + \frac{1}{c}\frac{bc - ad}{cz + d};$$

the first expression is linear, and the second consists of a linear transformation $(cz + d)$ followed by an inversion, expansion, and rotation by $(bc - ad)/c$, and a translation by a/c. As such, it is conformal everywhere except where the denominator vanishes, i.e., where $z \neq -d/c$.

Möbius or bilinear transformations are extremely important because of a peculiar property which they all share. Namely, if we define a "circle" (with quotation marks) to be either a circle or a straight line (which can be thought of as a circle with its center at infinity), then a bilinear transformation always takes a "circle" into a "circle." Let us proceed to prove this claim. This is because the equation of a "circle" in the z plane can be written

$$Az\bar{z} + (B + iC)z + (B - iC)\bar{z} + D = 0,$$

where A, B, C, and D are real constants; naturally if $A \neq 0$, this equation represents a circle (if it has any locus at all); if $A = 0$, then it represents a straight line. This is because

$$Az\bar{z} + (B + iC)z + (B - iC)\bar{z} + D$$
$$= A(x^2 + y^2) + 2Bx - 2Cy + D = 0;$$

obviously if $A = 0$ we have

$$2Bx - 2Cy + D = 0,$$

the equation of a straight line, while if $A \neq 0$, we can complete the square and obtain

$$\left(x + \frac{B}{A}\right)^2 + \left(y - \frac{C}{A}\right)^2 = \frac{B^2 + C^2 - DA}{A^2},$$

which is the equation of a circle with center $(-B/A, C/A)$ and radius squared $(B^2 + C^2 - DA)/A^2$, assuming that the latter is positive. Now, since $w = (az + b)/(cz + d)$, we have $cwz - az + dw - b = 0$ and, hence, solving for z, we get $z = -(dw - b)/(cw - a)$; since $ad - bc \neq 0$, note that z is also a bilinear function of w. Substituting this expression for z into the equation of the "circle," we find that

$$A\left(\frac{\bar{d}\bar{w} - \bar{b}}{\bar{c}\bar{w} - \bar{z}}\right)\left(\frac{dw - b}{cw - a}\right) - (B + iC)\left(\frac{dw - b}{cw - z}\right)$$
$$- (B - iC)\left(\frac{\bar{d}\bar{w} - \bar{b}}{\bar{c}\bar{w} - \bar{a}}\right) + D = 0,$$

assuming that $w \neq a/c$. By clearing the denominators, we obtain

$$A(\bar{d}\bar{w} - \bar{b})(dw - b) - (B + iC)(dw - b)(\bar{c}\bar{w} - \bar{a})$$
$$- (B - iC)(\bar{d}\bar{w} - \bar{b})(cw - a) + D(\bar{c}\bar{w} - \bar{a})(cw - a) = 0,$$

which, after algebraic manipulation, reduces to the form

$$(A|d|^2 - (B + iC)d\bar{c} - (B - iC)\bar{d}c + D|c|^2)w\bar{w}$$
$$+ (-A\bar{b}d - (B + iC)(-\bar{a}d) - (B - iC)(-c\bar{b}) - Dc\bar{a})w$$
$$+ (-Ab\bar{d} - (B + iC)(-b\bar{c}) - (B - iC)(-a\bar{d}) - D\bar{c}a)\bar{w}$$
$$+ A|b|^2 - (B + iC)b\bar{a} - (B - iC)\bar{b}a + D|a|^2 = 0.$$

The coefficient of $w\bar{w}$ and the constant term are both clearly real, and the coefficients of w and $\bar{w}$ are complex conjugates of one another, so this expression is recognizable as the equation of a "circle" in the w plane.

An alternative, and easier, proof would be to note that rotations, translations, and expansions preserve circles and straight lines (as we know from elementary analytic geometry), and the inversion $w = 1/z$ is a bilinear transformation with $a = d = 0$ and $b = c = 1$, so that under $w = 1/z$ the equation of the image of a "circle" becomes $Dw\bar{w} + (B - iC)w + (B + iC)\bar{w} + A = 0$, which is clearly a "circle" again. Thus, since a general bilinear transformation is always a composition of these four elementary transformations, it, too, must preserve "circles."

As an exercise in the use of the bilinear transformation, let us find the most general bilinear transformation $w = (az + b)/(cz + d)$ with $a \neq 0$ which maps the unit circle onto itself, i.e., which maps $\{z: |z| \leq 1\}$ conformally onto $\{w: |w| \leq 1\}$. We rewrite

$$w = \frac{z + b/a}{cz/a + d/a} = \frac{z + b/a}{(c/a)(z + d/c)},$$

and, by putting $\alpha = -b/a$, $\beta = c/a$, and $\delta = -d/c$, we have

$$w = \frac{z - \alpha}{\beta(z - \delta)}.$$

Clearly $|\alpha| < 1$, since α must go into the origin in the w plane, and hence it must be inside the unit circle in the z plane. Similarly, we must have $|\delta| > 1$, since δ goes into "the point at infinity," and, hence, it must lie outside the unit circle in the z plane. To find β and δ, we note that for any point $z = e^{i\theta}$ on the unit circle, we must have $|w| = 1$, so

$$\left|\frac{e^{i\theta} - \alpha}{\beta(e^{i\theta} - \delta)}\right| = 1,$$

and the point $z = 1$, in particular, must map into some point $e^{i\gamma}$ on the unit circle, so that

$$\frac{1 - \alpha}{\beta(1 - \delta)} = e^{i\gamma},$$

giving
$$\beta = \frac{1-\alpha}{1-\delta} e^{-i\gamma}.$$

By substituting for β in the first equation, we obtain
$$\left|\frac{1-\delta}{1-\alpha}\right| \left|\frac{e^{i\theta}-\alpha}{e^{i\theta}-\delta}\right| = 1 \quad \text{or} \quad \left|\frac{e^{i\theta}-\alpha}{1-\alpha}\right| = \left|\frac{e^{i\theta}-\delta}{1-\delta}\right|.$$

How do we solve an equation of this sort for δ? By inspection, $\delta = \alpha$ is a possibility; but since one of the two points δ and α must be inside the unit circle and the other outside, this possibility must be excluded. To discover another possibility, divide both numerator and denominator of the right-hand side by $|-\delta e^{i\theta}|$. Then
$$\left|\frac{e^{i\theta}-\alpha}{1-\alpha}\right| = \left|\frac{1/\delta - e^{-i\theta}}{(1/\delta)e^{-i\theta} - e^{-i\theta}}\right| = \left|\frac{e^{-i\theta} - 1/\delta}{1 - 1/\delta}\right|.$$

From this we see that $\delta = 1/\bar{\alpha}$ is a possibility, and this makes sense because now if α is inside the unit circle, δ is clearly outside. Thus
$$\beta = \frac{1-\alpha}{1-\delta} e^{-i\gamma} = \frac{1-\alpha}{1-1/\bar{\alpha}} e^{-i\gamma}$$

and, therefore,
$$w = e^{i\gamma}\left(\frac{1-1/\bar{\alpha}}{1-\alpha}\right)\left(\frac{z-\alpha}{z-1/\bar{\alpha}}\right) = e^{i\gamma}\left(\frac{\bar{\alpha}-1}{1-\alpha}\right)\left(\frac{z-\alpha}{\bar{\alpha}z-1}\right);$$

and this is the most general bilinear transformation possible which maps the unit circle into itself, α into the origin and 1 into $e^{i\gamma}$.

Let us now proceed to consider the effects of other functions of z on the complex plane. First, we consider the power function $w = z^{\alpha}$, where α is real, positive, and not equal to 1. If $\alpha > 1$, then this function is single-valued, analytic, and univalent in any sector $\beta < \arg z < \beta + 2\pi/\alpha$. $w' = \alpha z^{\alpha-1}$ is well-defined and nonzero in such a sector and, hence, $w = z^{\alpha}$ is a conformal mapping on the sector. Since $z^{\alpha} = r^{\alpha}e^{i\alpha\theta}$, w has the effect of multiplying the argument of z by the factor α. If $z_1 = r_1 e^{i\theta_1}$ and $z_2 = r_2 e^{i\theta_2}$ are two points in the sector, then the angle between lines from the origin to these two points is $\theta_2 - \theta_1$. After the mapping effected by $w = z^{\alpha}$, $\arg z_1^{\alpha} = \alpha\theta_1$, and $\arg z_2^{\alpha} = \alpha\theta_2$, so that the angle at the origin between the lines from $w = 0$ to z_1^{α} and z_2^{α} is $\alpha(\theta_2 - \theta_1)$ (Figure 4.9). Therefore, angles at the origin are multiplied by a factor of α, exactly as in the case of the power function z^n, where n is an integer. The case $\alpha < 1$ is quite similar, except that angles are decreased at the origin. We conclude that if we want a mapping which enlarges angles at the origin, we choose one with $\alpha > 1$, and if we want one which decreases angles at the origin, we choose one with $\alpha < 1$.

200 THE ELEMENTS OF CONFORMAL MAPPING

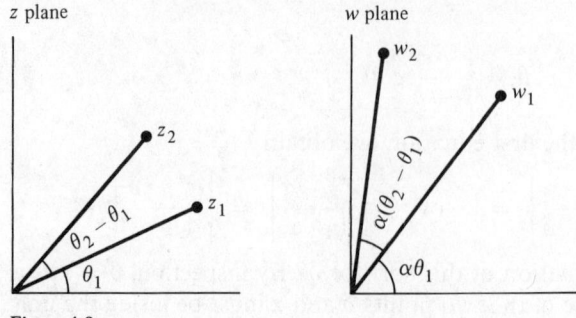

Figure 4.9

Finally, let us consider the effects of some of the transcendental functions on the complex plane. We have already noted that the exponential function $w = e^z$ is analytic, univalent, and, hence, conformal for $a < \text{Im } z < a + 2\pi$, where a is any real number. What sort of mapping does $w = e^z = e^x \cdot (\cos y + i \sin y)$ produce in one of these strips where it is conformal? Since $|e^z| = e^x$ and $\arg e^z = y$, we see that the absolute value and argument of the exponential function each depend on only one of the real variables x and y. Thus, a rectangle with sides parallel to the axes, with sides $x = x_1$, $x = x_2$, $y = y_1$, and $y = y_2$, will go into the portion of the w plane bounded by circles of radii e^{x_1} and e^{x_2}, and by lines through the origin inclined at angles y_1 and y_2, respectively (Figure 4.10). An infinite strip bounded by $y = y_1$ and $y = y_2$ will go into an entire sector in the w plane. A rectangle of height 2π will go into an annulus centered at the origin; and so on. Note that the exponential function effects a transformation from rectangular coordinates in the z plane to polar coordinates in the w plane. We might expect that the logarithm would do just the reverse; this question is, however, left to the reader in the exercises at the end of this section.

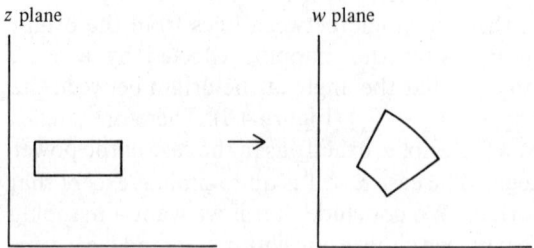

Figure 4.10

4.2. SOME EXAMPLES OF CONFORMAL MAPPINGS

Let us now consider the mapping produced by one further transcendental function, namely $w = \cos z$, which we recall is analytic and univalent (for instance) in the infinite strip $0 \leq \operatorname{Re} z < \pi$ and is conformal in this strip except at the point $z = 0$, where $w' = \sin z$ vanishes. We shall examine what $w = \cos z = \cos x \cosh y - i \sin x \sinh y$ does to a rectangle in the upper portion of this strip, bounded by segments parallel to the coordinate axes (Figure 4.11). For a line parallel to the x axis, y is a constant and, thus, $\sinh y = c$ and $\cosh y = \sqrt{1 + c^2}$, since $\cosh^2 y - \sinh^2 y = 1$. Therefore, in the w plane we have $u = \sqrt{1 + c^2} \cos x$ and $v = -c \sin x$, which yield

$$\frac{u^2}{1 + c^2} + \frac{v^2}{c^2} = 1,$$

the equation of an ellipse. Thus segments parallel to the x axis go into arcs of ellipses centered at the origin, with major axis along the u axis, in the w plane. On lines parallel to the y axis, x is a constant, so $\sin x = b$, and $\cos x = \sqrt{1 - b^2}$. Therefore, in the w plane we have $u = \sqrt{1 - b^2} \cosh y$, $v = -b \sinh y$, yielding

$$\frac{u^2}{1 - b^2} - \frac{v^2}{b^2} = 1,$$

the equation of a hyperbola. Hence, lines parallel to the y axis go into hyperbolas (which open horizontally). Thus, a rectangle as described above will be mapped into a region bounded by arcs of hyperbolas and ellipses, as shown in Figure 4.11; note that these arcs will be perpendicular, since they are the conformal images of perpendicular line segments.

Let us work one example now to illustrate the power of the methods we have learned so far, and to see how a number of different mappings can be combined to yield rather unusual results. Consider how we might map the interior of a lune (a domain bounded by two circular arcs) into the upper

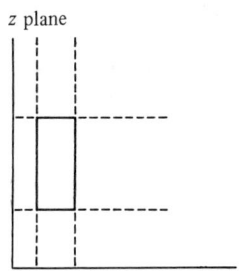

Figure 4.11

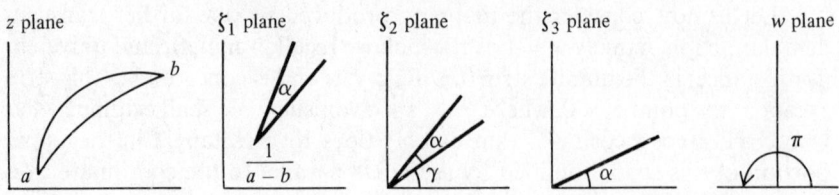

Figure 4.12

half plane (Figure 4.12). Suppose the two ends of the lune are at the points $z = a$ and $z = b$ and that the angle between the two arcs at a is α. The mapping $\zeta_1 = 1/(z - b)$, being a bilinear transformation which takes b into the point at infinity, takes the lune into an angular sector emanating from the point $1/(a - b)$. The angle at this point is the same as that at the point a, namely α. A further transformation of the form $\zeta_2 = \zeta_1 - 1/(a - b)$ takes this point to the origin. Then the transformation $\zeta_3 = e^{-i\gamma}\zeta_2$ will take one side of the sector into the positive real axis. Finally, $w = \zeta_3^{\pi/\alpha}$ opens the angle up to a full π radians. Thus, the total mapping required to take the lune into the upper half plane is

$$w = \zeta_3^{\pi/\alpha} = (e^{-i\gamma}\zeta_2)^{\pi/\alpha} = \left[e^{-i\gamma}\left(\zeta_1 - \frac{1}{a-b}\right)\right]^{\pi/\alpha}$$

$$= e^{-i\gamma\pi/\alpha}\left(\frac{1}{z-b} - \frac{1}{a-b}\right)^{\pi/\alpha}.$$

A. Exercises

1. Find the most general bilinear transformation which maps the unit circle into the upper half plane, taking a given point $e^{i\gamma}$ into the origin and taking the origin 0 into a given point a (Im $a > 0$) in the w plane.

2. Find the most general bilinear mapping which takes the upper half plane into the unit circle; in other words, find the inverse of the mapping in Exercise 1.

3. Find the most general bilinear mapping taking the upper half plane into itself.

4. Find a mapping which takes a lune into the unit circle.

5. Find a mapping which takes a given infinite strip parallel to the x axis conformally onto the upper half plane.

B. Problems

1 Describe the mapping induced by $w = \sin z$ in some detail, as was done for $\cos z$ in the text.

2 Do the same for $\log z$, after a suitable branch cut has been performed.

*3 Let $\tilde{F}(\zeta) = \tilde{u}(\zeta) + i\tilde{v}(\zeta)$ be an analytic function in the complex ζ plane, in a domain which includes the unit circle in its interior. Write out the Cauchy integral formula for $\tilde{F}$, with C being the unit circle, and evaluate for $\zeta = 0$; taking real parts of both sides, show that

$$\tilde{u}(0) = \frac{1}{2\pi} \int_0^{2\pi} \tilde{u}(e^{i\theta'}) \, d\theta' = \frac{1}{2\pi} \int_0^{2\pi} \tilde{f}(\theta') \, d\theta', \qquad \tilde{f}(\theta') = \tilde{u}(e^{i\theta'}).$$

This is the so-called *mean value principle* for harmonic functions, which states that the value of a harmonic function at the center of a circle is the mean of its values on the circumference.

*4 If $\zeta = e^{i\gamma}(z - a)/(\bar{a}z - 1)$ with $a = re^{i\phi}$ and $r < 1$, then $\tilde{F}, \tilde{u}$, and $\tilde{v}$ can be considered functions of z. Let us write $u(z) = \tilde{u}(\zeta(z)), f(\theta) = \tilde{f}(\theta'(\theta))$. Make the change of variables

$$e^{i\theta'} = e^{i\gamma} \frac{e^{i\theta} - a}{\bar{a}e^{i\theta} - 1}$$

in the mean value formula of Problem 3, and derive the fact that

$$u(a) = \frac{1}{2\pi} \int_0^{2\pi} \frac{(1 - r^2) f(\theta) \, d\theta}{1 + r^2 - 2r\cos(\theta - \phi)}, \qquad f(\theta) = u(e^{i\theta}).$$

This formula is called the *Poisson integral formula*. It gives the value of a harmonic function at any interior point of the unit circle in terms of its boundary values. It also turns out that if $f(\theta)$ is *any* continuous function on the unit circle, then $u(a)$ given by the above formula is a harmonic function of a, and

$$\lim_{a \to e^{i\theta}} u(a) = f(\theta).$$

How might one prove this?

C. Proofs

Show that the bilinear transformation which takes three given points z_1, z_2, z_3 into three prescribed points w_1, w_2, w_3 is given by the formula

$$\frac{w - w_1}{w - w_2} \frac{w_3 - w_2}{w_3 - w_1} = \frac{z - z_1}{z - z_2} \frac{z_3 - z_1}{z_3 - z_2}.$$

*Is used to denote problems of unusual difficulty.

4.3. THE SCHWARZ–CHRISTOFFEL TRANSFORMATION: MAPPING THE UPPER HALF PLANE CONFORMALLY ONTO A POLYGON

Is there any reason to assume that a function which maps some specified domain onto another even exists? It does not seem at all obvious that one necessarily should, and if it does not, then there is little sense in searching for it. The key result in this direction is the following important theorem, which we shall state without proof:

The Riemann Mapping Theorem Let D be a simply-connected domain other than the plane. Then there exists a univalent function f which maps D conformally onto the unit disk (or, equivalently, onto the upper half plane). Naturally this also implies the existence of a conformal mapping from the unit disk (or upper half plane) onto D, namely the inverse of the function f, which must exist since f is univalent.

A conformal mapping which will be indispensable to us in the applications in the next chapter is the one induced by the *Schwarz–Christoffel transformation*, which maps the upper half plane conformally onto the interior of a polygon. Let a polygon be given in the w plane, with vertices $w_1, w_2, \ldots, w_k$ and with interior angles $\gamma_1, \gamma_2, \ldots, \gamma_k$ at these vertices (Figure 4.13). If possible we wish to map the upper half z plane onto the interior of the polygon, such that the points $z_1, z_2, \ldots, z_k$ on the real axis go into the vertices of the polygon, and the real axis goes into the boundary of the polygon. The derivation which follows is not rigorous, but heuristic; however, it can be made rigorous with sufficient care.

Since the angle at z_i is π and the angle at w_i is γ_i, then near the point $z = z_i$ the function $w = f(z)$ performing the mapping must resemble a power

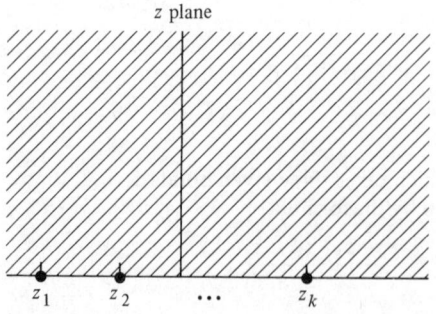
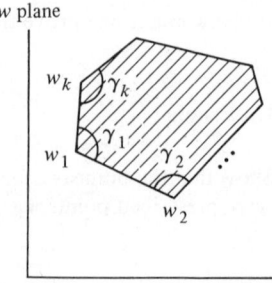

Figure 4.13

4.3. THE SCHWARZ–CHRISTOFFEL TRANSFORMATION

function. Thus, locally, near z_i, we must be able to expand $f(z)$ in the form

$$w - w_i = (z - z_i)^{\alpha_i}[a_0 + a_1(z - z_i) + \cdots], \quad a_0 \neq 0,$$

where the quantity in brackets is a power series. To determine the value of α_i, note that if $w - w_i = \rho e^{i\phi}$ and $z - z_i = re^{i\theta}$, then for very small values of r and ρ we have

$$\rho e^{i\phi} = (re^{i\theta})^{\alpha_i}(a_0 + a_1 re^{i\theta} + \cdots).$$

If we move clockwise around the polygon, approach the ith vertex, and proceed around a very small arc at w_i in the counter-clock-wise direction, then $\phi = \arg(w - w_i)$ changes by γ_i (Figure 4.14). θ changes by π at z_i; hence, the argument of the right-hand side of the above equation changes by $\theta\alpha_i = \pi\alpha_i$ plus the argument of the power series in parentheses; however, as r and ρ approach zero, the change in the argument of the power series likewise approaches zero. Hence $\gamma_i = \alpha_i \pi$ or $\alpha_i = \gamma_i/\pi$. Thus near z_i the expansion of $w - w_i$ has the form

$$w - w_i = (z - z_i)^{\gamma_i/\pi} H_i(z),$$

where $H_i(z)$ is a single-valued analytic function near z_i. Differentiating this with respect to z, we obtain

$$w' = \frac{\gamma_i}{\pi}(z - z_i)^{\gamma_i/\pi - 1} H_i(z) + (z - z_i)^{\gamma_i/\pi} H'_i(z) = \frac{\gamma_i}{\pi}(z - z_i)^{\gamma_i/\pi - 1} G_i(z),$$

where $G_i(z)$ is single-valued and analytic near z_i. Differentiating once again, we obtain

$$w'' = \frac{\gamma_i}{\pi}\left(\frac{\gamma_i}{\pi} - 1\right)(z - z_i)^{\gamma_i/\pi - 2} F_i(z),$$

where, again, $F_i(z)$ is single-valued and analytic near z_i. Putting all these facts together, we find the following differential equation for w near z_i:

$$\frac{w''}{w'} = \frac{(\gamma_i/\pi - 1)}{z - z_i} + h_i(z),$$

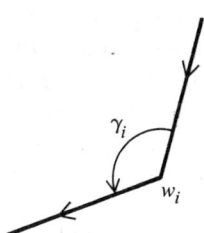

Figure 4.14

$h_i(z)$ being single-valued and analytic near z_i, but not necessarily near any of the other corners. If w''/w' looks like this locally (near z_i), then globally (over the whole polygon), it must be of the form

$$\frac{w''}{w'} = \frac{(\gamma_1/\pi - 1)}{z - z_1} + \frac{(\gamma_2/\pi - 1)}{z - z_2} + \cdots + \frac{(\gamma_k/\pi - 1)}{z - z_k} + h(z);$$

now $h(z)$ is a single-valued, analytic function in the upper half plane; and it is also bounded there, since the net effect of the mapping is to take the upper half plane into a polygon, and the only difficulties come at the vertices, which are taken care of by the first k terms. Thus, since $h(z)$ seems to be arbitrary, let us choose $h \equiv 0$ and see if we can obtain a reasonable result from this differential equation for the mapping. A first integral of this equation is

$$\log w' = \log(z - z_1)^{\gamma_1/\pi - 1} + \cdots + \log(z - z_k)^{\gamma_k/\pi - 1} + \log c,$$

where c is an arbitrary complex constant. We define a single-valued branch of each logarithm above, and of each power $(z - z_i)^{\gamma_i/\pi - 1}$ in the expressions below, by choosing $\arg(z - z_i)$ to be zero for real $z > z_i$, and π for real $z < z_i$. Likewise we choose $-\pi < \arg c \leq \pi$. Then, adopting the usual notation for products,

$$p_1 p_2 \cdots p_k = \prod_{i=1}^{k} p_i,$$

we have

$$\log w' = \log c \prod_{i=1}^{k} (z - z_i)^{\gamma_i/\pi - 1}$$

or finally

$$w' = c \prod_{i=1}^{k} (z - z_i)^{\gamma_i/\pi - 1}$$

and, thus,

$$w = c \int_{z_0}^{z} \prod_{i=1}^{k} (z - z_i)^{\gamma_i/\pi - 1} \, dz + d,$$

where d is a second arbitrary constant of integration. The integrand in this expression is single-valued and analytic in the upper half plane and on the real axis except at the points $z = z_i$; in fact, it is even analytic in the domain consisting of all z except those for which $\text{Re } z = z_i$ and $\text{Im } z \leq 0$ (the domain left after making a cut from each z_i vertically downward; see Figure 4.15). The point z_0 is arbitrarily chosen in the upper half plane or on the real axis where $z \neq z_i$, and its choice will naturally affect the value of d.

Now that we have arrived rather heuristically at this expression for w, we must ask ourselves the question: Does $w(z)$ as given by this formula

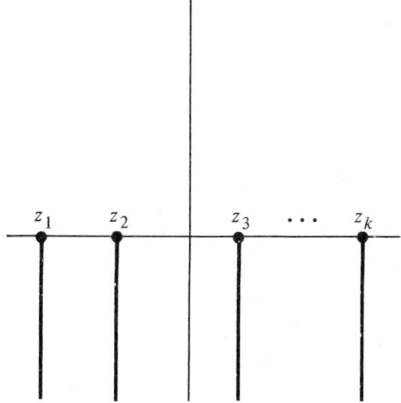

Figure 4.15

actually map the upper-half z plane onto a polygon in the w plane, and if so, can the constants be chosen so that the image polygon coincides with the given one? The answer to the first question* is yes; thus, we have the following:

4.3.1. Lemma The Schwarz–Christoffel transformation

$$w = c \int_{z_0}^{z} \prod_{i=1}^{k} (z - z_i)^{\gamma_i/\pi - 1} \, dz + d$$

maps the upper half plane conformally into a polygon with k vertices and with interior angles $\gamma_1, \gamma_2, \ldots, \gamma_k$ at these vertices.

Proof The function w has a derivative which is itself analytic in the upper half plane, namely

$$w' = c \prod_{i=1}^{k} (z - z_i)^{\gamma_i/\pi - 1},$$

and this derivative is not zero there; hence w is conformal on the upper half plane. Furthermore, $w(z_i)$ is finite for each z_i, since we can write

$$w'(z) = c(z - z_i)^{\gamma_i/\pi - 1} f(z)$$

---

* The second one will be answered shortly.

near z_i, where $f(z)$ is analytic. Hence if we integrate w' by parts we have

$$w(z_i) = c \int_{z_0}^{z_i} (z - z_i)^{\gamma_i/\pi - 1} f(z)\, dz + d$$

$$= \frac{c(z - z_i)^{\gamma_i/\pi}}{\gamma_i/\pi} f(z) \Big|_{z_0}^{z_i} - \frac{c\pi}{\gamma_i} \int_{z_0}^{z_i} (z - z_i)^{\gamma_i/\pi} f'(z)\, dz + d;$$

since $\gamma_i > 0$, the first term is finite (indeed, it vanishes) at z_i, and the second integral is likewise finite. Now, since

$$w = c \int_{z_0}^{z} w'\, dz + d \quad \text{and} \quad w' = |w'|e^{i \arg w'},$$

we have

$$w = c \int_{z_0}^{z} |w'| e^{i \arg w'}\, dz + d.$$

Now if z is to the right of z_i, then $\arg(z - z_i) = 0$ (Figure 4.16), while if z is to the left of z_i, then $\arg(z - z_i) = \pi$. Thus, for z on a segment joining any z_j and z_{j+1}, $\arg(z - z_i)$ for any i is constant. Hence

$$\arg w' = \arg c + (\gamma_1/\pi - 1)\arg(z - z_1) + \cdots + (\gamma_k/\pi - 1)\arg(z - z_k)$$

is a constant for any z on a segment between z_j and z_{j+1} on the real axis. For these values of z between z_j and z_{j+1} we have

$$w(z) = c \int_{z_0}^{z} |w'| e^{i \arg w'}\, dz + d = ce^{i \arg w'} \int_{z_0}^{z} |w'|\, dz + d;$$

$e^{i \arg w'}$ passes through the integral sign because it is constant on $z_j < z < z_{j+1}$.

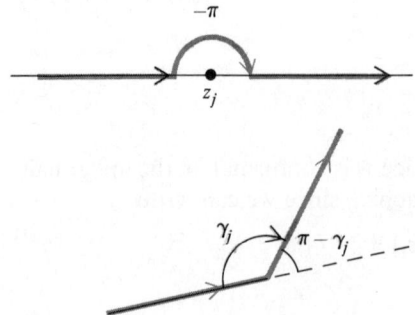

Figure 4.16

4.3. THE SCHWARZ–CHRISTOFFEL TRANSFORMATION

It is apparent from this formula that the argument of $w - w_j$ is also a constant as z ranges from z_j to z_{j+1}, since everything on the right has a constant argument for these values of z. Therefore, the image of the segment joining z_j to z_{j+1} is a straight line segment in the w plane, and from this we see that the image of the real axis must be a polygon in the w plane.* We must now study the shape and size of this polygon.

As z passes over a point z_j in the clockwise direction, the argument of $z - z_j$ changes by $-\pi$, so $\arg(z - z_j)^{\gamma_j/\pi - 1}$ changes by $-\pi(\gamma_j/\pi - 1) = -\gamma_j + \pi = \pi - \gamma_j$. Thus the inclination of the next side of the polygon differs by $\pi - \gamma_j$ from the inclination of the current side, so that the interior angle that the two sides form with each other is γ_j (Figure 4.16). This completes the proof of the lemma.

Thus w maps the upper-half z plane onto a polygon P with the proper interior angles. Let Q represent the given polygon. Then we have the following:

4.3.2. Lemma One can cause P to coincide with Q by proper choice of the constants $(c, d, z_1, z_2, \ldots, z_k)$.

What follows is a plausibility argument for the above lemma. In the expression

$$w = c \int_{z_0}^{z} \prod_{i=1}^{k} (z - z_1)^{\gamma_i/\pi - 1} \, dz + d,$$

there are $k + 2$ independent constants: c, d, and the z_i's.† As we know from 4.2, the effect of d is to translate the image polygon, and the effect of c is to rotate and/or change the scale of the coordinate system. Thus, if we can choose values for the z_i's which make the image polygon *similar* to the given one, then c and d will do the remaining work of bringing the two polygons into coincidence.

If the given polygon is a *triangle*, then, since the angles of the given triangle and of the image triangle are equal, the two figures are already similar and, hence, z_1, z_2, z_3 can be chosen completely arbitrarily: The two constants c and d are sufficient to bring the two triangles into coincidence.

In the general case, a polygon can always be decomposed into triangles. For instance, this can be done by means of drawing lines from one vertex to the others (Figure 4.17). In order for the two polygons to be similar, then,

---

* It might possibly, of course, be a closed polygonal curve which crosses itself one or more times. We shall see later that this can be prevented; at present, the crucial result is that the image of the real axis is a polygonal curve.
† Note that z_0 and d are not independent of one another. Assigning a value to one of them automatically determines the other.

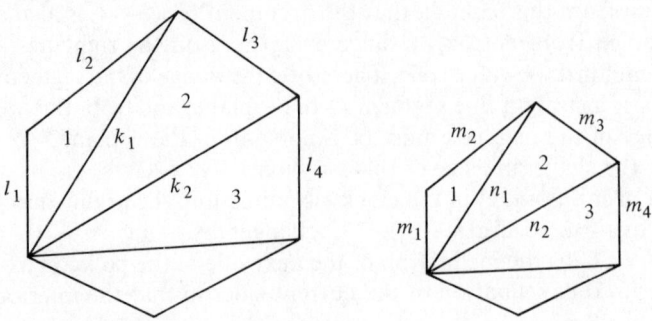

Figure 4.17

each of the triangles thus formed must be similar. Referring to the example in Figure 4.17, the two triangles numbered 1 will be similar if $l_1/m_1 = l_2/m_2$. It is not necessary to require any other ratios to be equal, since the angles between these pairs of corresponding sides are already equal (Lemma 4.3.1). The triangles numbered 2 will then be similar if $l_3/k_1 = m_3/n_1$. The triangles numbered 3 will be similar if $l_4/k_2 = m_4/n_2$. Once these three triangles are similar, the fourth ones are automatically similar since conformality plus similarity of the preceding three guarantees that their corresponding angles are equal. Thus, in this particular case, we have three relationships that must be satisfied and six constants $(z_1, z_2, z_3, z_4, z_5, z_6)$ to exploit. Clearly we can choose three of the z_i's arbitrarily and still have three available to be determined by the three ratios which must be satisfied (at least in theory).

It is easy to extend this argument to the general case, by noting that we will have $k - 3$ relationships to be satisfied by the k constants $z_1, z_2, \ldots, z_k$ so that three of the z_i's can always be chosen arbitrarily. This completes the argument.

In general, the integral involved in the Schwarz–Christoffel transformation may be difficult if not impossible to evaluate in closed form, and it may be quite a task to evaluate the $k - 3$ predetermined values of the z_i's, particularly if the polygon in question has many sides. Let us examine a very simple case and view in microcosm the difficulties which may arise.

EXAMPLE 6 Consider the problem of mapping the upper half plane conformally onto an isosceles right triangle with leg 1 (Figure 4.18). Here we take $w_1 = i$, $w_2 = 0$, $w_3 = 1$, and note that $\gamma_1 = \gamma_3 = \pi/4$, $\gamma_2 = \pi/2$. Since all the z_i's can be chosen arbitrarily, we take $z_1 = -1$, $z_2 = 0$, and $z_3 = 1$, for simplicity. Then

$$w' = c \prod_{i=1}^{3} (z - z_i)^{\gamma_i/\pi - 1} = c(z + 1)^{-\frac{3}{4}} (z - 0)^{-\frac{1}{2}} (z - 1)^{-\frac{3}{4}}$$

4.3. THE SCHWARZ–CHRISTOFFEL TRANSFORMATION

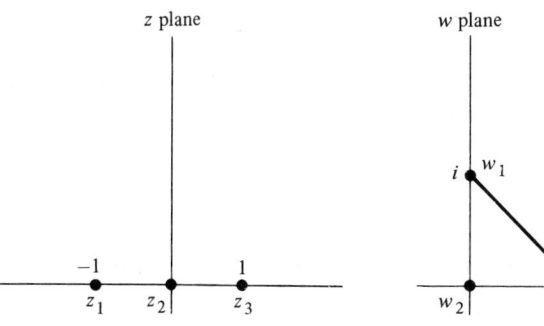

Figure 4.18

and, hence,

$$w' = \frac{c}{\sqrt{z}(z^2-1)^{\frac{3}{4}}} = \frac{c'}{\sqrt{z}(1-z^2)^{\frac{3}{4}}}.$$

We choose the branch of the multiple-valued function $\sqrt{z}(1-z^2)^{\frac{3}{4}}$, which is real for values of z on the real axis between zero and 1. This will force the constant c' to be real, since the segment $\overline{z_2 z_3}$ must go into the segment $\overline{w_2 w_3}$, which is also on the real axis. We then have

$$w = c' \int_{z_0}^{z} \frac{dz}{\sqrt{z}(1-z^2)^{\frac{3}{4}}} + d$$

and, if we choose $z_0 = 0$, this forces the constant d to be zero, so

$$w = c' \int_{0}^{z} \frac{dz}{\sqrt{z}(1-z^2)^{\frac{3}{4}}}.$$

To evaluate the constant c', we note that

$$w_3 = w(z_3) = 1 = c' \int_{0}^{1} \frac{dz}{\sqrt{z}(1-z^2)^{\frac{3}{4}}}.$$

The constant c', as it turns out, can be evaluated in terms of gamma functions. We shall turn this problem at the end of this section. Note that the integral for w cannot, of course, be evaluated in closed form, but must be evaluated in some other manner, for instance, by series expansion or numerical methods.

This example points out the difficulties one might have in mapping the upper half plane onto even a fairly simple polygon: Even when the given polygon was an isosceles right triangle very conveniently placed with respect to the axes, the evaluation of the constant c' will prove to be a nearly insuperable obstacle.

212 THE ELEMENTS OF CONFORMAL MAPPING

The primary utility of the Schwarz–Christoffel transformation in applications turns out to be in certain *degenerate* cases, however, where one either (1) chooses one of the points z_i to be at "infinity," or (2) takes one one of the w_i's at infinity, so that the image polygon is degenerate. Let us consider each of these problems in turn.

For the sake of definiteness, let us ask what happens when we let $z_k \to \infty$, which is the same as $1/z_k \to 0$. Then the mapping $t = -1/z = -1/(x + iy) = (-x + iy)/(x^2 + y^2)$ is real when z is real, and maps the upper half z plane conformally onto the upper half t plane, taking $z_k = \infty$ into $t_k = 0$. Let us assume further, for simplicity, that none of the z_i's is zero. Then, to map the upper half t plane into the given polygon in the w plane, we use the mapping

$$\frac{dw}{dt} = ct^{\gamma_k/\pi - 1} \prod_{i=1}^{k-1} (t - t_i)^{\gamma_i/\pi - 1}$$

since $t_k = 0$; and, substituting $t = -1/z$, we have

$$\frac{dw}{dt} = c(-1/z)^{\gamma_k/\pi - 1} \prod_{i=1}^{k-1} \left(-\frac{1}{z} + \frac{1}{z_i}\right)^{\gamma_i/\pi - 1}$$

$$= \frac{c(-1)^{\gamma_k/\pi - 1} \prod_{i=1}^{k-1} (z - z_i)^{\gamma_i/\pi - 1}}{z^{\gamma_k/\pi - 1} (z)^{(\gamma_1/\pi - 1) + \cdots + (\gamma_{k-1}/\pi - 1)} \prod_{i=1}^{k-1} z_i^{\gamma_i/\pi - 1}}$$

$$= \frac{c' \prod_{i=1}^{k-1} (z - z_i)^{\gamma_i/\pi - 1}}{z^{(\gamma_1/\pi - 1) + \cdots + (\gamma_k/\pi - 1)}},$$

where all the constant factors have been absorbed into the new constant c'; and, since for a polygon with k sides, $\gamma_1 + \cdots + \gamma_k = (k - 2)\pi$, the denominator of the last expression becomes z^{-2}. After use of the chain rule

$$\frac{dw}{dt} = \frac{dw}{dz}\frac{dz}{dt} = \frac{1}{t^2}\frac{dw}{dz} = z^2 \frac{dw}{dz},$$

we have

$$\frac{dw}{dz} = c' \prod_{i=1}^{k-1} (z - z_i)^{\gamma_i/\pi - 1}.$$

Thus, placing one of the vertices $z = z_k$ at infinity simply has the effect of eliminating the corresponding factor in the Schwarz–Christoffel formula. The reader is encouraged to try Example 6 with $z_1 = -1$, $z_2 = 1$, and $z_3 = \infty$

to see what sort of an expression for w emerges. Note also that choosing $z_k = \infty$ uses up one degree of freedom, so that only two of the remaining z_i's can be chosen arbitrarily.

When we use this simplification, and further, when the problem consists of mapping the upper half plane into a polygon which also has a vertex at infinity, then it is often possible to achieve a reasonably satisfactory solution. The technique used in this case is best illustrated by recourse to examples.

EXAMPLE 7 Map the upper half plane conformally onto the semi-infinite strip $-1 \leq Re\, w \leq 1$, $Im\, w \geq 0$ (Figure 4.19). The strip can be looked at as a triangle with one vertex (w_1) at infinity. With $w_2 = -1$ and $w_3 = 1$, the corresponding angles are $\gamma_1 = 0, \gamma_2 = \gamma_3 = \pi/2$. Let us choose the points in the z plane with z_1 at infinity, z_2 at $-a$ (real), and z_3 at a. Then

$$\frac{dw}{dz} = c(z-a)^{-1/2}(z+a)^{-1/2} = \frac{c'}{(a^2-z^2)^{1/2}},$$

the term for $z_1 = \infty$ having been replaced by 1 as the previous argument allows. Then, taking the lower limit of integration to be $z_0 = -a$, we have

$$w = c' \int_{-a}^{z} \frac{1}{\sqrt{a^2-t^2}} + d,$$

from which we immediately obtain

$$w(-a) = -1 = d \quad \text{so} \quad d = -1.$$

Thus

$$w = \int_{-a}^{z} \frac{c'\,dt}{\sqrt{a^2-t^2}} - 1.$$

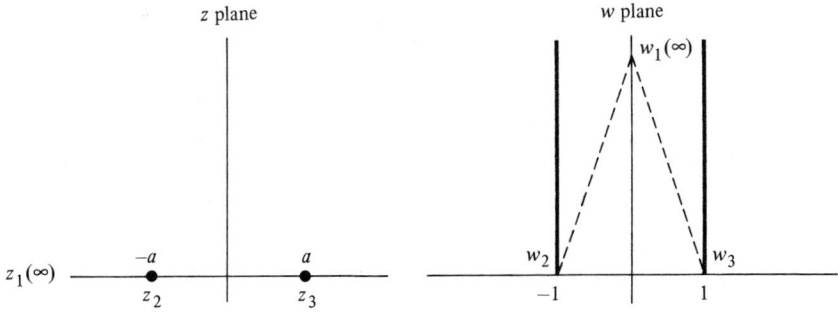

Figure 4.19

As z proceeds from $-a$ to a, w goes from -1 to 1, so

$$w(a) = 1 = \int_{-a}^{a} \frac{c'\,dt}{\sqrt{a^2 - t^2}} - 1,$$

whence

$$2 = c' \arcsin(t/a)\big|_{-a}^{a} = c'[\arcsin 1 - \arcsin(-1)] = c'[\pi/2 - (-\pi/2)] = c'\pi,$$

giving $c' = 2/\pi$, so that

$$w(z) = \frac{2}{\pi} \int_{-a}^{z} \frac{dt}{\sqrt{a^2 - t^2}} - 1,$$

which can be integrated in closed form, giving

$$w(z) = \frac{2}{\pi} \arcsin \frac{z}{a}.$$

The arcsine is a multiple-valued function, of course, but we have implicitly chosen which value it is to assume so that $w(z)$ is analytic and univalent for values of z; namely, we chose the value for the square root under the integral sign which is positive for real z between $-a$ and a. Hence for these values of z, $\arcsin(z/a)$ will assume real values between $-\pi/2$ and $\pi/2$.

EXAMPLE 8   Map the upper half plane conformally onto the infinite strip $0 \leq \operatorname{Im} w \leq a$ (Figure 4.20). To do this, we again look upon the figure in the w plane as the limiting position of a triangle, with $w_1 = w_2 = \infty$ and $w_3 = ia$, so that $\gamma_1 = \gamma_2 = 0$, $\gamma_3 = \pi$. In the z plane we take $z_1 = \infty$, $z_2 = 0$, $z_3 = 1$. Then

$$\frac{dw}{dz} = c(z - 0)^{-1}(z - 1)^0 = \frac{c}{z},$$

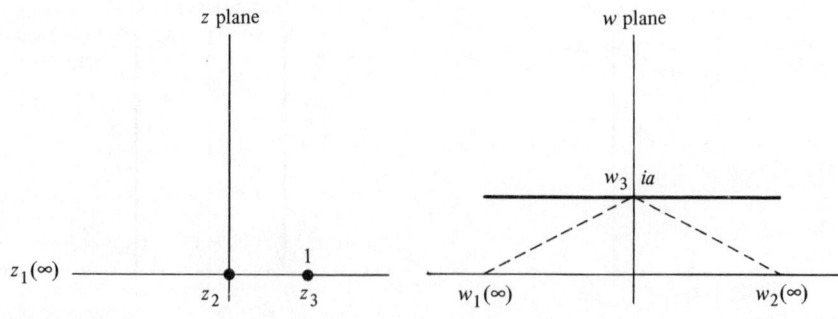

Figure 4.20

the term for $z_1 = \infty$ having been replaced by 1 as usual. Then the obvious solution of this differential equation is $w = c \log z + d$. For real z between 0 and 1, the imaginary part of w must be a, so we choose the branch of the logarithm which is real for $0 < z < 1$, which forces c to be real and $d = ia$. Since $z = 0$ is a branch point of the logarithm, if we proceed around the origin counterclockwise from a position on the positive real axis to one on the negative real axis, then arg z changes by π, so the logarithm changes by πi, and thus, for $z < 0$,

$$w = c \log|z| + ia + ic\pi.$$

This expression for w must be real, since the image of $\overline{z_1 z_2}$ is $\overline{w_1 w_2}$; hence $ia + ic\pi = 0$ and, therefore, $c = -a/\pi$. Thus, the final mapping is

$$w(z) = -\frac{a}{\pi} \log z + ia,$$

with the branch of the logarithm chosen to be real for real z between 0 and 1.

Of course, we can choose other values for z_1, z_2, and z_3, and thereby obtain a different mapping, one which also takes the upper half plane into the strip. For instance, if we choose $z_1 = -1$, $z_2 = 1$, and $z_3 = \infty$, then we obtain

$$\frac{dw}{dz} = c_0(z+1)^{-1}(z-1)^{-1} = \frac{c}{z+1} - \frac{c}{z-1} \quad \left(c = -\frac{c_0}{2} \right),$$

whence

$$w = c \log \left(\frac{z+1}{z-1} \right) + d.$$

For $-1 < z < 1$, w must be real, so we rewrite this as

$$w = c' \log \left(\frac{z+1}{1-z} \right) + d',$$

and choose $\arg[(z+1)/(1-z)]$ to be zero for $-1 < z < 1$, so that the logarithm is real for these values. This forces c' and d' to be real. Rewriting

$$w = c' \log(z+1) - c' \log(1-z) + d',$$

we note that if we proceed clockwise around $z_2 = 1$, then $\arg(z+1)$ is unchanged, while $\arg(1-z)$ changes by $-\pi$, so that for $z > 1$ we have

$$w = c' \log(z+1) - c' \log|1-z| + c'\pi i + d'.$$

We have entered the segment where Im $w = a$, so we must have

$$d' + c'\pi i = ia.$$

and, since d' is real, it must vanish, and $c' = a/\pi$. Thus
$$w = \frac{a}{\pi} \log \frac{z+1}{1-z},$$
where the branch of the logarithm is chosen so that it is real for $-1 < z < 1$.

Finally, let us evaluate c' in Example 6; we first define the *beta function*
$$B(\alpha, \beta) = \int_0^1 x^{\alpha-1}(1-x)^{\beta-1}\,dx,$$
for those real or complex values of α and β for which the integral exists. It is a fairly easy matter to derive a relationship between the beta and gamma functions. We do this by first writing
$$\Gamma(\alpha)\,\Gamma(\beta) = \int_0^\infty t^{\alpha-1} e^{-t}\,dt \int_0^\infty u^{\beta-1} e^{-u}\,du.$$
For $\alpha > 0$, $\beta > 0$ the two integrals on the right-hand side are absolutely convergent, so the order of integration can be rearranged as follows:
$$\Gamma(\alpha)\,\Gamma(\beta) = \int_0^\infty t^{\alpha-1}\,dt \int_0^\infty u^{\beta-1} e^{-(t+u)}\,du.$$
Next we change the variable of integration in the inner integral to $v = t + u$, whence $u = v - t$ and $du = dv$. As for the limits of integration, we have $0 \leq t \leq t + u = v < \infty$, so v runs from t to ∞, and we have
$$\Gamma(\alpha)\,\Gamma(\beta) = \int_0^\infty t^{\alpha-1}\,dt \int_t^\infty (v-t)^{\beta-1} e^{-v}\,dv$$
$$= \int_0^\infty t^{\alpha-1}\,dt \int_t^\infty v^{\beta-1} \left(1 - \frac{t}{v}\right)^{\beta-1} e^{-v}\,dv.$$
Now we again reverse the order of integration; the outer variable v will go from 0 to ∞, while t will go from 0 to v, since $0 \leq t \leq v < \infty$. Thus
$$\Gamma(\alpha)\,\Gamma(\beta) = \int_0^\infty e^{-v} v^{\beta-1}\,dv \int_0^v t^{\alpha-1} \left(1 - \frac{t}{v}\right)^{\beta-1}\,dt.$$
Now we make one further change of variables, this time putting $x = t/v$ in the inner integral, so that $dt = v\,dx$ and x now goes from 0 to 1:
$$\Gamma(\alpha)\,\Gamma(\beta) = \int_0^\infty e^{-v} v^{\beta-1}\,dv \int_0^1 (xv)^{\alpha-1}(1-x)^{\beta-1} v\,dx$$
$$= \int_0^\infty e^{-v} v^{\alpha+\beta-1}\,dv \int_0^1 x^{\alpha-1}(1-x)^{\beta-1}\,dx.$$
But this statement says that $\Gamma(\alpha)\,\Gamma(\beta) = \Gamma(\alpha+\beta)\,B(\alpha, \beta)$ or
$$B(\alpha, \beta) = \frac{\Gamma(\alpha)\,\Gamma(\beta)}{\Gamma(\alpha+\beta)}.$$

4.3. THE SCHWARZ–CHRISTOFFEL TRANSFORMATION

At first glance it may seem that deriving this formula was a rather artificial process with very little motivation behind it except "knowing in advance" that the formula was true. To a certain extent this criticism is justified; but one at least has a clue at the beginning as to where to go, since the same exponents occur in the product of the two gamma functions as in the beta function, and from this fact one is tempted strongly to change variables and orders of integration in an attempt to come up with something which looks like a beta function.

We can now use the relationship between these two functions to evaluate the constant c':

$$1 = c' \int_0^1 \frac{dz}{\sqrt{z}(1-z^2)^{\frac{3}{4}}}.$$

Let us change variables by putting $z = \sqrt{t}$, so that $dz = dt/(2\sqrt{t})$; then

$$1 = c' \int_0^1 \frac{dt}{2\sqrt{t}\, t^{1/4}(1-t)^{3/4}} = \frac{c'}{2} \int_0^1 \frac{dt}{t^{3/4}(1-t)^{3/4}}$$

$$= \frac{c'}{2} \int_0^1 t^{(1/4)-1}(1-t)^{(1/4)-1}\, dt.$$

Therefore

$$1 = \frac{c'}{2} \frac{\Gamma(1/4)\,\Gamma(1/4)}{\Gamma(1/2)} \quad \text{so} \quad c' = \frac{2\Gamma(1/2)}{[\Gamma(1/4)]^2} = \frac{2\sqrt{\pi}}{[\Gamma(1/4)]^2}.$$

A. Exercises

1 Find a Schwarz-Christoffel transformation which maps the upper half plane into (a) an equilateral triangle (b) any triangle.

2 Find a Schwarz-Christoffel transformation mapping the upper half plane into each of the following domains:

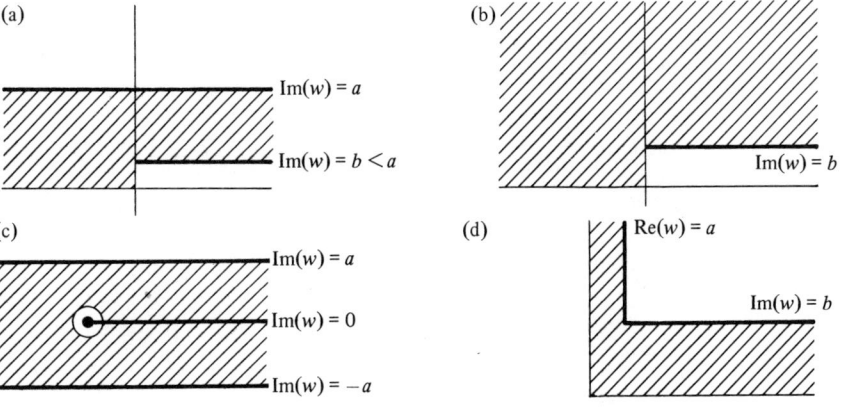

Figure 4.21

Chapter **5**

Two-Dimensional Incompressible Fluid Flow

We are about to begin our study of one of the most elegant applications of the complex variable theory and techniques which we have now learned: the two dimensional flow of an incompressible fluid. This application has the advantage of being easy to visualize, advanced enough to be of interest, and yet simple enough to be accessible to a student at this level. And, while fluid flow presents applications of our theory, it simultaneously aids us in more purely complex variables problems, such as in evaluating the constants which occur in the Schwarz-Christoffel transformation.

There are many great names associated with the study of fluid flows. A pioneer in the use of complex analysis in hydrodynamics was Riemann (1826–1866). He obtained many insights into the nature of complex functions by interpreting them as flows; like the great Einstein (who was inspired by Riemann's ideas about geometry), he made many of his discoveries not in the laboratory, but as a result of carefully conceived "Gedankenexperimente" (thought-experiments).

5.1. DERIVATION OF THE EQUATIONS AND THEIR IMPLICATIONS

In order both to simplify our study of fluid flow and to make it amenable to treatment by means of complex variables techniques, we are going to have to make a number of simplifying assumptions about our fluid and its flow. Once we have listed these assumptions, the reaction of many students may well be to the effect that we have emasculated the problem. In answer to this, we state that in building a mathematical model (i.e., a system of equations) to describe any physical situation, it is *absolutely essential* to make *some* simplifying assumptions. For example, it is simply impossible to consider individually each of the billions upon billions of molecules in a fluid, so we must treat it as if it were a continuum. Secondly, not all simplifying assumptions are totally unreasonable; many are such that only the finest degree of instrumentation could detect their truth or falsehood. Third, many assumptions are necessary in order to treat a problem by a given method; for instance, we must assume a two-dimensional flow if we are to use complex variable theory, since a complex vector is, in essence, two-dimensional. And fourth, it is remarkable how good the answers sometimes are!

For the sake of completeness, then, we list the assumptions we shall make:

1. We shall assume that the flow is *two dimensional*, in the sense that the velocity vectors at all points in the fluid are parallel to a given plane (which we shall take to be the z plane), and that any plane cross section of the fluid taken parallel to the z plane shows the same motion as any other.
2. The flow is a *steady-state* flow, meaning that at each point in the fluid the flow looks the same for all time, i.e., the velocity of the fluid at any point is constant and does not depend on time.
3. The fluid is *incompressible*, meaning that the density ρ (the mass per unit volume) is constant. Since the mass of any region of fluid of area A (and of unit height) is to be ρA however small A may be, this assumption also includes the assumption that the fluid is a continuum and not made up of a set of discrete molecules, however small and plentiful.
4. The flow is *irrotational*, meaning that it contains no "whirlpools," i.e., points of zero velocity around which fluid circulates in a closed path. (This concept will be formulated more precisely later.)
5. The flow is *source free*, meaning that nowhere in the flow region is fluid being created or destroyed.
6. And finally, the fluid is *nonviscous*, meaning that its motion is frictionless and that the fluid is unable to withstand any shear stress.

5.1. Derivation of the Equations and their Implications

Once we have made our assumptions, it is still possible to consider the flow physically in a number of different ways. One way would be to attempt to follow each point as it moves along in the flow; to do so, we would need to know the coordinates of the point at time $t(x(t), y(t))$ and also its velocity components in the x and y directions, $u(t)$ and $v(t)$. Another way would be to sit at the point (x, y) and watch the fluid as it goes by, with velocity components $u(x, y, t)$ and $v(x, y, t)$.* We shall use this latter approach, because our flow is steady state, and as a result u and v each depend on x and y alone, and not on t (Assumption 2).

Thus, in deriving our equations of motion, we shall use

$$\mathbf{q} = u + iv = \frac{dx}{dt} + i\frac{dy}{dt} = qe^{i\theta}$$

to represent the velocity vector with components u (in the x direction) and v (in the y direction). We shall use ρ to represent the (constant) density in units of mass per unit area (actually, assuming that the area is the base of a solid with unit height), and $p = p(x, y)$ to represent the pressure in units of force per unit length, at the point (x, y).

First, let us see what information we can glean from Assumptions 4 and 5. Assumption 4, that the flow is irrotational, tells us that there can be no net flow around any simple closed curve. In other words, if we consider the tangential component of the velocity of the flow around any simple closed contour, its average value around the contour must be zero; it cannot point in one direction more frequently than it does in any other.

Thus, suppose $C = \{z(s): z(s) = x(s) + iy(s), \ 0 \leq s \leq l\}$ is a simple closed contour, where the parameter s is arc length. Then the tangent vector to C is $z'(s) = x'(s) + iy'(s)$, and this tangent vector has unit length. Therefore the tangential component of the velocity vector is

$$(u + iv) \cdot [x'(s) + iy'(s)] = ux' + vy',$$

where "$\cdot$" symbolizes the usual dot product of vectors (not complex multiplication). The statement that the flow is irrotational, then, is the statement that

$$\oint_C (u + iv) \cdot (x' + iy')\, ds = \oint_C (ux' + vy')\, ds = \oint_C (u\, dx + v\, dy) = 0.$$

To repeat, this equation merely says that the average tangential component of $u + iv$ (the velocity vector) around any simple closed contour C is zero; in other words, there is no net *circulation* around C.

But now we can apply Green's Theorem to this last result, and we obtain

$$\oint_C (u\, dx + v\, dy) = \int\int_{I(C)} \left(\frac{\partial v}{\partial x} - \frac{\partial u}{\partial y} \right) dx\, dy = 0.$$

---

* These two approaches are due to Lagrange and Euler, respectively.

This result must hold in the interior of *any* simple closed contour C in the flow region whose interior also lies in the flow region; hence the integrand is identically zero

$$\frac{\partial v}{\partial x} - \frac{\partial u}{\partial y} \equiv 0$$

everywhere. For if this were not the case, then there would have to be a point where $v_x - u_y \neq 0$; for the sake of definiteness, say at this point that $v_x - u_y = \delta > 0$. Then because of continuity—an unspoken assumption throughout—we can surround this point by a small circle Σ of area A inside which $v_x - u_y > \delta/2$. But then

$$\iint_{I(\Sigma)} (v_x - u_y)\,dx\,dy > (\delta/2)\,A > 0,$$

a contradiction. Hence we conclude that throughout the flow region,

$$\frac{\partial v}{\partial x} - \frac{\partial u}{\partial y} = 0 \quad \text{or} \quad \frac{\partial u}{\partial y} = \frac{\partial v}{\partial x}.$$

Assumption 5 states that fluid is neither created or destroyed in the flow region. We can interpret this as saying that the net flow into a region bounded by a simple closed contour must equal the net flow out. How shall we put this fact mathematically? Let us examine a small portion of C (Figure 5.1) of length Δs. During a unit interval of time the amount of fluid leaving $I(C)$ over this arc is equal to* the area of the small rhombus indicated,

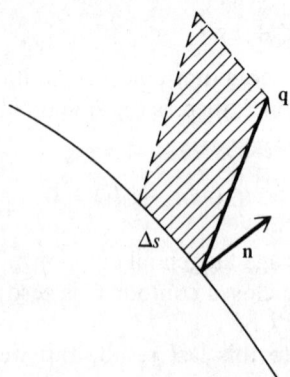

Figure 5.1

---

* In much of what follows, one should read "is closely approximated by" rather than "is equal to." We are talking like engineers!

5.1. DERIVATION OF THE EQUATIONS AND THEIR IMPLICATIONS 223

which has Δs as one side and the velocity vector $\mathbf{q} = u + iv$ as its other side. The area of this rhombus is the product of Δs and the component of $\mathbf{q}$ perpendicular (normal) to C. Now the vector $-y'(s) + ix'(s)$ is clearly a unit vector normal to C; hence the normal component of $\mathbf{q}$ is

$$(u + iv) \cdot (-y' + ix') = -uy' + vx'.$$

The area of the small rhombus is therefore $(-uy' + vx')\Delta s$; the total flux of fluid outwards through C is approximately the sum of all such small components or

$$\Sigma(-uy' + vx')\Delta s;$$

and in the limit as the number of such arcs on C increases without limit and the maximum Δs goes to zero, we obtain

$$\oint_C (-uy' + vx')\, ds = \oint_C (u\, dy - v\, dx) = 0.$$

This integral expresses the fact that no more, and no less, fluid enters $I(C)$ than leaves it.

By the same argument that we used on Assumption 4, we have by Green's Theorem that

$$\oint_C (u\, dy - v\, dx) = \iint_{I(C)} \left(\frac{\partial v}{\partial y} + \frac{\partial u}{\partial x} \right) dx\, dy = 0,$$

and, because C is an arbitrary simple closed curve, we must have

$$\frac{\partial u}{\partial x} + \frac{\partial v}{\partial y} = 0 \quad \text{or} \quad \frac{\partial u}{\partial x} = -\frac{\partial v}{\partial y}$$

throughout the region of flow. This last equation, together with the conclusion from Assumption 4 that

$$\frac{\partial u}{\partial y} = \frac{\partial v}{\partial x},$$

remind one of the Cauchy-Riemann equations except that the signs of the right-hand sides are different. We conclude, by Theorem 2.2.6, that the function

$$f(z) = u - iv,$$

which is the *conjugate* of the velocity vector

$$\bar{f}(z) = u + iv,$$

is an analytic function.

Before we pursue further implications of this, let us consider the principle of conservation of momentum. Mathematical statements of this principle arise from a consideration of Newton's law that any applied force

224 TWO-DIMENSIONAL INCOMPRESSIBLE FLUID FLOW

brings about a corresponding change in momentum (usually rendered $F = ma$). Let us examine a small rectangular element of the fluid as illustrated in Figure 5.2. Working in the x direction first, we see that the net force acting on the element is

$$(p(x, y) - p(x + \Delta x, y))\Delta y,$$

while the change of momentum is

$$ma_x = \rho \, \Delta x \, \Delta y \, \frac{du}{dt} = \rho \, \Delta x \, \Delta y \left(\frac{\partial u}{\partial x} \frac{dx}{dt} + \frac{\partial u}{\partial y} \frac{dy}{dt} \right) = \rho \, \Delta x \, \Delta y (u_x u + u_y v).$$

By equating the two, we obtain

$$\frac{p(x, y) - p(x + \Delta x, y)}{\Delta x} = \rho(u_x u + u_y v)$$

which, in the limit as Δx and Δy approach zero, becomes

$$-p_x = \rho(u_x u + u_y v)$$

or, finally,

$$u_x u + u_y v + \frac{p_x}{\rho} = 0.$$

If we now examine the y direction, we obtain similarly that

$$F_y = ma_y$$

or

$$[p(x, y) - p(x, y + \Delta y)] \Delta x = \rho \, \Delta x \, \Delta y \left(\frac{\partial v}{\partial x} \frac{dx}{dt} + \frac{\partial v}{\partial y} \frac{dy}{dt} \right)$$

$$= \rho \, \Delta x \, \Delta y (v_x u + v_y v).$$

Again dividing through by $\Delta x \, \Delta y$ and taking the limit, we obtain

$$v_x u + v_y v + \frac{p_y}{\rho} = 0.$$

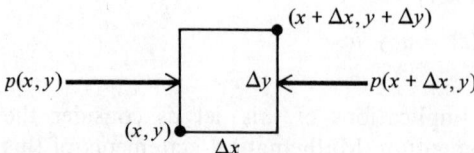

Figure 5.2

5.1. Derivation of the Equations and their Implications

By using the equations due to the source free and irrotational nature of the flow,

$$u_x = -v_y, \qquad u_y = v_x,$$

the two equations of conservation of momentum can be transformed into the two equations

$$u_x u + v_x v + \frac{p_x}{\rho} = 0, \qquad u_y u + v_y v + \frac{p_y}{\rho} = 0,$$

each of which contains differentiations with respect to just one of the independent variables x and y. Therefore, the two equations can be integrated, the first with respect to x (holding y constant) and the second with respect to y (holding x constant), yielding

$$\frac{1}{2}u^2 + \frac{1}{2}v^2 + \frac{p}{\rho} = c_1(y), \qquad \frac{1}{2}u^2 + \frac{1}{2}v^2 + \frac{p}{\rho} = c_2(x),$$

where c_1 is an arbitrary function of y and c_2 is an arbitrary function of x. If these two equations are to be consistent, i.e., if $c_1(y) = c_2(x)$ for all x and y, then c_1 and c_2 must, in fact, both be equal to the same constant c, and we have

$$\frac{1}{2}(u^2 + v^2) + \frac{p}{\rho} = c.$$

This equation is a mathematical statement of the so-called *Law of Bernoulli*, that pressure drops in a fluid as the speed increases. In fact, $u^2 + v^2$ is the square of the speed. We shall see a number of instances in which Bernoulli's law enables us to determine information about fairly difficult flows.

Now let us return for a moment to our consideration of the velocity vector $\bar{f}(z) = u + iv$, where, we recall, the function $f(z) = u - iv$ is an analytic function. We have shown that as a consequence of the fact that the flow is source free, we have

$$\oint_C \bar{f}_n \, ds = 0,$$

C being any simple closed curve and $\bar{f}_n$ the component of the velocity normal to C. Then it is also true that if we integrate $\bar{f}_n$ (or its negative) along any curve joining the points z_0 and z, then*

$$\int_{z_0}^{z} (-\bar{f}_n) \, ds = \psi(x, y) - \psi(x_0, y_0)$$

---

* The reason for using $-\bar{f}_n$ will become apparent shortly.

is independent of the path used to join the two points and is, hence, a well-defined (real-valued) function. This function ψ obtained by integrating $-\bar{f}_n$, or, what is the same,

$$\psi(x, y) - \psi(x_0, y_0) = \int_{z_0}^{z} (u\,dy - v\,dx)$$

is called the *stream function* of the flow. If c is any constant, then the curves whose equations are $\psi(x, y) = c$ are called the *streamlines* of the flow. There is no net flux of fluid across a streamline, for if z and z_0 are on the same streamline, then

$$\int_{z_0}^{z} (u\,dy - v\,dx) = \psi(x, y) - \psi(x_0, y_0) = c - c = 0.$$

Hence *the tangent to a streamline always points in the direction of flow.*

The advantage of studying a flow through its streamlines is that once they are known, the velocity field is also known. The stream function has further properties that make its determination more straightforward and which link it even more closely to u and v. For, since the function $f(z) = u(x, y) - iv(x, y)$ is analytic, it follows that the integral

$$\int_{z_0}^{z} f(z)\,dz = F(z) - F(z_0)$$

is independent of the path joining z_0 to z; therefore, the function $F(z)$ is also analytic. But

$$F(z) - F(z_0) = \int_{z_0}^{z} (u - iv)(dx + i\,dy) = \int_{z_0}^{z} (u\,dx + v\,dy)$$

$$+ i \int_{z_0}^{z} (u\,dy - v\,dx).$$

And now we see why we chose the sign as we did in the definition of ψ, for we see from this equation that

$$\operatorname{Im}\left[F(z) - F(z_0)\right] = \psi(x, y) - \psi(x_0, y_0)$$

or simply

$$\operatorname{Im}(F(z)) = \psi(x, y).$$

Hence the stream function ψ is the imaginary part of an analytic function. Therefore, by Theorem 2.5.4, ψ is harmonic throughout the region of flow,

$$\frac{\partial^2 \psi}{\partial x^2} + \frac{\partial^2 \psi}{\partial y^2} = 0.$$

If the fluid under consideration flows past an obstacle or along a boundary

5.1. DERIVATION OF THE EQUATIONS AND THEIR IMPLICATIONS

(Figure 5.3), then each portion of the boundary and of the obstacle must be a streamline or a portion of a streamline, i.e., a curve along which ψ is constant. In the example illustrated, one can arbitrarily choose the value of ψ on one boundary, the value 0 often being chosen for simplicity. Then, since the total flux Φ of the fluid between the boundaries is normally given, one can determine ψ on the top boundary since $\psi(\text{top}) - \psi(\text{bottom}) = \Phi - 0 = \Phi$. Considerations of symmetry often determine ψ on the obstacle as well; then the combination of Laplace's equation in the flow region together with known values of ψ on the boundary lead to what is called a *boundary value problem* for ψ. The study of such problems forms a large and important branch of mathematics, and most such problems are difficult or impossible to solve explicitly. Needless to say, we shall confine ourselves to relatively simple cases.

To get back to the function $F(z)$, the real part of F is usually denoted by $\phi(x, y)$, thus $F(z) = \phi(x, y) + i\psi(x, y)$. The function $\phi(x, y)$ is also harmonic,

$$\phi_{xx} + \phi_{yy} = 0,$$

and ϕ and ψ satisfy the Cauchy-Riemann equations:

$$\phi_x = \psi_y, \qquad \phi_y = -\psi_x.$$

Therefore $F'(z) = \phi_x + i\psi_x = \psi_y - i\phi_y$, or, if we wish,

$$F'(z) = \phi_x - i\phi_y$$

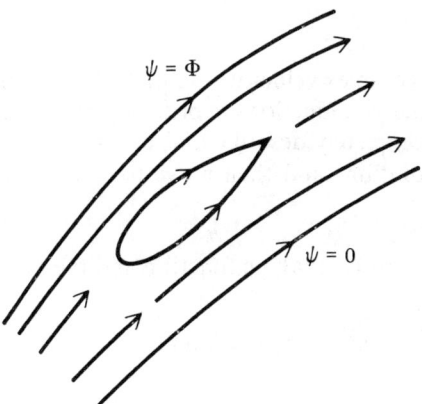

Figure 5.3

228 TWO-DIMENSIONAL INCOMPRESSIBLE FLUID FLOW

in terms of ϕ alone. But by its definition, F is an indefinite integral of the conjugate f of the velocity field, so that

$$F'(z) = f(z) = u - iv = \phi_x - i\phi_y.$$

Hence the velocity components of our flow are given by the respective partial derivatives of ϕ:

$$\phi_x = u, \quad \phi_y = v.$$

A harmonic function such as ϕ whose partial derivatives give the velocity components of a fluid flow is called a *velocity potential*.

The function $F(z)$ is called the *complex velocity potential* of the flow. Once we know the function $F(z)$ we know all there is to know about the flow through the velocity potential and the stream function, the real and imaginary parts of F. Because ϕ and ψ are related by the Cauchy-Riemann equations $\phi_x = \psi_y$, $\phi_y = -\psi_x$, then once we know ψ we can determine ϕ to within an arbitrary additive constant (or vice versa). The value of this constant is immaterial to the flow, since it will not affect the velocity field, which is given by the partial derivatives of ϕ. Hence it is usually customary to choose the value of ϕ to be zero at some convenient point in the flow (just as we chose ψ to be zero along one boundary—an act which does not affect the values of the derivatives of ψ).

In summary:

1. $F(z) = \phi(x, y) + i\psi(x, y)$, analytic, the complex velocity potential;
2. $\phi(x, y)$, the velocity potential;
3. $u = \phi_x$, $v = \phi_y$, the velocity components of the flow;
4. $\psi(x, y)$, the stream function, constant on streamlines;
5. $\bar{f}(z) = u(x, y) + iv(x, y)$, the velocity vector of the flow;
6. $f(z) = u - iv = F'(z)$, the conjugate of the velocity vector, an analytic function.

Any analytic function $F(z)$ is the complex velocity potential of a certain fluid flow. One can, therefore, study analytic functions in order to determine a large class of flows which can be completely described, or else start with simpler flows which have obvious solutions and then move toward more complicated ones.

EXAMPLE 1 As an example of a fairly simple analytic function and the flow it describes, let us consider

$$F(z) = z^2 = (x + iy)^2 = (x^2 - y^2) + i(2xy).$$

Here the velocity potential is

$$\phi = x^2 - y^2$$

5.1. DERIVATION OF THE EQUATIONS AND THEIR IMPLICATIONS

and the stream function is

$$\psi = 2xy$$

Thus the streamlines are the curves given by the equations $2xy =$ constant. If the constant is zero, then $x = 0$ or $y = 0$ or both, i.e., the x and y axes are streamlines. In each of the four quadrants of the plane, the streamlines are hyperbolas having the axes as asymptotes; for the first quadrant, see Figure 5.4. The components of the velocity of the flow are $u = \phi_x = 2x$, $v = \phi_y = -2y$. In the first quadrant, we see that all velocity vectors $\bar{f}(x) = 2x - 2iy$ are directed toward the lower right, and the direction of flow is as indicated in the figure. Thus the complex velocity potential $F(z) = z^2$ describes flow around a 90° corner in the first quadrant.

Notice that the velocity is zero at the origin. Any point along the boundary at which the velocity is zero is called a *stagnation point*. It turns our that any flow around a corner of less than 180° will produce a stagnation point at the corner; the complex potential for such a flow is $F(z) = z^\alpha$ where $\alpha > 1$, or in polar coordinates, $F(z) = r^\alpha(\cos\alpha\theta + i\sin\alpha\theta)$. Thus we have the velocity potential $\phi(x, y) = r^\alpha \cos\alpha\theta$. Now because $r = \sqrt{x^2 + y^2}$, we have

$$\frac{\partial r}{\partial x} = \frac{1}{2}(x^2 + y^2)^{-1/2}(2x) = \frac{x}{r}, \qquad \text{likewise} \quad \frac{\partial r}{\partial y} = \frac{y}{r}.$$

And, since $\tan\theta = y/x$, we have

$$\frac{\partial \theta}{\partial x} = \frac{-y/x^2}{1 + y^2/x^2} = \frac{y}{r^2}; \qquad \frac{\partial \theta}{\partial y} = \frac{1/x}{1 + y^2/x^2} = \frac{x}{r^2}.$$

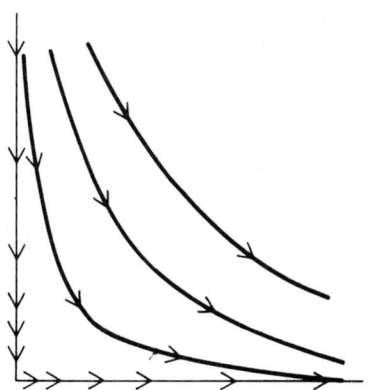

Figure 5.4

Therefore, the velocity components are

$$\frac{\partial \phi}{\partial x} = \alpha r^{\alpha-1} \frac{x}{r} \cos \alpha\theta - \alpha r^\alpha \sin \alpha\theta \cdot \left(\frac{-y}{r^2}\right)$$

$$= \alpha r^{\alpha-2}(x \cos \alpha\theta + y \sin \alpha\theta),$$

Similarly

$$\frac{\partial \phi}{\partial y} = \alpha r^{\alpha-2}(y \cos \alpha\theta - x \sin \alpha\theta).$$

If $1 < \alpha < 2$, then x and y approach zero at the origin more strongly than than $\alpha r^{\alpha-2}$ approaches infinity there, and if $\alpha > 2$ then the velocity components are obviously zero at the origin. We have already considered the case $\alpha = 2$. Thus, all cases having been covered, we see that an inside corner with an angle of less than 180° will always be a stagnation point.

EXAMPLE 2 Now let us consider the function $F(z) = \log z = \log r + i\theta$, where as usual $r = |z| = \sqrt{x^2 + y^2}$, and $\theta = \arg z$. Because of the ambiguity in the argument, let us restrict our consideration to the upper half plane, as given by $0 \leq \theta \leq \pi$. In this domain and with this determination of the argument, F is analytic and single-valued. Then the stream function is $\psi = \arg z$, which is constant along radial lines emanating from the origin (Figure 5.5). The velocity potential is $\phi = \log |z| = \frac{1}{2} \log(x^2 + y^2)$; hence the velocity vector is

$$\bar{f}(z) = \phi_x + i\phi_y = \frac{1}{2}\left(\frac{2x}{x^2+y^2} + i\frac{2y}{x^2+y^2}\right) = \frac{x+iy}{x^2+y^2}.$$

This vector has the same direction as $x + iy$ (away from the origin), but a magnitude that decreases inversely with r. Figure 5.5 illustrates the flow and a few of the streamlines.

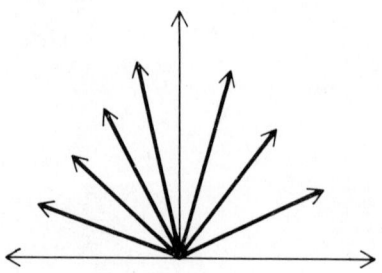

Figure 5.5

We shall consider some further flow problems in the exercises, i.e., given F, determine the flow described; and in subsequent sections of this chapter, we shall solve much more complicated flow problems from scratch, using the methods of conformal mapping.

A. Exercises

Describe the flows given by each of the following complex velocity potentials. If convenient for the problem, you may restrict your consideration to a portion of the plane; for example, to make the function single-valued as in Example 2 or to consider one of a number of congruent slices of the plane, as in Example 1.

1 $F(z) = z^3$
2 $F(z) = z^{3/2}$
3 $F(z) = 1/z$
4 $F(z) = z^2 + 1$
5 $F(z) = (z + 1)^2$
6 $F(z) = iz^2$
7 If $F(z) = \phi + i\psi$, then what flow is described by $-\psi + i\phi$?

5.2. CERTAIN BASIC FLOWS

In this section we shall consider how to approach certain very basic flow problems and solve them from the beginning. We shall build a basic "library" of flows which we can solve in this way; in subsequent sections we shall use conformal mapping techniques to reduce much more difficult problems to one of the basic ones which we know how to solve.

First we consider the following problem:

EXAMPLE 3 Describe the flow through an infinite channel, one side of which is the real axis and the other side of which is $y = b$ (Figure 5.6).*
We assume that somewhere 'way out on the left—we normally say at infinity—the fluid flows into the channel with uniform horizontal velocity $\mathscr{U}$.

Both the x axis and the line $y = b$ are streamlines, because they form the boundaries of the flow region (see Section 5.1). Hence along these two lines we know that the stream function ψ is a constant; for simplicity we choose ψ to be equal to 0 on the x axis. The total flux into the channel is

---

* When we speak of a flow, we tacitly assume that all of the conditions mentioned in Section 5.1 hold for the flow.

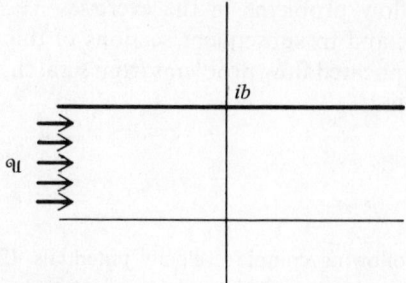

Figure 5.6

simply the velocity multiplied by the width, $\mathscr{U}b$; hence the value of ψ for $y = b$ must be $\mathscr{U}b$. Thus we have the following boundary value problem for ψ:

$$\psi_{xx} + \psi_{yy} = 0, \qquad \psi(x, 0) = 0, \qquad \psi(x, b) = \mathscr{U}b.$$

It would appear that ψ must be a function of y alone, since the boundary conditions (on curves where y does not vary) do not depend on x. If we assume that this is indeed true, then we have

$$\psi''(y) = 0, \qquad \psi(0) = 0, \qquad \psi(b) = \mathscr{U}b.$$

By integrating once, we have $\psi'(y) = c_1$, and a second time, we obtain $\psi(y) = c_1 y + c_2$; utilizing the boundary conditions,

$$\psi(0) = 0 = c_2, \qquad \psi(b) = c_1 b = \mathscr{U}b,$$

so that $c_1 = \mathscr{U}$ and, therefore, $\psi = \mathscr{U}y$.

The reader who has some experience with partial differential equations may perceive that the above boundary value problem also has as solution any function of the form

$$\sin \frac{n\pi y}{b} (c_n e^{-n\pi x/b} + d_n e^{n\pi x/b}) + \mathscr{U}y,$$

where n is any positive integer $(1, 2, 3, \ldots)$ and c_n and d_n are arbitrary constants. A neophyte may well ask how we dare to disregard such obviously pertinent solutions. In answer to this question, we note that the Cauchy-Riemann equations imply that the partial derivatives of ϕ, i.e., the velocity components, are

$$u = \phi_x = \psi_y = \frac{n\pi}{b} \cos \frac{n\pi y}{b} (c_n e^{-n\pi x/b} + d_n e^{n\pi x/b}) + \mathscr{U}$$

$$v = \phi_y = -\psi_x = -\frac{n\pi}{b} \sin \frac{n\pi y}{b} (-c_n e^{-n\pi x/b} + d_n e^{n\pi x/b}).$$

As $x \to +\infty$, v must approach zero, which is not possible unless $d_n = 0$. But then as $x \to -\infty$, u must approach $\mathscr{U}$, which forces c_n to be zero. Hence this solution is indeed spurious, and $\psi = \mathscr{U}y$ must be the only one possible.

Let us determine the potential function $\phi(x, y)$; since it is the harmonic conjugate of ψ, it satisfies the Cauchy-Riemann equations with ψ:

$$\phi_x = \psi_y = \mathscr{U}, \qquad \phi_y = -\psi_x = 0;$$

The integration of these two equations leads us to the two relationships

$$\phi = \mathscr{U}x + c_1(y), \qquad \phi = c_2(x),$$

whence we have $\phi = \mathscr{U}x + c$, where c is a constant. Nothing is lost by taking c to be zero, since we obtain the velocity components by differentiating ϕ; c does not affect the value of the derivative of ϕ. Hence $\phi = \mathscr{U}x$ and, therefore, $F(z) = \phi + i\psi = \mathscr{U}x + i\mathscr{U}y = \mathscr{U}z$. The velocity vector is $\bar{f}(z) = F'(z) = \phi_x + i\phi_y = \mathscr{U} + i0 = \mathscr{U}$, which indicates quite realistically that the velocity remains uniformly horizontal towards the right with magnitude $\mathscr{U}$, in the absence of external forces (frictional or gravitational).

Therefore, flow from left to right through a channel of width b, with velocity in from the left equal to $\mathscr{U}$, is described by the complex velocity potential $F(z) = \mathscr{U}z$. Interestingly enough, we need not restrict our considerations to a channel of finite height: Horizontal flow to the right through the upper half plane, or indeed, through the entire plane, is described by the function $F(z) = \mathscr{U}z$, as long as $\mathscr{U}$ is the horizontal velocity where the flow enters out at "infinity."

The flow through a channel can be thought of as a flow with a source of strength $\mathscr{U}b$ at the left (at "infinity") and a sink, i.e., negative source, at the right. It might be interesting to bring this so-called "source at infinity" up a bit closer, say to the origin, where it could be studied. In order to do this, we shall use, for the first time, the techniques of conformal mapping

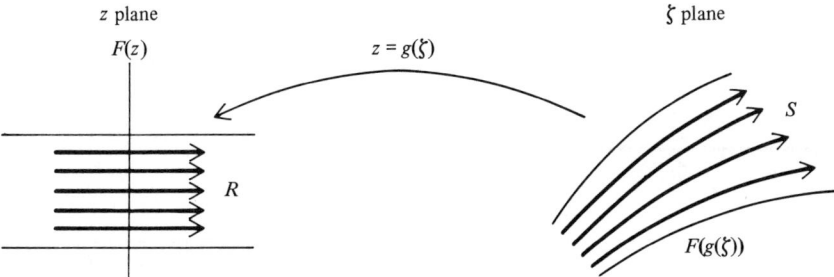

Figure 5.7

234 TWO-DIMENSIONAL INCOMPRESSIBLE FLUID FLOW

as applied to hydrodynamic problems. Before we do this, let us discuss the general rationale involved.

Suppose (as we have just done) that we have obtained a complex velocity potential $F(z)$ which describes a flow in some region R of the z plane. Suppose in some other complex plane, say the ζ plane, we have another flow through a region S of the ζ plane. And finally, suppose that there exists an analytic function $z = g(\zeta)$ which maps S conformally onto R, the boundary onto the boundary and the interior onto the interior (Figure 5.7). Then it turns out that $F(z) = F(g(\zeta))$ is the complex velocity potential for the flow in the ζ plane, except for a possible constant factor related to the strength of the flow.

Considering the care with which we have developed our concepts up to this point, this contention is not remarkably hard to prove. We do so in the following steps:

1. Since $g(\zeta)$ is analytic in S, and since $F(z)$ is analytic in R (the range of g), the composite function $F(g(\zeta))$ is analytic in S (Theorem 2.2.5).
2. If $z = x + iy$ and $\zeta = \xi + i\eta$, then the functions $\phi(x, y) = \phi(x(\xi, \eta), y(\xi, \eta))$ and $\psi(x, y) = \psi(x(\xi, \eta), y(\xi, \eta))$ are harmonic functions of ξ and η throughout S (Proof 5, Section 2.5).
3. The stream function ψ (considered as a function of ξ and η) is constant along the boundaries of S [since $z = g(\zeta)$ maps the boundaries of S into the boundaries of R, by hypothesis, and ψ is constant on the boundaries of R]. Therefore $\psi(x(\xi, \eta), y(\xi, \eta))$ is the stream function of the flow in the ζ plane, aside from a possible constant factor depending on the strength of the flow. Its conjugate harmonic function $\phi(x(\xi, \eta), y(\xi, \eta))$ must be proportional to the velocity potential and, hence, $F(g(\zeta))$ is proportional to the complex potential of the flow through S.

EXAMPLE 4 In order to make this clearer, let us first find a conformal mapping (using the Schwarz-Christoffel transformation), which will map the

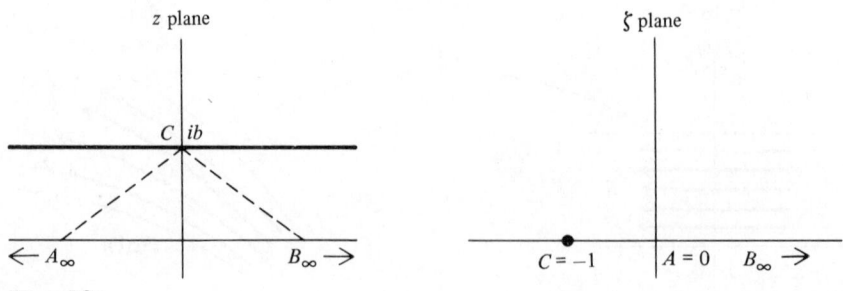

Figure 5.8

upper half ζ plane into a horizontal strip of width b in the z plane, as illustrated in Figure 5.8. We shall consider the strip in the z plane to be a degenerate triangle with the angles at A and B at infinity. Thus the angles are $\gamma_A = \gamma_B = 0$, $\gamma_C = \pi$. In the ζ plane, we are at liberty to choose all three points arbitrarily; we choose B at infinity (so that its factor will be missing from the derivative of the Schwarz-Christoffel transformation), $A = 0$ and $C = -1$. This will move the source at A to the origin in the ζ plane. Then the derivative of this mapping is

$$\frac{dz}{d\zeta} = c(\zeta - 0)^{\gamma_A/\pi - 1}(\zeta + 1)^{\gamma_C/\pi - 1} = c\zeta^{-1},$$

where c is a constant to be determined. Then

$$z(\zeta) = g(\zeta) = c \log \zeta + d.^*$$

The constants c and d must be chosen in such a fashion that for $0 < \zeta < \infty$ (real), z is real, since the segments AB in the two planes must correspond. If we choose the argument of ζ to be zero on the positive real axis, then $\log \zeta$ is real and this means that $d = 0$ and c is real. Next, the two points C must correspond, meaning that $ib = c \log(-1)$; since we chose $\arg \zeta = 0$ on the positive real axis, it follows that as we pass through the upper half plane to $\zeta = -1$, the argument of ζ increases to π; hence $\log(-1) = \log|-1| + \pi i = \pi i$. We, thus, have $ib = c\pi i$, so that $c = b/\pi$. Therefore

$$z = g(\zeta) = \frac{b}{\pi} \log \zeta,$$

with the argument of ζ chosen as noted, maps the upper half ζ plane conformally into the interior of the given strip in the z plane. This should not be surprising, since we saw previously that the exponential does the reverse.

Before continuing this example, let us pause for a moment to consider what we have done so far. In Example 3, we found that the complex velocity potential $F(z) = \mathcal{U}z$ described the flow through a horizontal channel of width b, with flux $\mathcal{U}b$. The source of this flow was at the point we have called "A_∞" in Figure 5.8, the sink at "B_∞."

Then, in Example 4, we discovered that the function

$$z = g(\zeta) = \frac{b}{\pi} \log \zeta$$

mapped the upper half plane conformally onto the strip of Example 3, in such a way that the source point "A" now corresponds to the origin in the ζ plane. The discussion between the examples has made it clear that $F(g(\zeta))$ must then be the complex velocity potential of the flow in the ζ plane:

---

* Of course, from earlier parts of the text, this could have been written down immediately; but the calculation is an instructive application of the Schwarz-Christoffel transformation.

EXAMPLE 4 (Continuation) In this case, then, the complex velocity potential

$$F(z) = \mathcal{U}z = \mathcal{U}\frac{b}{\pi}\log \zeta$$

must describe the flow in the upper half ζ plane from a source at the origin with strength $\mathcal{U}b$. In fact, in Example 2 we discovered that this type of complex potential describes a flow directed radially outwards from the origin (Figure 5.9). The velocity vector (in the present case) is

$$\frac{\mathcal{U}b}{\pi} \frac{\xi + i\eta}{\xi^2 + \eta^2},$$

which has the same direction as the radius vector $\xi + i\eta$, and has magnitude

$$\frac{\mathcal{U}b}{\pi} \frac{\sqrt{\xi^2 + \eta^2}}{\xi^2 + \eta^2} = \frac{\mathcal{U}b}{\pi\rho},$$

where $\rho = \sqrt{\xi^2 + \eta^2}$. Thus, since the velocity vector is normal to a circle of radius ρ centered at the origin, the total flux is given by

$$\int_0^\pi \frac{\mathcal{U}b}{\pi\rho} \rho \, d\phi,$$

ϕ being the polar angle in the ζ plane. But the value of this integral is precisely $\mathcal{U}b$, which was the total flux in the z plane as well. Therefore, we have solved the flow problem in the upper half plane, with a source at the origin, by the use of conformal mapping plus our previous example of basic flow through a channel.

One further example is in order now: Without any more work, it is also possible to consider the case of a sink at the origin. If the potential

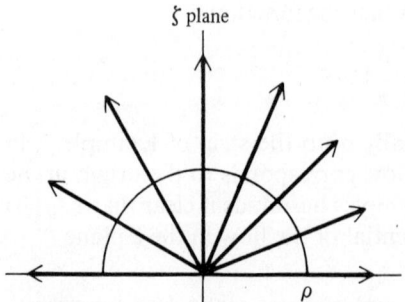

Figure 5.9

function

$$F(z) = \frac{\mathscr{U}b}{\pi} \log z$$

describes a flow with source at the origin, then obviously

$$G(z) = -F(z) = -\frac{\mathscr{U}b}{\pi} \log z$$

describes a flow with a sink at the origin. Changing the sign of F has no effect on the streamlines, but it reverses the direction of the velocity field.

The reader may at this point notice that we appear to be ignoring one of our assumptions of Section 5.1, that the fluid flow be source-free, in bringing a source to the origin as we have just done. This is not really the case, however: Our assumption was to the effect that there be no sources in *the flow region itself*; we said nothing about sources on the *boundary*. In fact, in either of the flows discussed above, the net flow over any simple closed curve *within* the flow region is zero; when the source is on the boundary, no simple closed curve can encircle it and still remain within the flow region, assuming, of course, that the flow region is simply connected.

EXAMPLE 5   Let us consider one more basic flow, the so-called "dipole" flow. We begin by mapping the upper half plane into the strip as before, with the difference that this time we place B at $\zeta = -h$ and A at $\zeta = h$ (Figure 5.10). We may as well place C at zero in the ζ plane, although since $\gamma_C = \pi$, and, thus, $\gamma_C/\pi - 1 = 0$, the factor for C will not enter in any event. Naturally C must be between B and A for the mapping to be conformal. The order of the points B, C, A as we proceed along the real axis of the ζ plane with the region to be mapped on our left must be the same as the "counterclockwise" order in the degenerate triangle BCA in the z plane. The flow region must always be on the left as we traverse the boundary

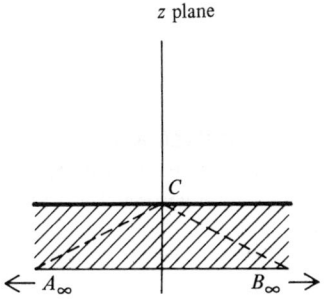 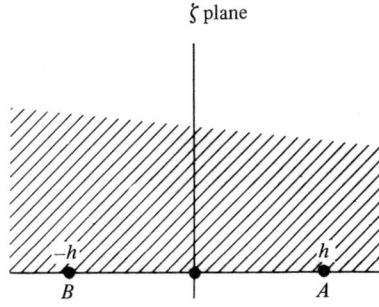

Figure 5.10

points in order, in both planes, because a reflection is *not* a conformal mapping (Section 4.2). Thus, in this case the Schwarz-Christoffel formula gives

$$\frac{dz}{d\zeta} = \frac{dg}{d\zeta} = c(\zeta + h)^{-1}(\zeta - h)^{-1} = \frac{c}{2h}\left(\frac{1}{\zeta - h} - \frac{1}{\zeta + h}\right)$$

by partial fractions. By integrating, we have

$$z = g(\zeta) = \frac{c}{2h}[\log(\zeta - h) - \log(\zeta + h)] + d.$$

One can see readily that this mapping indeed corresponds to a source at $\zeta = h$ [the term $\log(\zeta - h)$] and a sink at $-h$ [the term $-\log(\zeta + h)$]. Now, if we choose the branches of both logarithms which are real for real $\zeta > h$, then clearly c is real and $d = 0$, and we have

$$z = g(\zeta) = \frac{c}{2h}[\log(\zeta - h) - \log(\zeta + h)].$$

Then, since $F(z) = \mathscr{U}z$ is the complex velocity potential in the z plane, the potential in the ζ plane is

$$F(z(\zeta)) = F_h(\zeta) = K\frac{\log(\zeta - h) - \log(\zeta + h)}{2h}.$$

The exact value of the positive constant K is not important at the moment, for we are more interested in what happens to this flow as $h \to 0$. As the source and sink approach each other, they do not cancel, but rather complement one another. For, the dipole at the origin will emit fluid in one direction and absorb it in another. Indeed,

$$F(\zeta) = \lim_{h \to 0} F_h(\zeta) = \lim_{h \to 0} K\frac{\log(\zeta - h) - \log(\zeta + h)}{2h} = -\frac{K}{\zeta},$$

and thus $F(\zeta) = -K/\zeta$ describes dipole flow in the upper half plane.

For simplicity, let us consider dipole flow in the upper-half z plane, for which the complex velocity potential is $F(z) = -K/z$. Then in terms of its real and imaginary parts, we have

$$F = \phi + i\psi = -\frac{K}{r}(\cos\theta - i\sin\theta).$$

The streamlines, i.e., those curves along which ψ is constant, are given by $(K/r)\cos\theta = c$ or $r = (K/c)\cos\theta$. These curves are circles through the origin with centers on the positive imaginary axis. To determine the direction of flow along these streamlines, we first need

$$F'(z) = \frac{K}{z^2} = \frac{K}{r^2}(\cos 2\theta - i\sin 2\theta),$$

which is the conjugate of the velocity vector

$$\bar{f}(z) = \frac{K}{r^2}(\cos 2\theta + i \sin 2\theta).$$

For $0 < \theta < \pi/4$, the velocity vector is directed upwards to the right; for $\pi/4 < \theta < \pi/2$, upwards to the left, and so on. Therefore the dipole is as shown in Figure 5.11.

To summarize this section thus far, we have discovered the complex velocity potentials for a number of basic flows; we shall use these, plus conformal mapping techniques, to solve many problems of greater complexity in future sections. Our basic flows then, are as follows:

Flow	Complex Potential
Upper half plane or channel	$F(z) = \mathcal{U}z$ ($\mathcal{U}$ real)
Source at origin	$F(z) = K \log z$ (K real)
Dipole at origin	$F(z) = K/z$ (K real)

EXAMPLE 6 (Flow through an aperture) Some interesting flows can also be described by potential functions F which are defined *implicitly*. For example, consider F in the equation

$$z = c \cosh F, \quad c > 0.$$

F is not uniquely defined by this equation, since as we saw earlier in the book, the inverse hyperbolic cosine is infinitely multi-valued. However, we can choose a single-valued branch of F with no trouble by restricting F to a horizontal strip of width $2\pi i$ in the F plane. We then have

$$z = x + iy = c \cosh \phi \cos \psi + ic \sinh \phi \sin \psi;$$

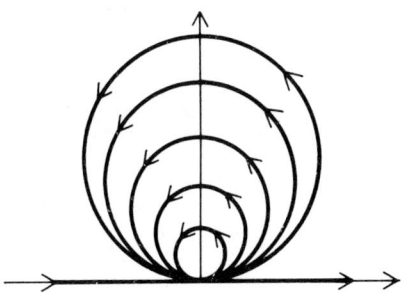

Figure 5.11

240 TWO-DIMENSIONAL INCOMPRESSIBLE FLUID FLOW

and since from this it follows that

$$\frac{x}{c \cos \psi} = \cosh \phi, \qquad \frac{y}{c \sin \psi} = \sinh \phi,$$

we can eliminate ϕ between the two equations and obtain

$$\frac{x^2}{c^2 \cos^2 \psi} - \frac{y^2}{c^2 \sin^2 \psi} = 1.$$

Thus $\psi =$ constant, i.e., the streamlines, correspond to hyperbolas with foci at $(\pm c, 0)$ (see Figure 5.12). In the limit as $\psi \to \pi/2$, we see that the y axis is a streamline; and as ψ approaches 0 or π we obtain the rays along the real axis from $-c$ to $-\infty$ and $+c$ to $+\infty$.

Thus it would appear that F describes a flow through an aperture $2c$ units wide in a flat plate. However, we shall see in a future section that this implies infinite velocities at $(\pm c, 0)$, an untenable situation since Bernoulli's law then states that the pressure becomes negative infinity! So instead, to salvage the situation, one could allow one of the hyperbolic streamlines to be the boundary of the flow region (Figure 5.13), in which case F describes flow through an aperture which is hyperbolic in cross section.

EXAMPLE 7 (Flow around an elliptical cylinder) In this example we consider the potential function implicitly defined by

$$z = c \cos F \qquad (c > 0).$$

As in the previous example, we choose a single-valued branch of F. We then note that

$$z = x + iy = c \cosh \psi \cos \phi - ic \sinh \psi \sin \phi,$$

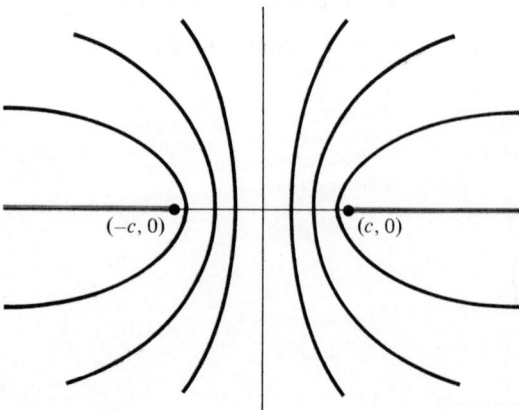

Figure 5.12

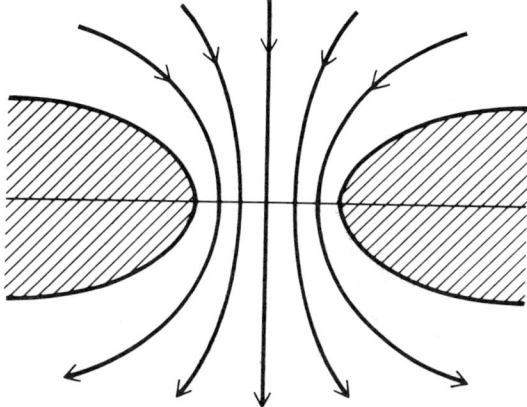

Figure 5.13

so that

$$\frac{x}{c\cosh\psi} = \cos\phi, \qquad \frac{y}{c\sinh\psi} = \sin\phi$$

and, hence,

$$\frac{x^2}{c^2\cosh^2\psi} + \frac{y^2}{c^2\sinh^2\psi} = 1.$$

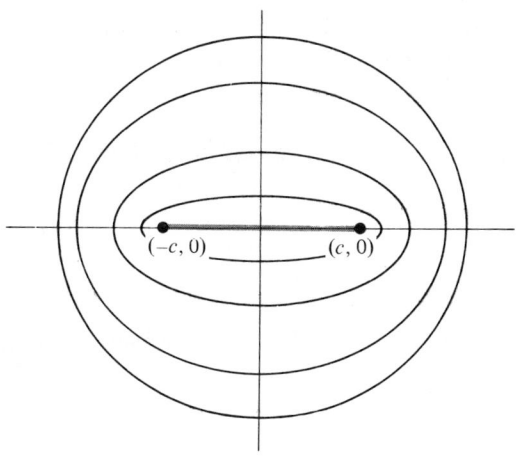

Figure 5.14

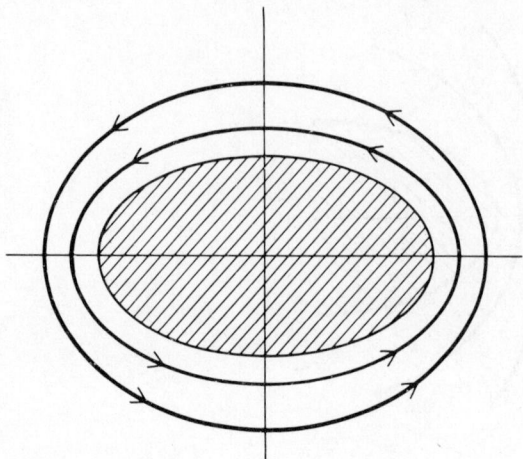

Figure 5.15

Thus, the streamlines $\psi =$ constant correspond to ellipses in the z plane with foci at $(\pm c, 0)$. As $\psi \to 0$, these ellipses collapse to form the segment joining the foci (Figure 5.14). As in Example 6, our first thought might be to consider this as flow around a rod; however, as we shall see later, this would imply an infinite velocity at $z = \pm c$. Hence, we do as before: We consider one of the streamlines to be the boundary of an elliptical cylinder; then the values of ψ larger than this value describe flow around this cylinder (Figure 5.15). Note that in this case the flow region is multiply-connected since points on the boundary *can* be enclosed by a simple closed curve (for instance, one of the streamlines). There is a nonzero net flow around any one of the streamlines; however, this flow is still irrotational since the net flow around a simple closed curve *not* containing a boundary point is zero.

We shall consider flow *past* a cylinder (as opposed to this flow *around* one) in a future Section 5.4.

A. Exercises

By mapping the flow regions into the upper half plane, using the techniques of this section, solve the following flow problems (note that it may take a series of mappings to solve a problem):

5.2. Certain Basic Flows

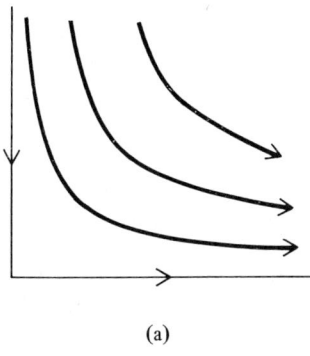

(a)

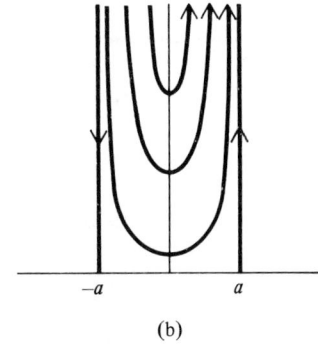

(b)

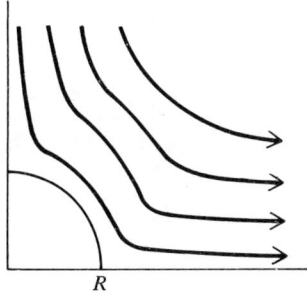

(Hint: first use $\log(z/R)$ to map into a region like that in Problem 4.)

(c)

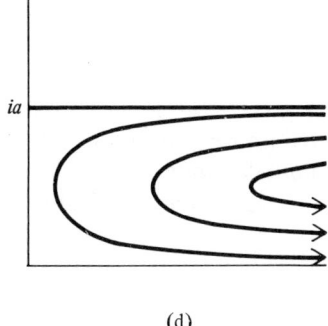

(d)

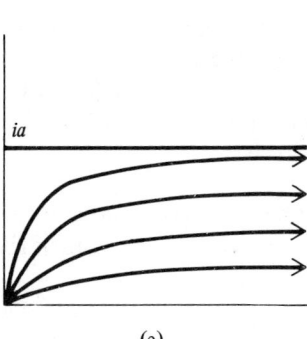

(e)

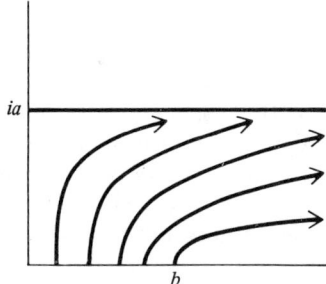

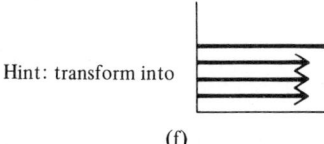

Hint: transform into

(f)

Figure 5.16

5.3. PRINCIPAL VALUES; HYDRODYNAMICS AS AN AID TO CONFORMAL MAPPING

Before we proceed with the main content of this section, let us consider a topic which is extremely useful in assigning a value to certain otherwise divergent integrals which have considerable importance in this aspect of complex variables.

The reader will recall from the calculus that the integral of the function $y = 1/x$ over any interval containing the origin is a divergent improper integral; this is because (formally)

$$\int_{-k}^{k} \frac{dx}{x} = \int_{-k}^{0} \frac{dx}{x} + \int_{0}^{k} \frac{dx}{x} = \lim_{\delta \to 0^+} \int_{-k}^{-\delta} \frac{dx}{x} + \lim_{\eta \to 0^+} \int_{\eta}^{k} \frac{dx}{x}.$$

δ and η must go to zero independently of one another, and if they do, then the above expression is an "indeterminate form of the type $\infty - \infty$." The limit does not exist.

However, it is clear that the limit

$$\lim_{\delta \to 0^+} \left(\int_{-k}^{-\delta} \frac{dx}{x} + \int_{\delta}^{k} \frac{dx}{x} \right)$$

does exist and, in fact, is zero, because it is equal to

$$\lim_{\delta \to 0^+} (\log|x|\big|_{-k}^{-\delta} + \log|x|\big|_{\delta}^{k}) = \lim_{\delta \to 0^+} (\log \delta - \log k + \log k - \log \delta) = 0.$$

By this means one can assign a value to divergent integrals which behave like $1/(x - c)$ near a point c by demanding that the *symmetric* interval $(c - \delta, c + \delta)$ be used before taking the limit to evaluate the improper integrals. The value thus obtained is called the *Cauchy principal value* of the integral under consideration. To be more precise:

5.3.1. Definition Suppose $y = f(x)$ is defined and continuous on an interval $[a, b]$. If we define $g(x) = f(x)/(x - c)$, where $a < c < b$, then the number given by the following limit (if the limit exists):

$$\lim_{\delta \to 0^+} \left(\int_{a}^{c-\delta} g(x)\, dx + \int_{c+\delta}^{b} g(x)\, dx \right)$$

is called the *Cauchy principal value* of the integral of g from a to b, and is denoted by

$$P \int_{a}^{b} g(x)\, dx.$$

As we saw above, for instance,

$$P \int_{-k}^{k} \frac{dx}{x} = 0.$$

We often can evaluate rather difficult principal value integrals by use of the techniques of the calculus of residues quite similar to those studied in Chapter 3.

EXAMPLE 8   Let us consider the evaluation of

$$P \int_{-\infty}^{\infty} \frac{dx}{(x^2 + 1)(x - 1)}.$$

The contour which we use must run along a portion of the real axis, but it must avoid passing through the singularity at $x = 1$. We shall use the contour C in Figure 5.17; we proceed along the real axis from $-R$ to $1 - \delta$, around the semicircle $1 + \delta e^{i\theta}$ from $\theta = \pi$ to $\theta = 0$, then along the real axis from $1 + \delta$ to R, and finally around the circle $z = Re^{i\theta}$ from $\theta = 0$ to $\theta = \pi$. We then take the limit as $R \to \infty$ and as $\delta \to 0$; note that the integral over the large semicircle will go to zero, since the degree of the denominator exceeds the degree of the numerator by *three*.

The integral we are seeking to evaluate over C preparatory to taking the limits is

$$\oint_C \frac{dz}{(z^2 + 1)(z - 1)};$$

there is one pole of the integrand inside C, namely at $z = i$; and the residue there is

$$\frac{1}{(z + i)(z - 1)}\bigg|_{z=i} = \frac{1}{2i(i - 1)} = \frac{1}{2(-i - 1)} = \frac{1}{2}\left(\frac{-1 + i}{2}\right) = \frac{-1 + i}{4}.$$

Therefore

$$\oint_C \frac{dz}{(z^2 + 1)(z - 1)} = 2\pi i \left(\frac{-1 + i}{4}\right) = \frac{\pi}{2}(-i - 1).$$

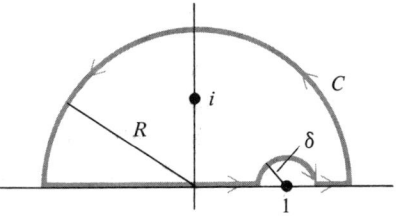

Figure 5.17

But the integral around C is the sum of the four integrals around each of the curves making up C:

$$\int_{-R}^{1-\delta} \frac{dx}{(x^2+1)(x-1)} + \int_{1+\delta}^{R} \frac{dx}{(x^2+1)(x-1)}$$

$$+ \int_{0}^{\pi} \frac{iRe^{i\theta}\, d\theta}{(R^2 e^{2i\theta}+1)(Re^{i\theta}-1)} + \int_{\pi}^{0} \frac{i\delta e^{i\theta}\, d\theta}{[(\delta e^{i\theta}+1)^2+1]\delta e^{i\theta}};$$

the limit of the first two integrals as $\delta \to 0$ and $R \to \infty$ is the principal value that we are seeking, and the third goes to zero. The fourth is equal to

$$i \int_{\pi}^{0} \frac{d\theta}{2 + 2\delta e^{i\theta} + \delta^2 e^{2i\theta}},$$

which clearly approaches $-\pi i/2$ as $\delta \to 0$. Hence

$$P \int_{-\infty}^{\infty} \frac{dx}{(x^2+1)(x-1)} - \frac{\pi i}{2} = \frac{\pi}{2}(-i-1)$$

and, therefore,

$$P \int_{-\infty}^{\infty} \frac{dx}{(x^2+1)(x-1)} = -\frac{\pi}{2}.$$

The next example illustrates that it is sometimes necessary to transform the integral so that a finite path of integration becomes an infinite path, one which is more amenable to the process of contour integration.

EXAMPLE 9 Let us find

$$P \int_{0}^{k} \sqrt{\frac{k-x}{x}} \frac{dx}{x-1}, \qquad k > 1.$$

(We assume the positive real value of the square root). Note that although the integral is improper at $x = 0$, it is finite there since the integrand behaves like $x^{-1/2}$. The principal value is to be taken at $x = 1$. In order to stand a chance with contour integration, we need a transformation which will take the interval $[0, k]$ into the positive real axis $[0, \infty)$, taking 0 into 0 and k into ∞.* One such transformation is

$$\zeta = \frac{x}{k-x} \qquad \text{or} \qquad x = \frac{k\zeta}{1+\zeta}.$$

---

* We assume that a change of variables of this type takes a principal value integral into another principal value integral (which it does, as the reader can verify). This is not always true, but it is for all the applications which we shall consider.

Then

$$x - 1 = \frac{k\zeta - (1 + \zeta)}{1 + \zeta} = \frac{\zeta(k - 1) - 1}{1 + \zeta}$$

and

$$dx = \frac{(1 + \zeta)k - k\zeta}{(1 + \zeta)^2} d\zeta = \frac{k \, d\zeta}{(1 + \zeta)^2},$$

Hence, upon substitution, we have

$$P \int_0^k \sqrt{\frac{k - x}{x}} \frac{dx}{x - 1} = P \int_0^\infty \frac{k}{\sqrt{\zeta}} \frac{d\zeta}{(1 + \zeta)^2} \frac{1 + \zeta}{\zeta(k - 1) - 1}$$

$$= P \int_0^\infty \frac{k \, d\zeta}{\sqrt{\zeta}(1 + \zeta)[\zeta(k - 1) - 1]},$$

where the principal value is of course taken at $\zeta = 1/(k - 1)$. To find the integral, we use a contour C which completely encircles the origin (Figure 5.18). Because of the multiple-valued $\sqrt{\zeta}$, the two integrals along the real axis will not cancel each other.

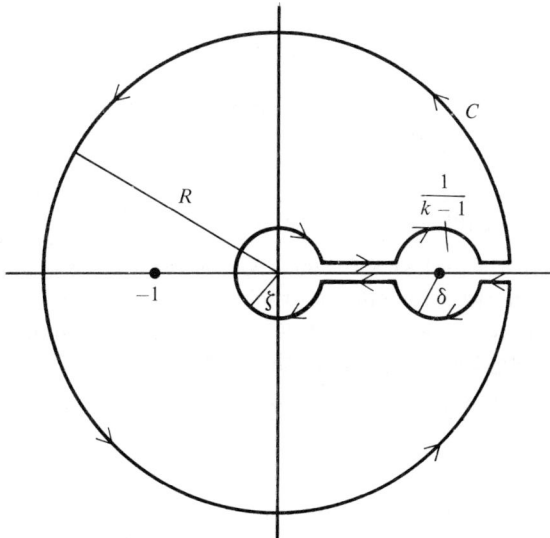

Figure 5.18

The integrand has a pole inside C at $\zeta = -1$, and the residue there is

$$\lim_{\zeta \to -1} (\zeta + 1) \frac{k}{\sqrt{\zeta}(1+\zeta)[\zeta(k-1)-1]} = \frac{k}{i(-k)} = i.$$

Note that we have chosen $\arg \zeta = 0$ for real $\zeta > 0$ and, hence, at $\zeta = -1$, $\arg \zeta = \pi$, meaning that $-1 = e^{\pi i}$, and hence $\sqrt{-1} = e^{\pi i/2} = i$.

As $R \to \infty$, the integral around the portion of C which is the circle of radius R will go to zero, since the degree of the denominator exceeds the degree of the numerator by more than 1, in fact by $\frac{5}{2}$. The integral around the circle $\zeta = \eta e^{i\theta}$ is

$$\int_{2\pi}^{0} \frac{k i \eta e^{i\theta}\, d\theta}{\sqrt{\eta e^{i\theta/2}} (1 + \eta e^{i\theta})[\eta e^{i\theta}(k-1)-1]};$$

since the integrand behaves like $\sqrt{\eta}$ as $\eta \to 0$, this integral also goes to zero. Looking at the integral around the two halves of the circle $\zeta = 1/(k-1) + \delta e^{i\theta}$, since we chose $\arg \zeta = 0$ on the top half of the real axis, we must have $\arg \zeta = 2\pi$ on the bottom half; and hence on the bottom, we obtain just the negatives of the values for $\sqrt{\zeta}$ that we have on the top half. As a result, the two integrals cancel. To see this, note that the two integrals are

$$\int_{\pi}^{0} \frac{k i \delta e^{i\theta}\, d\theta}{\sqrt{1/(k-1) + \delta e^{i\theta}} \exp(i \arg \zeta /2)\, s(\delta)}$$

$$+ \int_{2\pi}^{\pi} \frac{k i \delta e^{i\theta}\, d\theta}{\sqrt{1/(k-1) + \delta e^{i\theta}} \exp(i \arg \zeta /2)\, s(\delta)},$$

where $s(\delta) = [1/(k-1) + 1 + \delta e^{i\theta}](k-1)\delta e^{i\theta}$, a single-valued function. As $\delta \to 0$, the two integrands approach

$$\frac{i}{\sqrt{1/(k-1)}\exp(0)}, \quad \frac{i}{\sqrt{1/(k-1)}\exp(\pi i)},$$

constants which are negatives of each other. Since both integrals are over intervals of the same length (π), we conclude that their values cancel. Thus, in evaluating our principal value, we are left with the integrals along portions of the real axis. Along the top portion, these are

$$\left(\int_{\eta}^{1/(k-1)-\delta} + \int_{1/(k-1)+\delta}^{R} \right) \left(\frac{k\, d\zeta}{\sqrt{\zeta}(1+\zeta)[\zeta(k-1)-1]} \right).$$

Along the bottom portion, the limits of integration are reversed from the above; but the integrand is just the negative of the one above, as was the case along the small circular arcs. Hence these two integrals, in the limit, simply give us twice the principal value

$$2P \int_0^\infty \frac{k\,d\zeta}{\sqrt{\zeta}(1+\zeta)[\zeta(k-1)-1]} = -2\pi,$$

and therefore, going back to the original integral, we have

$$P \int_0^k \sqrt{\frac{k-x}{x}} \, \frac{dx}{x-1} = -\pi.$$

Using these results on principal values, we can now illustrate two different approaches to the problem of hydrodynamic flow through a "bent channel."

EXAMPLE 10 Consider the problem of describing the flow through the "bent channel" in the z plane as illustrated in Figure 5.19. The velocity in at the top is $\mathscr{U}$; hence the total flux into the channel is $\mathscr{U}a$. Because of the conservation of mass, the flux out must equal the flux in, giving $\mathscr{U}_\infty b = \mathscr{U}a$, or finally, $\mathscr{U}_\infty = \mathscr{U}a/b$.

In order to solve this problem, it will be necessary to map the upper half ζ plane (Figure 5.20) into the bent channel. In the ζ plane, we shall choose A at ∞, B at 0, and C at 1; this forces D to be at some point $k > 1$, to be determined. (Recall that only three points can be chosen at will.) Note that the point C is a "sink" (off to the right at "infinity" in the z plane), so the point C represents a sink in the ζ plane as well; according to our careful groundwork in Section 5.2, then, the complex velocity potential

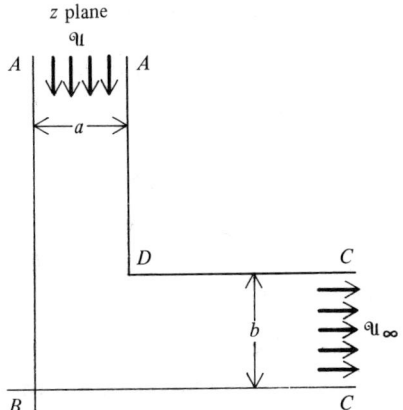

Figure 5.19

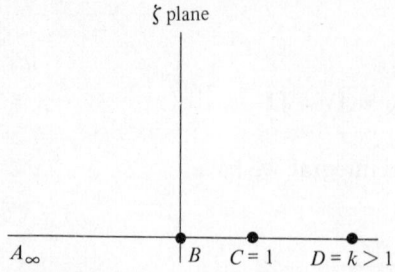

Figure 5.20

for the flow in the ζ plane is

$$F(\zeta) = -\frac{\mathcal{U}_\infty b}{\pi} \log(\zeta - 1) = -\frac{\mathcal{U}a}{\pi} \log(\zeta - 1),$$

the minus sign having been chosen precisely because $C = 1$ is a sink.

To map the ζ plane into the z plane, we shall use a Schwarz-Christoffel transformation, which gives

$$\frac{dz}{d\zeta} = c(\zeta - 0)^{-1/2} (\zeta - 1)^{-1} (\zeta - k)^{1/2} = c'\sqrt{\frac{k-\zeta}{\zeta}} \frac{1}{\zeta - 1},$$

where $ic = c'$. Notice that we shall want the segments BC in the two planes to correspond; if we choose the branch of the square root which is real and positive for $0 < \zeta < 1$ and write

$$z(\zeta) = c' \int_{\zeta_0}^{\zeta} \sqrt{\frac{k-\zeta}{\zeta}} \frac{d\zeta}{\zeta - 1} + d,$$

then we know that c' must be real. Next, if we choose $\zeta_0 = 0$, then $d = 0$. It still remains to determine the constants c' and k. We can do this relatively easily by using the few facts that we know about hydrodynamics. For instance, we know that the conjugate of the velocity vector in the z plane is given by dF/dz; and by the chain rule,

$$\frac{dF}{dz} = \frac{dF}{d\zeta}\frac{d\zeta}{dz} = \frac{dF}{d\zeta} \cdot 1 \bigg/ \left(\frac{dz}{d\zeta}\right),$$

so that

$$\frac{dF}{dz} = -\frac{\mathcal{U}a}{\pi(\zeta-1)} \cdot \frac{\sqrt{\zeta(\zeta-1)}}{c'\sqrt{(k-\zeta)}} = -\frac{\mathcal{U}a}{\pi c'}\sqrt{\frac{\zeta}{k-\zeta}}.$$

Using this expression for the conjugate of the velocity in the z plane in terms of ζ, we shall observe the velocity at points where it is known, in an attempt to determine c' and k.

Recall that we have chosen the value of the square root to be real and positive when $0 < \zeta < 1$; thus, for these values of ζ, we have chosen $\arg[\zeta/(k - \zeta)] = 0$. If we now leave the segment BC in the ζ plane and proceed on a small semicircle counterclockwise around the origin in the upper half plane, then $\arg(k - \zeta)$ does not change, while $\arg \zeta$ changes by π. Thus on the segment BA_∞, we have

$$\frac{dF}{dz} = -\frac{\mathcal{U}a}{\pi c'}\sqrt{\frac{|\zeta|e^{\pi i}}{k - \zeta}} = -\frac{\mathcal{U}ae^{\pi i/2}}{\pi c'}\sqrt{\frac{|\zeta|}{k - \zeta}} = -\frac{\mathcal{U}ai}{\pi c'}\sqrt{\frac{\zeta}{\zeta - k}}.$$

In the limit as $\zeta \to \infty$, we have

$$\frac{dF}{dz} = -\frac{\mathcal{U}ai}{\pi c'}.$$

On the other hand, we are given (in the z plane; Figure 5.19) that at A

$$\frac{dF}{dz} = 0 + i\mathcal{U} \qquad [\text{note } \bar{f}(z) = 0 - i\mathcal{U}]$$

and, therefore,

$$-i\mathcal{U} = -\frac{\mathcal{U}ai}{\pi c'},$$

so that $c' = -a/\pi$. On the other hand, as $\zeta \to 1$ (the point corresponding to C) we get

$$\frac{dF}{dz} \to -\frac{\mathcal{U}a}{\pi c'}\frac{1}{\sqrt{(k-1)}} = \frac{\mathcal{U}}{\sqrt{(k-1)}}.$$

Hence $\mathcal{U}/\sqrt{k-1} = \mathcal{U}a/b$, so that $k - 1 = (b/a)^2$ or $k = (a^2 + b^2)/a^2$. Therefore, for the Schwarz-Christoffel transformation we have

$$\frac{dz}{d\zeta} = -\frac{1}{\pi}\frac{\sqrt{(a^2 + b^2) - a^2\zeta}}{\sqrt{\zeta(\zeta - 1)}}.$$

All constants have been evaluated by recourse to hydrodynamics; it is very interesting indeed to note that one can solve this geometric problem so neatly by appealing to an application.

Alternatively, one can evaluate the constants k and c' as follows: Since $\zeta = k$ corresponds to $z = a + ib$ (the point D in the z plane), we must have

$$\int_C \frac{dz}{d\zeta}d\zeta = c'\int_C \sqrt{\frac{k - \zeta}{\zeta}}\frac{d\zeta}{\zeta - 1} = a + ib,$$

where C is any simple arc in the upper half ζ plane joining 0 and k. For

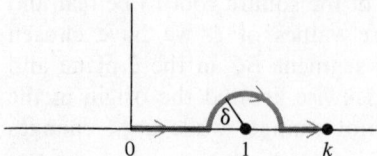

Figure 5.21

example, we can choose for C the real axis from 0 to $1 - \delta$ and from $1 + \delta$ to k, plus the semicircle $\zeta = 1 + \delta e^{i\theta}$ for θ from π to 0 (Figure 5.21). Then we have

$$a + ib = c'\left(\int_0^{1-\delta} + \int_{1+\delta}^k\right)\left(\sqrt{\frac{k-\zeta}{\zeta}}\,\frac{d\zeta}{\zeta - 1}\right)$$

$$+ c'\int_\pi^0 \sqrt{\frac{k - 1 - \delta e^{i\theta}}{1 + \delta e^{i\theta}}}\,\frac{i\delta e^{i\theta}\,d\theta}{\delta e^{i\theta}}.$$

In the limit as $\delta \to 0$, the first two integrals approach the principal value obtained in example 5.19, which was $-\pi$. The third integral approaches

$$\int_\pi^0 \sqrt{(k-1)}\,i\,d\theta = -\pi i\sqrt{k-1}.$$

Hence we obtain

$$a + ib = -c'\pi - c'\pi i\sqrt{k-1}.$$

Therefore, by equating real parts, we have $c' = -a/\pi$, and by equating imaginary parts, $b = -c'\pi\sqrt{k-1} = a\sqrt{k-1}$, whence we get the same value for k as before.

The above simply serves to show that there is more than one way to skin a mathematical cat. In this case, however, the purely geometric way of evaluating the constants, namely the principal value method, required significantly more work than did the method which used hydrodynamics purely as a help in evaluating the constants.

A. Exercises

Evaluate the following principal value integrals:

1 $P\displaystyle\int_{-\infty}^\infty \frac{dx}{x(x^2 - 2x + 2)}$ 2 $P\displaystyle\int_{-\infty}^\infty \frac{dx}{x^3 - 1}$

3 $P \int_0^\infty \dfrac{dx}{x^2 - 1}$

4 $P \int_{-\infty}^\infty \dfrac{dx}{x(x - 1)(x^2 + 1)}$

5 $P \int_0^\infty \dfrac{\sqrt{x}\, dx}{(x - 1)(x^2 + 1)}$

5.4. FLOW PAST AN OBSTACLE; THE KUTTA-JOUKOWSKI AIRFOIL THEORY

In Example 7, we considered flow around a cylinder of elliptical cross section. In this section we propose to discuss flow around ("circulation") and/or flow past ("streaming") obstacles of various cross sections. First, to get a more precise idea of what "circulation" means, we consider flow around a circular cylinder.

EXAMPLE 11   The complex potential function

$$F(z) = ik \log \dfrac{z}{a} \qquad (k > 0, a > 0),$$

describes flow around the circular cylinder $|z| = a$ (Figure 5.22). First, note that when $|z| = a$, we have $F(z) = -k \arg z$, which is real, and, thus,

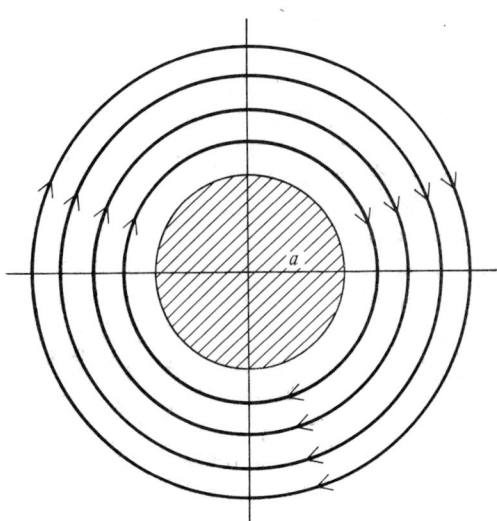

Figure 5.22

the stream function ψ vanishes on $|z| = a$, so the circle is certainly a streamline. Likewise, so is any larger circle centered at the origin. Obviously F is a multiple-valued function because every time we encircle $|z| = a$, the imaginary part of the logarithm increases by 2π. In fact,

$$F(z) = -k\arg\frac{z}{a} + ik\ln\left|\frac{z}{a}\right| = \phi(x, y) + i\psi(x, y),$$

so we see that the stream function ψ is single-valued, while the potential decreases by $2\pi k$ each time we encircle the cylinder. F has a single-valued derivative, however, which gives the conjugate of the velocity field of the flow, so the multiple-valuedness of the velocity potential ϕ seems relatively harmless. Thus, having chosen a particular value of $\theta = \arg z$ and of ϕ, a change of θ to the value $\theta + 2\pi$ produces the new value of the velocity potential $\phi - 2k\pi$. The decrease in the value of ϕ in a circuit about the cylinder, namely $2k\pi$, is called the *circulation*, and the constant k is called the *strength of the circulation*. Note that the circulation about any simple closed curve in the flow region which does not enclose $|z| = a$ is zero, and thus this flow is irrotational. In fact, nonzero circulation about a simple closed curve C only occurs when C encloses boundary points, namely points on $|z| = a$. As we have noted before, this can only occur when, as in this example and in Example 7, the flow region is multiply connected.

The next interesting result describes what happens when a cylinder is inserted into a flow for which the potential function F is already known; we shall first need the following:

5.4.1. Lemma *(Reflection principle)* Suppose $F(z) = \phi(x, y) + i\psi(x, y)$ is analytic in a domain D which includes a portion l of the real axis and that F is real on l. Then $F(\bar{z}) = \overline{F(z)}$ for all z in D, that is, the value of F at the conjugate of a point is the conjugate of the value at the given point.

Proof An equivalent statement of the theorem is that $F(z) \equiv \overline{F(\bar{z})}$. We shall define a function G by

$$G(z) = \overline{F(\bar{z})}.$$

First we shall show that G is analytic. Let

$$G = \xi(x, y) + i\eta(x, y).$$

Then, since

$$\overline{F(\bar{z})} = \phi(x, -y) - i\psi(x, -y),$$

we have

$$\xi(x, y) = \phi(x, -y), \qquad \eta(x, y) = -\psi(x, -y)$$

and, hence,
$$\xi_x(x, y) = \phi_x(x, -y) \quad , \quad \xi_y(x, y) = -\phi_y(x, -y).$$
Furthermore,
$$\eta_x(x, y) = -\psi_x(x, -y), \quad \eta_y(x, y) = \psi_y(x, -y).$$
Since F is analytic, we have
$$\phi_x(x, -y) = \psi_y(x, -y) \quad , \quad \phi_y(x, -y) = -\psi_x(x, -y),$$
and, therefore, we obtain
$$\xi_x(x, y) = \eta_y(x, y) \quad , \quad \xi_y(x, y) = -\eta_x(x, y).$$
Thus, the real and imaginary parts of G satisfy the Cauchy-Riemann equations, so G is analytic.

To show that F and G are identical, we note that they are both analytic in D and are identical on l, since by hypothesis F is real on l, meaning that $\psi(x, 0) = 0$, and hence
$$G(x) = \xi(x, 0) + i\eta(x, 0) = \phi(x, 0) - i\psi(x, 0)$$
$$= \phi(x, 0) = F(x).$$
Thus the analytic function $F(z) - G(z)$ is zero on the segment l. Hence $F - G$ has nonisolated zeros; so by Lemma 4.1.2, F is identically equal to G throughout D. QED.

5.4.2. Theorem Let F be the complex potential of a flow and assume that its singularities (if any) are all at a distance more than a units away from the origin. We further assume that the real axis is a streamline (without loss of generality, the streamline given by $\psi = 0$). Then if the circular cylinder $|z| = a$ is inserted in the flow, the complex potential of the new flow is
$$G(z) = F(z) + F\left(\frac{a^2}{z}\right).$$

Proof Let $G = \xi(x, y) + i\eta(x, y)$. We shall need to show first that $|z| = a$ is a streamline and, second, that the function G is, in fact, analytic (at least wherever F is) so that it is a *bona fide* potential.

1. $|z| = a$ is a streamline. Note that on $|z| = a$, we have $\bar{z} = a^2/z$, and hence on $|z| = a$ we obtain
$$G(z) = F(z) + F(\bar{z}).$$
If we can show that G is real on $|z| = a$, we shall have shown that the circle is part of the streamline given by $\eta(x, y) = 0$. But, according to hypothesis, F is real on the real axis and, therefore, $F(\bar{z}) = \bar{F}(z)$ by the reflection prin-

ciple; hence

$$G(z) = F(z) + \bar{F}(z) = 2 \operatorname{Re} F(z)$$

on $|z| = a$; hence G is real and $|z| = a$ is part of the streamline $\eta(x, y) = 0$.

2. G is analytic wherever F is: Since

$$G(z) = F(z) + F\left(\frac{a^2}{z}\right)$$

and $F(z)$ has singularities farther than a units from the origin, $F(a^2/z)$ has all its singularities inside the circle of radius a and none outside. Furthermore, $F(a^2/z)$ is an analytic function of an analytic function, so is itself analytic; hence G, being the sum of two analytic functions, is itself analytic, thus completing the proof.

A particular application of the theorem is the following:

EXAMPLE 12 If we have uniform horizontal flow $F(z) = \mathcal{U}z$, then the hypotheses of the theorem are satisfied; hence insertion of a cylinder of radius a in the flow yields a potential of

$$G(z) = \mathcal{U}\left(z + \frac{a^2}{z}\right).$$

This flow is illustrated in Figure 5.23. This can be modified to give uniform flow in the direction given by angle α simply by multiplying G by $e^{i\alpha}$, a

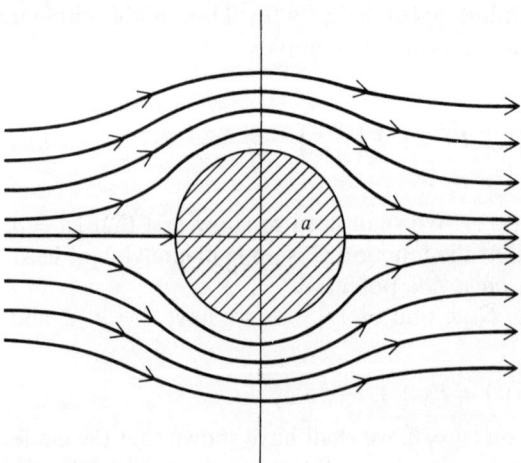

Figure 5.23

rotation:
$$\tilde{G}(z) = \mathcal{U}e^{i\alpha}\left(z + \frac{a^2}{z}\right).$$

Thus it is relatively easy to modify the potential of the flow in case the straight streamline $\psi = 0$ of the original flow is not along the real axis.

We obtain a highly interesting flow by combining circulation (Example 11) with streaming (Example 12).

EXAMPLE 13 By adding the two potentials for circulation and horizontal streaming, we obtain the potential

$$H(z) = \mathcal{U}\left(z + \frac{a^2}{z}\right) + ik\log\frac{z}{a},$$

which describes the flow when both phenomena are present. Note that

$$H'(z) = \mathcal{U}\left(1 - \frac{a^2}{z^2}\right) + \frac{ik}{z} = \frac{\mathcal{U}z^2 + ikz - \mathcal{U}a^2}{z^2}.$$

Since H' is the conjugate of the velocity field of the flow, the points where H' vanishes are stagnation points of the flow; these occur at

$$z = \frac{-ik \pm \sqrt{4a^2\mathcal{U}^2 - k^2}}{2\mathcal{U}},$$

which can be written in the form

$$z = a\left(\frac{ik}{2a\mathcal{U}} \pm \sqrt{1 - \frac{k^2}{4a^2\mathcal{U}^2}}\right).$$

If $k < 2a\mathcal{U}$ then we can write

$$\frac{k}{2a\mathcal{U}} = \sin\beta,$$

in which case we have

$$z = a(-i\sin\beta \pm \cos\beta),$$

so both stagnation points are on the circle and the flow looks as in Figure 5.24. If $k = 2a$, then there is a single stagnation point at $z = -a$ (Figure 5.25). Finally, if $k > 2a\mathcal{U}$, we can write

$$\frac{k}{2a\mathcal{U}} = \cosh\beta,$$

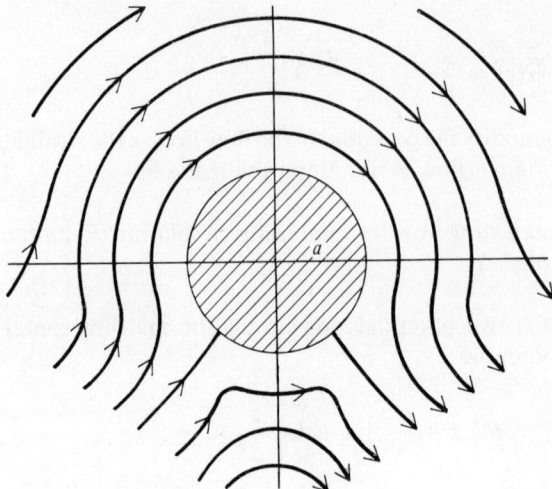

Figure 5.24

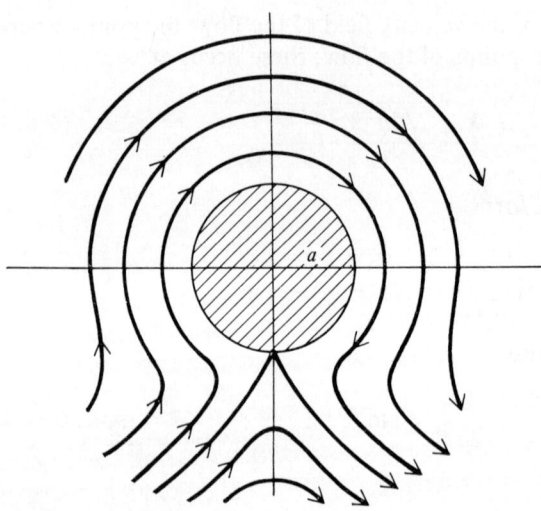

Figure 5.25

whereupon we obtain
$$z = ai(-\cosh\beta \pm \sinh\beta),$$
which gives only one root outside the circle, namely
$$z = aie^{\beta}.$$
This case (Figure 5.26) illustrates what happens when the circulation is very high with respect to the streaming speed. There is some fluid which eddies completely around the cylinder and never runs off downstream.

Everything that we have said thus far about the flow around or past a circular cylinder can be said about a cylinder of any cross section, provided that we can map the circle onto the curve of the more general shape. For instance, let us consider the Joukowski transformation
$$w = z + \frac{b^2}{z} \quad (b > 0)$$
(which we shall discuss further at the end of this section). In terms of x and y,
$$w = \left(x + \frac{b^2 x}{x^2 + y^2}\right) + i\left(y - \frac{b^2 y}{x^2 + y^2}\right)$$
$$= \left(\frac{x(x^2 + y^2 + b^2) + iy(x^2 + y^2 - b^2)}{x^2 + y^2}\right).$$

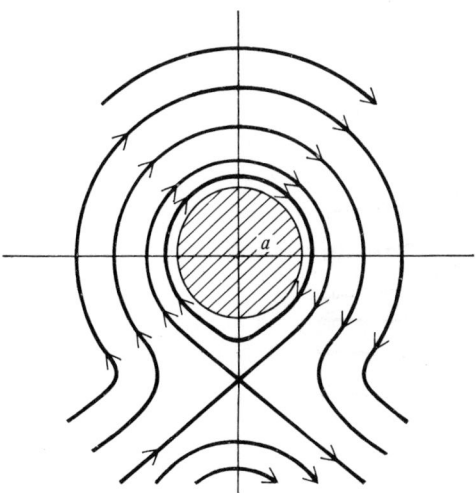

Figure 5.26

This transformation takes the circle of radius a given by $z = ae^{i\theta}$ ($0 \leq \theta \leq 2\pi$) into

$$w = \frac{a^2 + b^2}{a} \cos\theta + i\frac{a^2 - b^2}{a} \sin\theta.$$

Thus, if $w = u + iv$, we have

$$\frac{u^2}{[(a^2 + b^2)/a^2]^2} + \frac{v^2}{[(a^2 - b^2)/a^2]^2} = 1,$$

which is an ellipse provided that $b < a$. Thus, whenever $b < a$, the Joukowski transformation maps the circle of radius a into an ellipse. Thus we can consider examples such as the following:

EXAMPLE 14 (Horizontal streaming past an ellipse whose major axis is on the real axis.) Suppose an ellipse is given in the z plane with semi-major axis A, semiminor axis B, symmetric with respect to the origin (Figure 5.27). Then one can determine two constants a and b from the pair of equations

$$A = a + \frac{b^2}{a}, \quad B = a - \frac{b^2}{a}.$$

In the complex ζ plane, horizontal streaming past a circular cylinder of radius a, centered at the origin, is given by

$$F(\zeta) = \mathscr{U}\left(\zeta + \frac{a^2}{\zeta}\right).$$

The mapping of the flow region from the ζ plane to the z plane is given by

$$z = \zeta + \frac{b^2}{\zeta}.$$

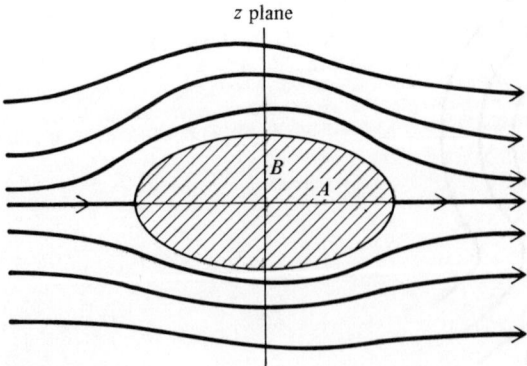

Figure 5.27

Thus, the potential function describing the flow in the z plane is

$$G(z) = F(\zeta(z)),$$

where z is the point which comes from some given point ζ under the Joukowski transformation.

One can in quite a similar way discuss oblique streaming past an elliptic cylinder (see problems at the end of the section).

Finally, in this section, we shall consider the effect that streaming and circulation past a cylinder have on the pressure exerted on the cylinder and on the moment about the origin. Even though we are considering incompressible flow—and air is obviously compressible—this theory goes a long way toward explaining lift on an airfoil. First, we shall need the following remarkable result. [Here, as before, $p(x, y)$ represents the pressure at a given point, and ρ the density; we shall let p_x and p_y represent the total force per unit length on the cylinder in the x and y directions, respectively, and M the moment about the origin of the cylinder caused by the forces on it.]

5.4.3. Theorem *(Blasius)* Let C be a simple closed curve (the boundary of a cylinder) and suppose F is the complex potential of a flow past C. Then the following two equations hold:

$$p_x - ip_y = \frac{i}{2} i\rho \oint_C \left(\frac{dF}{dz}\right)^2 dz,$$

$$M = \mathrm{Re}\left[-\frac{1}{2}\rho \oint_C z \left(\frac{dF}{dz}\right)^2 dz \right].$$

Proof Basically, this theorem is a result of Bernoulli's equation. First, we note that the incremental force on a small portion of C in the x and y directions (respectively) is approximately

$$\Delta p_x = -p\Delta y, \qquad \Delta p_y = p\Delta x$$

(see Figure 5.28).* The incremental moment about the origin on this same small segment is approximately

$$\Delta M = x\Delta x \cdot y\Delta y \cdot p.$$

Hence we have

$$\Delta p_x - i\Delta p_y = \Delta(p_x - ip_y) = -p\Delta y - ip\Delta x = -ip\Delta \bar{z}.$$

---

* As in other derivations in this chapter, we are using approximations; the discussion can be made rigorous with sufficient care.

262 TWO-DIMENSIONAL INCOMPRESSIBLE FLUID FLOW

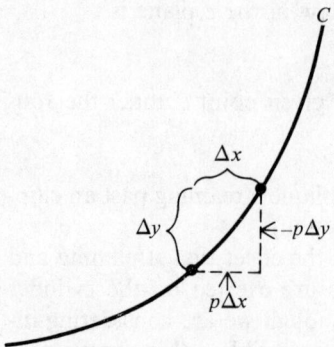

Figure 5.28

Likewise

$$\Delta M = \operatorname{Re}(pz\Delta \bar{z}).$$

Now Bernoulli's law states that at any point

$$p = c - \frac{1}{2}\rho q^2 = c - \frac{1}{2}\rho \frac{dF}{dz}\frac{d\bar{F}}{d\bar{z}};$$

when we integrate around C, the constant c will integrate to zero and will have no resultant effect, so it can safely be eliminated. In the small, then, we have approximately

$$p = -\frac{1}{2}\rho \frac{\Delta F}{\Delta z}\frac{\Delta \bar{F}}{\Delta \bar{z}},$$

or, finally,

$$\Delta \bar{z} = -\frac{1}{2}\frac{\rho}{p}\frac{\Delta F}{\Delta z}\Delta \bar{F}.$$

Putting this into the expressions for incremental pressure and moment, we have

$$\Delta(p_x - ip_y) = \frac{i}{2}\rho \frac{\Delta F}{\Delta z}\Delta \bar{F},$$

$$\Delta M = \operatorname{Re}\left(-\frac{\rho}{2}z\frac{\Delta F}{\Delta z}\Delta \bar{F}\right).$$

Now along C, which is a streamline, F is real and, hence, $F = \bar{F}$ and $\Delta F =$

$\Delta \bar{F}$. From this we obtain that

$$\Delta(p_x - ip_y) = \frac{i}{2}\rho\left(\frac{\Delta F}{\Delta z}\right)^2 \Delta z,$$

$$\Delta M = \operatorname{Re}\left[-\frac{\rho}{2}z\left(\frac{\Delta F}{\Delta z}\right)^2 \Delta z\right].$$

Adding up all such small increments around C gives us an approximation to Blasius's expressions, and in the limit as $\max|\Delta z| \to 0$ we obtain precisely the desired formulas.

By using the Blasius formulas as a basis for our proof, we are now able to prove the elegant theorem on which most of aerodynamics is based:

5.4.4. Theorem *(Kutta-Joukowski)* Suppose that an airfoil of arbitrary cross section is at rest in a uniform flow with velocity $\mathscr{U}$, angle of incidence α, and circulation k (Figure 5.29). Then the airfoil undergoes a lift of $k\rho\mathscr{U}$ perpendicular to the direction of the flow, in the clockwise sense.

Proof By our previous work, for large z, the complex potential must look as follows:

$$F(z) = \mathscr{U}ze^{i\alpha} + \frac{ik}{2\pi}\log z + \frac{b}{z} + \cdots,$$

where the first term is due to the uniform flow, the second to the circulation, and all subsequent terms go to zero as $|z| \to \infty$. Then

$$\frac{dF}{dz} = \mathscr{U}e^{i\alpha} + \frac{ik}{2\pi z} - \frac{b}{z^2} + \cdots;$$

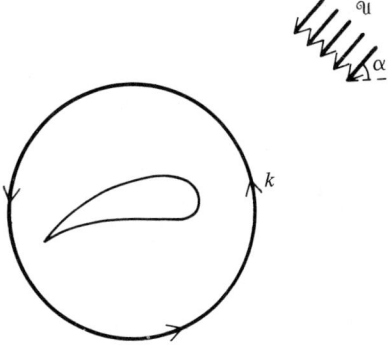

Figure 5.29

therefore, we have

$$\left(\frac{dF}{dz}\right)^2 = \mathcal{U}^2 e^{2i\alpha} + \frac{ik\mathcal{U}e^{i\alpha}}{\pi z} - \frac{k^2 + 8\pi^2 b \mathcal{U} e^{i\alpha}}{4\pi^2 z^2} + \cdots.$$

Thus, according to Blasius's theorem, if we integrate around a very large circle (large enough to contain the airfoil, of course, and large enough for the above to hold), we obtain

$$p_x - p_y = \frac{1}{2}i\rho \oint \left(\frac{dF}{dz}\right)^2 dz = \frac{1}{2}i\rho \cdot 2\pi i\left(\frac{ik\mathcal{U}e^{i\alpha}}{\pi}\right) = -ik\rho\mathcal{U}e^{i\alpha};$$

hence, the total pressure on the airfoil is

$$p_x + ip_y = ik\rho\mathcal{U}e^{-i\alpha} = k\rho\mathcal{U}e^{i(\pi/2 - \alpha)}$$

as claimed. Incidentally, by the same techniques, one can show that the moment about the origin is

$$M = \operatorname{Re}(2\pi i\rho b \mathcal{U} e^{i\alpha}).$$

Note that the above results are completely independent of the *shape* of the airfoil. All that is necessary to produce lift in a uniform flow is circulation about the airfoil!

A. Exercises

Establish the last result of Theorem 5.4.4.

B. Problems

1 Discuss horizontal streaming and circulation about an ellipse.
2 Discuss streaming past an ellipse at an oblique angle.
3 Combine the previous flow with circulation and discuss.

C. Proofs

1 Show that the Joukowski transformation

$$w = z + \frac{b^2}{z}$$

can also be written in the form

$$\frac{w + 2b}{w - 2b} = \left(\frac{z + b}{z - b}\right)^2.$$

2 Let C be a circle centered in the first quadrant, through the point $z = -b$. Show that the mapping

$$w = z + \frac{b^2}{z}$$

maps the exterior of this circle conformally onto the exterior of its image in the w plane and that, in fact, the mapping is conformal even on the circle except at $z = -b$.

3 Show, numerically or otherwise, that the image of the circle described in Problem 2 is a streamlined airfoil shape, and, hence, discuss how one can find the complex potential for the streaming and circulation past an airfoil.

5.5. FREE STREAMLINES: CONTRACTION COEFFICIENTS

In order to solve a flow problem completely, it is sometimes necessary to do more than just map the flow region onto the upper half plane. For example, in Example 10, we used the problem of hydrodynamic flow through a "bent channel" in order to evaluate the constants in a Schwarz-Christoffel transformation mapping the upper half plane onto the bent channel. While in the process, we computed dF/dz in terms of ζ, namely

$$\frac{dF}{dz} = \mathcal{U}\sqrt{\frac{\zeta}{k-\zeta}}.$$

The point $\zeta = k$ corresponds to the point $a + ib$ in the z plane (see Figures 5.19 and 5.20), which is the inside (270°) corner of the channel. Since dF/dz is the conjugate of the velocity field $\bar{f}(z) = u + iv$, it follows that the velocity becomes infinite at the point $\zeta = k$ or $z = a + ib$.

This is likely to prove troublesome, for according to Bernoulli's Law (see Section 5.1), we have

$$\frac{1}{2}(u^2 + v^2) + \frac{p}{\rho} = c,$$

relating the pressure and the speed. At the point A at the top entrance of the channel, $u^2 + v^2$ is given (it is $\mathcal{U}^2$), and the pressure also has some preassigned value, say p_∞. Thus we have

$$c = \frac{1}{2}\mathcal{U}^2 + \frac{P_\infty}{\rho},$$

so that

$$\frac{1}{2}(u^2 + v^2) + \frac{p}{\rho} = \frac{1}{2}\mathcal{U}^2 + \frac{P_\infty}{\rho}.$$

266 TWO-DIMENSIONAL INCOMPRESSIBLE FLUID FLOW

If the speed $q = \sqrt{u^2 + v^2}$ becomes infinite as we approach $a + ib$, then the pressure p would have to approach $-\infty$ for the equation to balance, which is a physical impossibility. Before we concentrate further on the bent channel problem, however, let us consider the following related but somewhat simpler problem.

EXAMPLE 15 Let us examine the flow problem illustrated in Figure 5.30; fluid is to flow in at the left at a constant horizontal speed $\mathscr{U}$ and up over the step illustrated, which is a units high. We map the upper half ζ plane (Figure 5.31) onto the flow region in the z plane by means of the transformation

$$\frac{dz}{d\zeta} = c(\zeta - 0)^{-\frac{1}{2}}(\zeta - 1)^{\frac{1}{2}}.$$

Then we have

$$z = c \int_0^\zeta \sqrt{\frac{t-1}{t}}\, dt,$$

the additive constant d being zero since we have begun our integration at the origin, which goes into the origin in the z plane. Now for real ζ between 0 and 1, we are on the segment BC: hence, z must be pure imaginary. Therefore, let us rewrite the expression for z as

$$z = ic \int_0^\zeta \sqrt{\frac{1-t}{t}}\, dt,$$

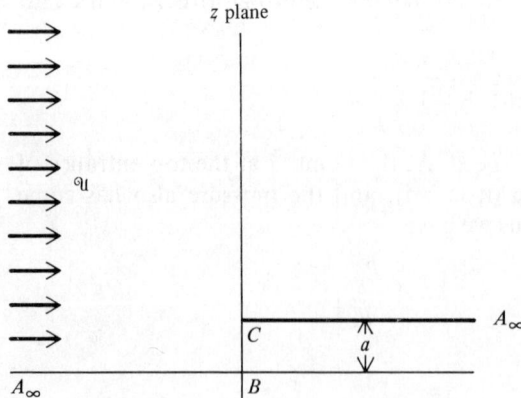

Figure 5.30

5.5. Free Streamlines: Contraction Coefficients

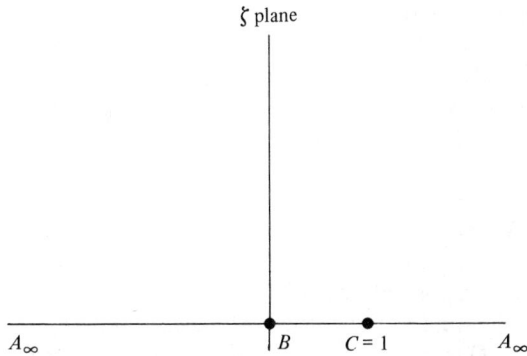

Figure 5.31

taking the real, positive branch of the square root for $0 < t < 1$. Then, since the point $\zeta = 1$ corresponds to $z = ia$, we have

$$ia = ic \int_0^1 \sqrt{\frac{1-t}{t}}\, dt = ic \int_0^1 t^{(1/2)-1}(1-t)^{(3/2)-1}\, dt;$$

we recognize the last integral to be a beta function (see Section 4.3), namely

$$B\left(\frac{1}{2}, \frac{3}{2}\right) = \frac{\Gamma(\frac{1}{2})\Gamma(\frac{3}{2})}{\Gamma(2)} = \frac{\sqrt{\pi}\frac{1}{2}\sqrt{\pi}}{1} = \frac{\pi}{2}.$$

Hence, $ia = ic\pi/2$, giving $c = 2a/\pi$.

Now in the ζ plane, the complex velocity potential is of the form $F(\zeta) = k\zeta$, since both source and sink are off at infinity. In the z plane, the conjugate of the velocity vector $(u - iv = qe^{-i\theta})$ is

$$\frac{dF}{dz} = \frac{dF}{d\zeta}\frac{d\zeta}{dz} = k\frac{\pi}{2a}\sqrt{\frac{\zeta}{\zeta - 1}}.$$

As $\zeta \to -\infty$, dF/dz must approach $\mathscr{U} + i0$, the given horizontal velocity off at the left. Now

$$\lim_{\zeta \to -\infty} \frac{dF}{dz} = \lim_{\zeta \to -\infty} \frac{k\pi}{2a}\sqrt{\frac{\zeta}{\zeta - 1}} = \frac{k\pi}{2a}$$

and, hence, $k\pi/2a = \mathscr{U}$ or $k = 2a\mathscr{U}/\pi$. Thus we have the expression

$$\frac{dF}{dz} = \mathscr{U}\sqrt{\frac{\zeta}{\zeta - 1}},$$

an expression strikingly similar to the one for bent channel flow. Note here that as $\zeta \to -1$ or $z \to ia$ (the top edge of the step), the velocity becomes infinite. But, as we have already pointed out, this is an impossibility, because Bernoulli's law requires that the pressure approach $-\infty$, which is impossible. What can we do to salvage this problem?

The answer, a surprisingly elegant and simple one, is due to Helmholtz, who reasoned as follows: Whenever a fluid flows past a corner which opens wider than 180°, the velocity there should become "infinite" and the pressure should decrease to "minus infinity." However, long before the situation reaches this point, the fluid at the corner will begin to boil due to the greatly reduced pressure. This boiling will continue until the flow has reached a steady state, when the fluid will be flowing up over the step along a "free streamline" (Figure 5.32), below which is vapor, above which is liquid.* The streamline will come into the line BC tangent to it (since, if the angle at which the streamline met BC exceeded 180°, the velocity would again increase until further boiling took place, and the flow would thus not be steady). Likewise, the pressure on both sides of the streamline must be the same, on account of the steady state of the flow. This pressure must be p_∞, which implies by Bernoulli's Law

$$\frac{1}{2}q^2 + \frac{p_\infty}{\rho} = \frac{1}{2}\mathscr{U}^2 + \frac{p_\infty}{\rho}$$

that $q^2 = \mathscr{U}^2$ or $q = \mathscr{U}$, that is, that the speed of the flow along the stream-

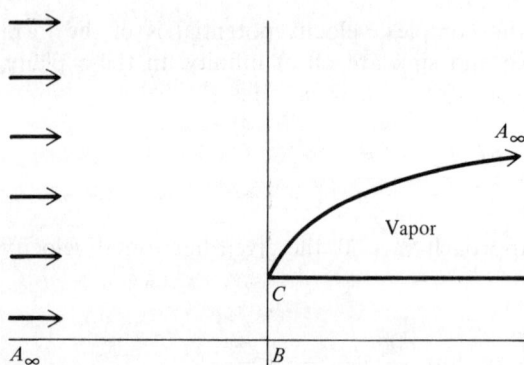

Figure 5.32

---

* The student may wish to consider why the streamline should necessarily begin at C and not elsewhere on the segment BC. We shall assume without argument that it starts at C.

5.5. FREE STREAMLINES: CONTRACTION COEFFICIENTS

line CA_∞ is constant and equals $\mathscr{U}$. This will give us a boundary condition which will help us solve the flow problem.

In order to solve this problem, we need not only the complex velocity potential $F(z) = \phi + i\psi$, but also the equation of the free streamline. Now, since ψ is constant along streamlines, and $A_\infty BCA_\infty$ is a streamline, clearly ψ is constant there, and we may as well choose $\psi = 0$ since this is the boundary of the flow region. Likewise, we are free to choose the value of ϕ at some convenient point; we shall take $\phi(ia) = 0$.

Our procedure to solve the problem will be as follows: First we are going to look at the complex F plane (real axis ϕ, imaginary axis ψ). Next we shall look at the so-called *reciprocal hodograph plane*, with complex variable v, where $v = \mathscr{U}/F'(z)$.* (We shall see why this weird choice shortly.) Now, if we can map a suitable portion of the plane conformally onto a suitable portion of the F plane, we will have a *differential equation* relating F and F', which we can try to solve.

The F plane looks as illustrated in Figure 5.33. Because $\psi = 0$ on the lower boundary in the z plane, it goes into the real axis in the F plane. The point C goes into the origin, since we chose $\phi(ia) = 0$. B goes into some point $-k$ to the left of the origin, which we shall try to determine later. The rest of the flow region goes into the upper half plane; it is clear that the value of ψ at any point D in the flow region is positive, since this value is precisely the measure of the flux $\mathscr{F}$ across any curve joining D to C (Figure 5.34).

Now let us consider the complex variable $F'(z) = f(z) = u - iv = qe^{i\theta}$. This variable maps the flow region in the z plane into a finite region in the F' plane, which might be rather difficult to map into the upper half F plane. For this reason we examine a complex variable given essentially

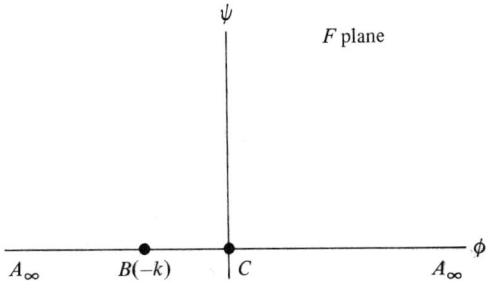

Figure 5.33

---

* The hodograph plane is given by the velocity vector itself, i.e., it is the $\bar{f}(z) = u + iv$ plane.

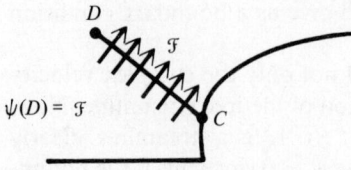

Figure 5.34

by the reciprocal of F, namely

$$v = \frac{\mathcal{U}}{F'(z)} = \frac{\mathcal{U}}{qe^{-i\theta}} = \frac{\mathcal{U}}{q}e^{i\theta}.$$

In the v plane, or *reciprocal hodograph* plane, the flow region is mapped into an infinite region (Figure 5.35). We arrive at this region as follows: Along the free streamline CA, the speed is constantly equal to $\mathcal{U}$, so v is simply $e^{i\theta}$. At C, the velocity is vertical, so $\theta = \pi/2$ and, hence, $v = i$. As we progress outwards along the free streamline toward A, θ approaches 0 and the velocity becomes horizontal, so that the point A corresponds to 1. Now if we move down from C to B, the velocity stays vertical ($\theta = \pi/2$), but q approaches 0. Recall that $z = 0$ is a stagnation point; the speed always approaches 0 at an inside corner with angle less than 180°. Similarly, if we approach B from A_∞ at the left, the velocity is horizontal ($\theta = 0$), and q approaches zero from the value $\mathcal{U}$ (out at A). Clearly the image of the flow region is the shaded portion of the v plane, since no velocity vector has argument less than zero or greater than $\pi/2$.

It is now our goal to map the image of the flow region in the reciprocal hodograph plane (Figure 5.35) into the upper half F plane (Figure 5.33). It will be necessary to do this in two steps; it is rather difficult *a priori* to see a mapping which in one step will perform this service. If we consider

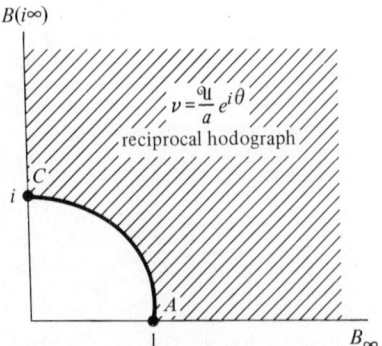

Figure 5.35

the complex t plane, where $t = \log v$, we see that the image of the flow region is a degenerate polygon which can be mapped into the upper half F plane by means of a Schwarz-Christoffel transformation (Figure 5.36). To verify that the flow region has this image, note that $\log v$ has constant real part (0) on the arc of the unit circle AC; constant imaginary part ($\pi/2$) on the imaginary axis CB; and is real and positive for real v on the portion of the real axis AB. To map the upper half F plane onto this semi-infinite strip in the t plane, we use the Schwarz-Christoffel transformation

$$\frac{dt}{dF} = cF^{-1/2}(F+k)^{-1}$$

or

$$t = \frac{\pi i}{2} + c \int_0^F \frac{d\sigma}{(\sigma+k)\sqrt{\sigma}},$$

where the additive constant of integration must be $\pi i/2$ in order that the origin in the F plane correspond to $t = \pi i/2$. To find the value of c, note that if we integrate to $A(\infty)$ in the F plane, then we must obtain

$$0 = \frac{\pi i}{2} + c \int_0^\infty \frac{d\sigma}{(\sigma+k)\sqrt{\sigma}}.$$

We integrate by making the substitution $\sigma = \tau^2$, whence $d\sigma = 2\tau d\tau$, and we have

$$0 = \frac{\pi i}{2} + c \int_0^\infty \frac{2d\tau}{\tau^2 + k} = \frac{\pi i}{2} + \frac{2c}{\sqrt{k}} \arctan \frac{\tau}{\sqrt{k}}\Big|_0^\infty = \frac{\pi i}{2} + \frac{c\pi}{\sqrt{k}},$$

and thus $c = -i\sqrt{k}/2$. In particular, for real values of $F > 0$, we have

$$t = \frac{\pi i}{2} - i \arctan \sqrt{\frac{F}{k}}.$$

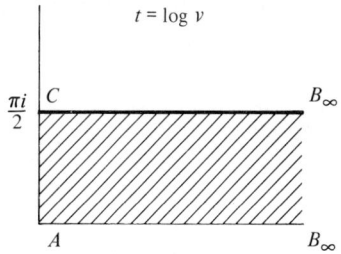

Figure 5.36

272 TWO-DIMENSIONAL INCOMPRESSIBLE FLUID FLOW

Hence
$$v = e^t = e^{\pi i/2} e^{-i \arctan \sqrt{F/k}} = i[\cos(\arctan\sqrt{F/k}) - i\sin(\arctan\sqrt{F/k})].$$
According to Figure 5.37, if $\arctan\sqrt{F/k} = \theta$, then $\cos\theta = \sqrt{k}/\sqrt{F+k}$, and $\sin\theta = \sqrt{F}/\sqrt{F+k}$. Hence

$$v = \sqrt{\frac{F}{F+k}} + i\sqrt{\frac{k}{F+k}}.$$

Now $v = \mathcal{U}/(dF/dz) = \mathcal{U}dz/dF$ and, therefore,

$$\mathcal{U}dz = \frac{\sqrt{F} + i\sqrt{k}}{\sqrt{F+k}} dF$$

or, finally,

$$\mathcal{U}z = \mathcal{U}ia + \int_0^F \frac{\sqrt{\sigma} + i\sqrt{k}}{\sqrt{\sigma+k}} d\sigma,$$

the constant of integration being as it is because $F = 0$ and $z = ia$ must correspond. If we take real and imaginary parts, we have

$$x = \frac{1}{\mathcal{U}} \int_0^F \frac{\sqrt{\sigma}}{\sqrt{\sigma+k}} d\sigma, \qquad y = a + \frac{1}{\mathcal{U}} \int_0^F \frac{\sqrt{k}}{\sqrt{\sigma+k}} d\sigma,$$

which are parametric equations of the free streamline in terms of the parameter F, $0 \leq F < \infty$. Note that for very large values of F, x behaves like F and y behaves like $\sqrt{F}$, so that the curve is much like a parabola ($y \sim \sqrt{x}$).

The reader may have noticed that we have not yet evaluated the constant k. We can try to do this by going back to the expression for t in terms of F

$$t = \frac{\pi i}{2} - \frac{i\sqrt{k}}{2} \int_0^F \frac{d\sigma}{(\sigma+k)\sqrt{\sigma}}$$

and integrating in the negative direction from C. To get to the segment $A_\infty B$ in the F plane, we must bypass $-k$ by using a small semicircle (Figure 5.38); then we note that on the segment AB in the t plane (Figure 5.36),

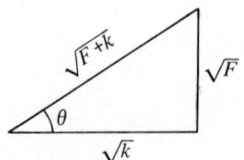

Figure 5.37

5.5. FREE STREAMLINES: CONTRACTION COEFFICIENTS

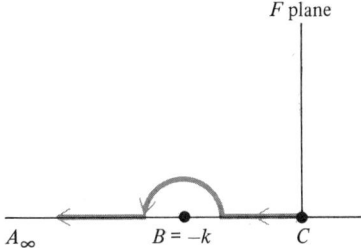

Figure 5.38

t must be real. Thus we have

$$\text{real} = \frac{\pi i}{2} - \frac{i\sqrt{k}}{2}\left(\int_0^{-k+\delta} + \int_{-k-\delta}^F\right)\left(\frac{d\sigma}{(\sigma+k)\sqrt{\sigma}}\right)$$

$$- \frac{i\sqrt{k}}{2}\int_0^\pi \frac{i\delta e^{i\theta}\,d\theta}{\delta e^{i\theta}\sqrt{-k+\delta e^{i\theta}}}.$$

In the limit as $\delta \to 0$, we obtain

$$\text{real} = \frac{\pi i}{2} - \frac{i\sqrt{k}}{2} P\int_0^F \frac{d\sigma}{(\sigma+k)\sqrt{\sigma}} - \frac{i\sqrt{k}}{2}\cdot\frac{i}{\sqrt{-k}}\pi.$$

The principal value integral is imaginary, since $\sigma < 0$; taking account the factor i, we see that the term involving the principal value integral is real. The remainder of the expression is imaginary, and hence it must vanish:

$$0 = \frac{\pi i}{2} - \frac{i\sqrt{k}}{2}\frac{i}{\sqrt{-k}}\pi = \frac{\pi i}{2} - \frac{\pi i}{2} = 0.$$

Well! Our quest for the elusive k has ended in a tautology, and it would appear that the value of k is irrelevant, or at least, that it cannot be obtained by this method. We have carried out this fruitless effort primarily to remind the reader that there are many blind alleys in this work. However, discouragement need not automatically follow such attempts, since there are many ways to attack such problems. Hydrodynamics calculations *are* difficult, and complicated. It is not always possible to solve a problem completely; but we can often glean considerable information from even incomplete solutions.

To illustrate further the difficulties which often arise, let us derive the differential equation for F, the complex velocity potential. Since $t = \log v = \log(\mathscr{U}/F')$, we have

$$\log\left(\frac{\mathscr{U}}{F'}\right) = \frac{\pi i}{2} - \frac{i\sqrt{k}}{2}\int_0^F \frac{d\sigma}{(\sigma+k)\sqrt{\sigma}};$$

274 TWO-DIMENSIONAL INCOMPRESSIBLE FLUID FLOW

By exponentiating both sides and solving for F', we obtain

$$F' = -\mathcal{U}i \exp\left(\frac{i\sqrt{k}}{2}\int_0^F \frac{d\sigma}{(\sigma+k)\sqrt{\sigma}}\right).$$

Needless to say, this equation is highly nonlinear and cannot be solved in closed form. It is comforting, however, to realize that we have come even this far.

We shall devote the remainder of the chapter to two other problems in which we shall concentrate on the data which we *can* obtain. The first of these has a much nicer answer than the second one, the bent channel problem, which we have worked on before and will save till last.

EXAMPLE 16 This problem is called the *Borda Mouthpiece*. (Figure 5.39). Fluid flows out from the z plane horizontally through a semi-infinite channel as illustrated.* Velocity out at the right end is $\mathcal{U}$. It is as if a pipe to the outside were inserted into a very large container of fluid, and the fluid then drained out through the pipe. The width of the channel is $2a$ while the width of the column of fluid at the right is $2\sigma a$, where $\sigma(0 < \sigma < 1)$ is called the *contraction coefficient*. We shall content ourselves in this problem with determining the value of σ, given the flow as illustrated. Because of the symmetry of the problem about the central dotted line, we choose $\psi = 0$ along this line of symmetry. We can then choose ϕ at a point; hence we choose $\phi(\pm ia) = 0$, that is, ϕ is to vanish at BB'. Then the F plane will look as illustrated in Figure 5.40; the image of the flow region is an infinite strip bounded above by the streamline $\psi = \mathcal{U}\sigma a$, and below by the streamline $\psi = -\mathcal{U}\sigma a$. (The total flux out the mouthpiece is, of course, $2\mathcal{U}\sigma a$.)

Since the F plane is not the upper half plane in this case, we shall have to map it onto the upper half plane in some other variable, say ζ, which we do by means of the Schwarz-Christoffel mapping $dF/d\zeta = c\zeta^{-1}$; since

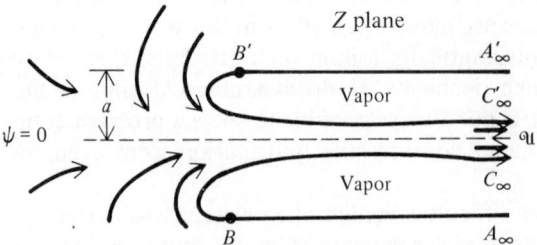

Figure 5.39

---

* Obviously we are neglecting the effects of gravity.

5.5. FREE STREAMLINES: CONTRACTION COEFFICIENTS 275

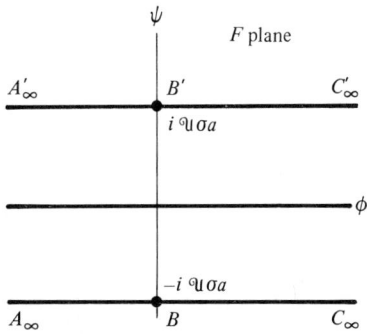

Figure 5.40

A goes into ∞, no factor appears for it; and since the angles at B and B' are π, no terms appear for them either. The result is shown in Figure 5.41. The total flux out through the sink at CC', as remarked before, is $-2\mathcal{U}\sigma a$; hence our previous work yields

$$\frac{dF}{d\zeta} = -\frac{2\mathcal{U}\sigma a}{\pi\zeta}.$$

Now for the moment let us leave the ζ plane and consider the reciprocal hodograph plane, that is, the complex v plane where $v = \mathcal{U}/F' = (\mathcal{U}/q)\,e^{i\theta}$. The reciprocal plane appears as in Figure 5.42. As in the previous example, the speed is constant ($\mathcal{U}$) along both streamlines; hence $v = e^{i\theta}$ there. On the top streamline, θ is $-\pi$ at B' and increases to 0 at C'. On the bottom, θ is $-\pi$ at B and decreases to 0 as we proceed to C. Along $B'A'_\infty, \theta$ is constant at $-\pi$ while q decreases to zero; similarly along BA_∞.

Next we consider the t plane, where $t = \log v$ (Figure 5.43). The image is the semi-infinite strip illustrated; the circumference of the unit circle

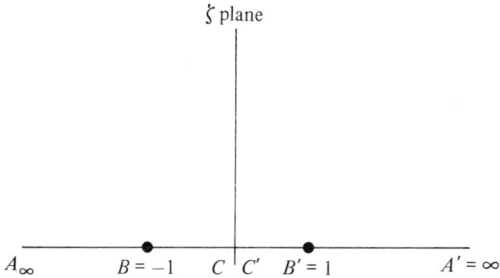

Figure 5.41

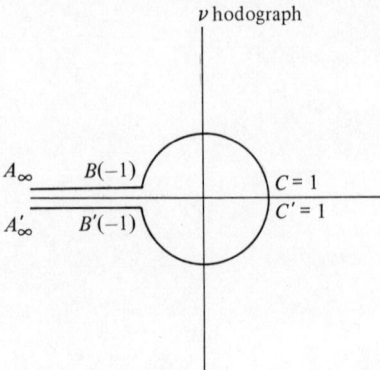

Figure 5.42

goes into a segment of the imaginary axis, while the two sides of the negative real axis go into lines parallel to the real axis in the t plane.

We are now ready to perform our final mapping, from the upper half ζ plane into this degenerate polygon in the t plane. The Schwarz-Christoffel transformation which does this is

$$\frac{dt}{d\zeta} = c'(\zeta + 1)^{-1/2}(\zeta - 1)^{-1/2}$$

or

$$t = c \int_0^\zeta \frac{d\zeta}{\sqrt{1-\zeta^2}},$$

for real ζ between -1 and 1 with the positive square root. In particular, $\zeta = 1$ corresponds to $t = -\pi i$; consequently,

$$-\pi i = c \int_0^1 \frac{d\zeta}{\sqrt{1-\zeta^2}} = c \arcsin \zeta \Big|_0^1 = \frac{c\pi}{2}.$$

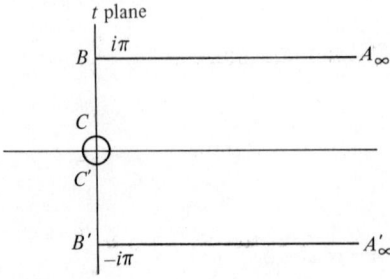

Figure 5.43

5.5. Free Streamlines: Contraction Coefficients

Hence $c = -2i$, and for this range of ζ, we have $t = -2i \arcsin \zeta$. Therefore we can find v in terms of ζ:

$$v = e^t = e^{-2i \arcsin \zeta} = \cos(2 \arcsin \zeta) - i \sin(2 \arcsin \zeta)$$
$$= [\cos^2(\arcsin \zeta) - \sin^2(\arcsin \zeta)] - i[2 \sin(\arcsin \zeta) \cos(\arcsin \zeta)]$$
$$= (1 - \zeta^2 - \zeta^2) - i(2\zeta \sqrt{1 - \zeta^2}) = (1 - 2\zeta^2) - 2i\sqrt{1 - \zeta^2}$$

(Figure 5.44).

But in terms of our original function F, v is

$$v = \frac{\mathscr{U}}{dF/dz} = \frac{\mathscr{U}}{(dF/d\zeta)(d\zeta/dz)} = \frac{dz}{d\zeta}\left(-\frac{\pi \zeta}{2\sigma a}\right).$$

Therefore,

$$\mathscr{U} dz = v\left(-\frac{2\sigma \mathscr{U} a}{\pi \zeta}\right) d\zeta = \left(-\frac{2\sigma \mathscr{U} a}{\pi \zeta}\right)[(1 - 2\zeta^2) - 2i\zeta \sqrt{1 - \zeta^2}] d\zeta.$$

We have so far not actually chosen the position of the real and imaginary axes in the z plane, and our positioning of the origin will not affect the solution of the problem. If we choose the origin at B, then $z = 0$ will correspond to $\zeta = -1$, and we shall have

$$\mathscr{U} z = \int_0^z \mathscr{U} dz = \int_{-1}^{\zeta} (1 - 2\zeta^2 - 2i\zeta \sqrt{1 - \zeta^2})\left(-\frac{2\sigma \mathscr{U} a}{\pi \zeta} d\zeta\right).$$

We are primarily interested in the y coordinate, since its value as $\zeta \to 0$ will give the height to the streamline of the fluid flowing out through the mouthpiece, an essential if we are to find σ. Hence we take imaginary parts of both sides, obtaining

$$\mathscr{U} y = \int_{-1}^{\zeta} (-2\zeta \sqrt{1 - \zeta^2})\left(-\frac{2\sigma \mathscr{U} a}{\pi \zeta}\right) d\zeta = \frac{4\sigma \mathscr{U} a}{\pi} \int_{-1}^{\zeta} \sqrt{1 - \zeta^2} \, d\zeta.$$

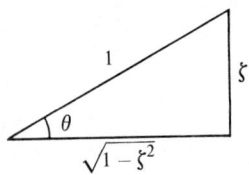

Figure 5.44

278 TWO-DIMENSIONAL INCOMPRESSIBLE FLUID FLOW

To find y_∞, the height to the streamline BC from the real axis, we evaluate

$$y_\infty = \frac{4\sigma a}{\pi} \int_{-1}^{0} \sqrt{1 - \zeta^2}\, d\zeta.$$

Making the substitution $\zeta = \sin\theta$ in this integral, the limits of integration become $-\pi/2$ to 0, and $d\zeta = \cos\theta\, d\theta$, so that

$$y_\infty = \frac{4\sigma a}{\pi} \int_{-\pi/2}^{0} \cos^2\theta\, d\theta = \frac{4\sigma a}{\pi} \int_{-\pi/2}^{0} \left(\frac{1}{2} + \frac{1}{2}\cos 2\theta\right) d\theta = \frac{4\sigma a}{\pi} \frac{\pi}{4} = \sigma a.$$

Therefore $y_\infty = \sigma a$; on the other hand (Figure 5.45), $y_\infty = a - \sigma a = (1 - \sigma)a$; hence, $\sigma = 1 - \sigma$ and, therefore, the contraction coefficient is $\sigma = \frac{1}{2}$. Incidentally, the equation

$$z = -\frac{2\sigma a}{\pi} \int_{-1}^{\zeta} \left(\frac{1 - 2\zeta^2 - 2i\zeta\sqrt{1 - \zeta^2}}{\zeta}\right) d\zeta$$

also gives the parametric equation of the free streamline BC for real ζ between -1 and 0.

As a final example we shall return to the problem of the bent channel and attempt to determine the contraction coefficient of this flow.

EXAMPLE 17 The entire sequence of maps used in this problem is illustrated in Figure 5.46. The reader will recall that we have fluid flowing into the top of the channel at velocity $\mathscr{U}$; as it passes around the corner at D, we now know that a free streamline forms (Figure 5.46a). As the fluid flows out at the right (at speed $\mathscr{U}_\infty$), it is contracted by a factor of σ. Since the flux in equals the flux out, we have $\mathscr{U}_\infty \sigma b = \mathscr{U} a$ and, hence, $\mathscr{U}_\infty = \mathscr{U} a/\sigma b$. To obtain the F plane (Figure 5.46b), we take $\psi = 0$ along the positive real and imaginary axes in the z plane; thus, $\psi = \mathscr{U} a$ along the other boundary and the free streamline. We choose ϕ to be zero at $a + ib$ so that the point D corresponds to $F = ia$. The point B goes into some point on the real F axis; but the location of this point is unimportant since in all of the other planes B is at infinity. Note that since the image of the flow region in the

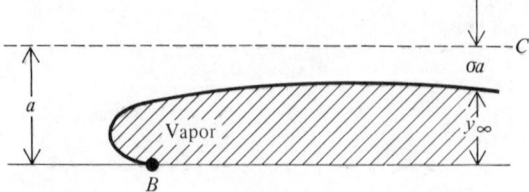

Figure 5.45

5.5. FREE STREAMLINES: CONTRACTION COEFFICIENTS

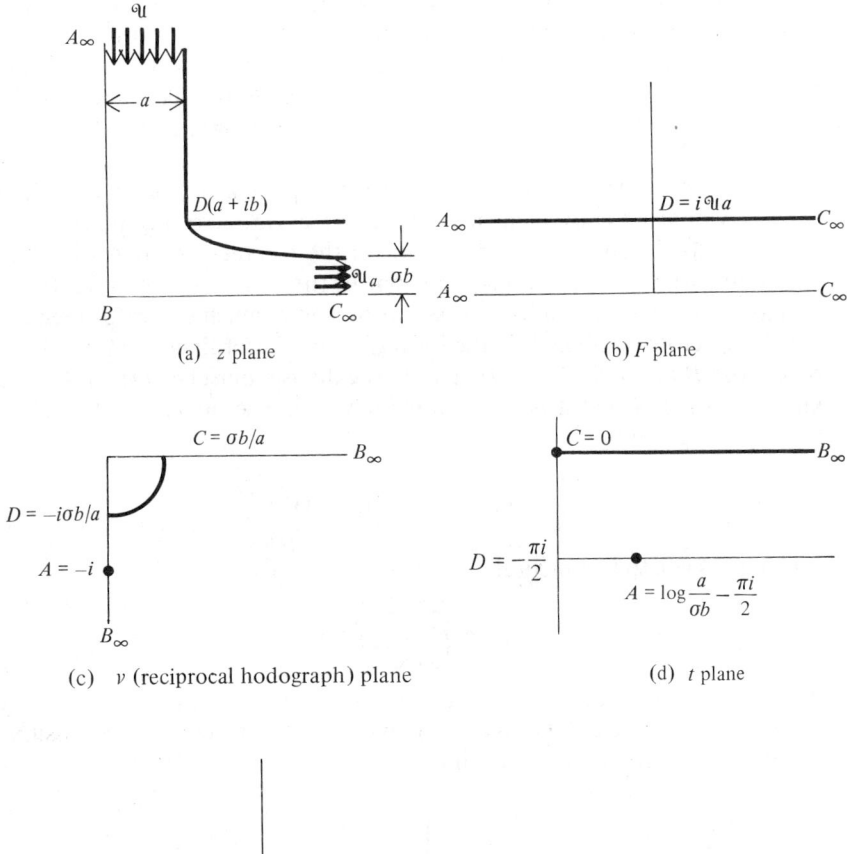

Figure 5.46

F plane is not the upper half plane, we shall have to use the ζ plane as in the last example. Before we examine the configuration of the upper half ζ plane, however, let us move over to the reciprocal hodograph and t planes. In the reciprocal hodograph plane (Figure 5.46c), the mapping $v = (\mathcal{U}/q) e^{i\theta}$ maps the flow region into the portion of the fourth quadrant outside the circle of radius $\sigma b/a$ (and not the unit circle as in the previous examples). This is because the speed along the free streamline is $\mathcal{U}_\infty = \mathcal{U}a/\sigma b$ and not $\mathcal{U}$ as before.

In obtaining the t plane (Figure 5.46d), instead of $t = \log v$ we use $t = \log(av/\sigma b)$, so that the circle mentioned above goes into the portion of the imaginary t axis shown. Note the location of A in the two planes. Note also that B is at ∞ in the reciprocal hodograph plane because the speed at B is 0; being a corner with angle less than 180°, it is a stagnation point.

We now see that if we wish to map the upper half ζ plane into the t plane it would be convenient if C and D were symmetrically placed with respect to the origin in the ζ plane; for in the t plane, the angle of $\pi/2$ at each point will give rise to a square root in the Schwarz-Christoffel transformation. A quantity under a square root sign is much easier to integrate if its two factors combine into the form $\zeta^2 - a^2$. Accordingly, in the ζ plane we choose B at ∞, C at -1, D at 1. This exhausts our choices; all that we know about A is that it is at some point $k > 1$. The mapping which does this, then, is given by

$$\frac{dt}{d\zeta} = c(\zeta + 1)^{-1/2}(\zeta - 1)^{-1/2},$$

which gives rise to the integral

$$t = c \int_{-1}^{\zeta} \frac{d\zeta}{\sqrt{\zeta^2 - 1}},$$

the constant of integration being zero, so that $t = 0$ and $\zeta = -1$ may correspond. For real ζ between -1 and 1, we shall choose the positive imaginary branch of $\sqrt{\zeta^2 - 1}$. Thus $\sqrt{\zeta^2 - 1} = i\sqrt{1 - \zeta^2}$ and

$$t = -ic \int_{-1}^{\zeta} \frac{d\zeta}{\sqrt{1 - \zeta^2}}.$$

In particular, $\zeta = 1$ corresponds to $t = -\pi i/2$ and, therefore,

$$-\frac{\pi i}{2} = -ic \int_{-1}^{1} \frac{d\zeta}{\sqrt{1-\zeta^2}} = -ic \arcsin \zeta \Big|_{-1}^{1} = -ic\pi,$$

giving $c = \frac{1}{2}$. Therefore, we have

$$t = \frac{1}{2}\int_{-1}^{\zeta} \frac{d\zeta}{\sqrt{\zeta^2 - 1}} = -\frac{i}{2}\int_{-1}^{\zeta} \frac{d\zeta}{\sqrt{1 - \zeta^2}}.$$

To find the value of k, we need to extend this integral beyond D to the point A. We shall then have

$$t = \log\left(\frac{a}{\sigma b}\right) - \frac{\pi i}{2} = \frac{1}{2}\int_{-1}^{1} \frac{d\zeta}{\sqrt{\zeta^2 - 1}} + \frac{1}{2}\int_{1}^{k} \frac{d\zeta}{\sqrt{\zeta^2 - 1}}.$$

In the first integral, as remarked above, we take the positive imaginary branch of the square root, which corresponds to choosing $\zeta + 1$ to have argument 0 and $\zeta - 1$ to have argument π. In the second integral, then, $\zeta - 1$ will have argument 0, so that we are taking the positive *real* branch of the square root. Thus

$$\log\left(\frac{a}{\sigma b}\right) - \frac{\pi i}{2} = -\frac{\pi i}{2} + \frac{1}{2}\log(\zeta + \sqrt{\zeta^2 - 1})\Big|_1^k = -\frac{\pi i}{2}$$

$$= -\frac{\pi i}{2} + \frac{1}{2}\log(k + \sqrt{k^2 + 1}).$$

Therefore we have $\log(a/\sigma b) = \frac{1}{2}\log(k + \sqrt{k^2 - 1})$, which can be solved for k in a straightforward manner, yielding

$$k = \frac{1}{2}\left[\left(\frac{a}{\sigma b}\right)^2 + \left(\frac{\sigma b}{a}\right)^2\right].$$

Let us now map the upper half ζ plane onto the flow region in the F plane and see if we can evaluate σ, the only remaining unknown constant. For this mapping we use

$$\frac{dF}{d\zeta} = c(\zeta + 1)^{-1}(\zeta - 1)^0(\zeta - k)^{-1} = \frac{c}{(\zeta + 1)(\zeta - k)}.$$

Integrating in the ζ plane beginning at the point $D = 1$ we have

$$F = c\int_1^\zeta \frac{d\zeta}{(\zeta + 1)(\zeta - k)} + i\mathcal{U}a.$$

The constant c must be real and positive in order for these values to fall on the line $\operatorname{Im} F = \mathcal{U}a$ for real ζ between -1 and k. To find the value of c, we note that if we integrate along a path P as shown in Figure 5.47 to some point to the right of A, we have entered the real axis in the F plane

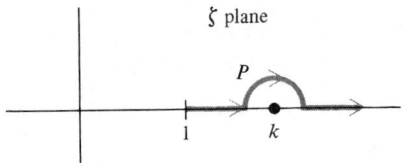

Figure 5.47

on the left. Thus the values of the expression integrated along P,

$$c \int_1^\zeta \frac{d\zeta}{(\zeta+1)(\zeta-k)} + i\mathcal{U}a = c\left(\int_1^{k-\delta} + \int_{k+\delta}^\zeta\right)\left(\frac{d\zeta}{(\zeta+1)(\zeta-k)}\right)$$

$$+ c \int_\pi^0 \frac{i\delta e^{i\theta}\, d\theta}{(\delta e^{i\theta}+k+1)\delta e^{i\theta}} + i\mathcal{U}a,$$

must be real. This must still hold in the limit as $\delta \to 0$, when we have

$$F = \text{real} = cP \int_1^\zeta \frac{d\zeta}{(\zeta+1)(\zeta-k)} - c\frac{i\pi}{k+1} + i\mathcal{U}a.$$

The principal value is real; hence the two imaginary constants must cancel and we obtain

$$c = \frac{(k+1)\mathcal{U}a}{\pi},$$

k having been determined previously. Therefore,

$$\frac{dF}{d\zeta} = \frac{(k+1)\mathcal{U}a}{\pi} \frac{1}{(\zeta+1)(\zeta-k)}.$$

To find the parametric equation of the free streamline, we set

$$\frac{\partial v}{b\sigma} = e^t = \frac{\mathcal{U}a}{\sigma b}\left(\frac{dF}{dz}\right)^{-1} = \frac{\mathcal{U}a}{\sigma b}\left(\frac{dF}{d\zeta}\frac{d\zeta}{dz}\right)^{-1},$$

as before (except for the factor multiplying v), so that

$$dz = \frac{\sigma b}{\mathcal{U}a} e^t \frac{dF}{d\zeta} d\zeta.$$

We shall have to express the right hand side (e^t, in particular) in terms of ζ. Since the free streamline DC corresponds to the interval from 1 to -1 in the ζ plane, we rewrite t as

$$t = -\frac{i}{2}\int_{-1}^\zeta \frac{d\zeta}{\sqrt{1-\zeta^2}} = -\frac{i}{2}\left(\int_{-1}^1 \frac{d\zeta}{\sqrt{1-\zeta^2}} - \int_\zeta^1 \frac{d\zeta}{\sqrt{1-\zeta^2}}\right)$$

$$= -\frac{i}{2}\left(\pi - \arcsin\zeta\Big|_\zeta^1\right) = -\frac{i}{2}\left(\pi - \frac{\pi}{2} + \arcsin\zeta\right)$$

$$= -\frac{i}{2}\left(\arcsin\zeta + \frac{\pi}{2}\right).$$

5.5. FREE STREAMLINES: CONTRACTION COEFFICIENTS

We next need to use this result in order to compute e^t in terms of ζ. Now

$$e^t = \exp\left[-\frac{i}{2}\left(\arcsin \zeta + \frac{\pi}{2}\right)\right]$$

$$= \cos\left[\frac{1}{2}\left(\arcsin \zeta + \frac{\pi}{2}\right)\right] - i\sin\left[\frac{1}{2}\left(\arcsin \zeta + \frac{\pi}{2}\right)\right].$$

We next need the half-angle formulas to simplify these further:

$$\cos\left[\frac{1}{2}\left(\arcsin \zeta + \frac{\pi}{2}\right)\right] = \sqrt{\frac{1}{2} + \frac{1}{2}\cos\left(\arcsin \zeta + \frac{\pi}{2}\right)} = \sqrt{\frac{1}{2} - \frac{1}{2}\zeta},$$

recalling also that $\cos(\theta + \pi/2) = -\sin\theta$. Likewise,

$$\sin\left[\tfrac{1}{2}(\arcsin \zeta + \pi/2)\right] = \sqrt{\tfrac{1}{2} + \tfrac{1}{2}\zeta}.$$

Hence

$$e^t = \sqrt{\tfrac{1}{2} - \tfrac{1}{2}\zeta} - i\sqrt{\tfrac{1}{2} + \tfrac{1}{2}\zeta}.$$

By putting this into the formula for dz, we have

$$dz = \frac{\sigma b}{\mathcal{U}a} e^t \frac{dF}{d\zeta} d\zeta$$

$$= \frac{\sigma b}{a}\left(\sqrt{\frac{1}{2} - \frac{1}{2}\zeta} - i\sqrt{\frac{1}{2} + \frac{1}{2}\zeta}\right)\frac{(k+1)a}{\pi}\frac{d\zeta}{(\zeta + 1)(\zeta - k)}.$$

By extracting a factor of $\sqrt{\frac{1}{2}}$, and putting all the constants together, then integrating, we have

$$\int_{a+ib}^{z} dz = \frac{\sigma b(k+1)}{\sqrt{2\pi}} \int_{1}^{\zeta}\left(\sqrt{1 - \zeta} - i\sqrt{1 + \zeta}\right)\frac{d\zeta}{(\zeta + 1)(\zeta - k)}.$$

Integration in the z plane is from $a + ib$ (the point D) along the free streamline to z; in the ζ plane it is from 1 to $\zeta < 1$ along the real axis. We are really only interested in imaginary parts, since $y_\infty = \sigma b$ will give us the contraction coefficient. Thus we examine

$$\int_{b}^{\sigma b} dy = \sigma b - b = -\frac{\sigma b(k+1)}{\sqrt{2\pi}} \int_{1}^{-1} \frac{d\zeta}{\sqrt{\zeta + 1}(\zeta - k)}.$$

We shall have solved our problem if we can evaluate this last integral. We do this by means of two substitutions: first, $v = \zeta + 1$, whence the integral becomes

$$-\int_{2}^{0} \frac{dv}{\sqrt{v}(v - 1 - k)};$$

then we set $v = u^2$, so that $dv = 2u\,du$ and we obtain

$$-\int_{\sqrt{2}}^{0} \frac{2u\,du}{u(u^2 - 1 - k)} = -2\int_{\sqrt{2}}^{0} \frac{du}{(u - \sqrt{1+k})(u + \sqrt{1+k})}$$

$$= -\frac{1}{\sqrt{k+1}} \int_{0}^{\sqrt{2}} \left(\frac{1}{\sqrt{1+k} - u} + \frac{1}{\sqrt{1+k} + u} \right) du$$

$$= \frac{1}{\sqrt{k+1}} \log\left(\frac{\sqrt{1+k} - u}{\sqrt{1+k} + u} \right) \bigg|_{0}^{\sqrt{2}} = \frac{1}{\sqrt{1+k}} \log \frac{\sqrt{1+k} - \sqrt{2}}{\sqrt{1+k} + \sqrt{2}}.$$

Hence

$$\sigma b - b = \frac{\sigma b(k+1)}{\sqrt{2\pi}} \frac{1}{\sqrt{k+1}} \log \frac{\sqrt{1+k} - \sqrt{2}}{\sqrt{1+k} + \sqrt{2}}.$$

Solving for σ, we finally have

$$\sigma = \left(1 - \frac{1}{\pi} \sqrt{\frac{k+1}{2}} \log \frac{\sqrt{1+k} - \sqrt{2}}{\sqrt{1+k} + \sqrt{2}} \right)^{-1}.$$

Previously, we have calculated a formula for k in terms of σ. This formula gives us σ in terms of k. In theory, at least, we have solved our problem. However, in practice, this example serves to illustrate that problems cannot always be solved as nicely as the last example! For, upon replacing k in the preceding equation with the expression derived previously, we obtain a transcendental equation for σ, which presumably determines σ uniquely, but which cannot be solved except by numerical methods.

A. Exercises

Discuss each of the following flow problems, finding all applicable constants and the contraction coefficient where possible.

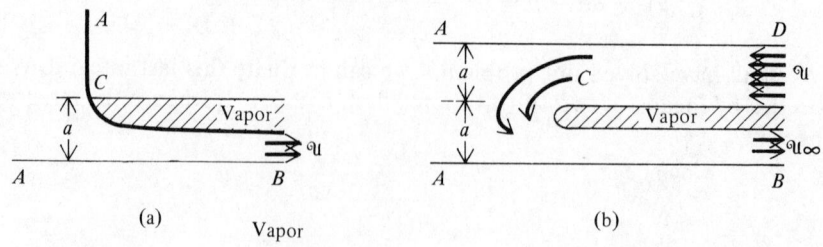

(a) Vapor

(b)

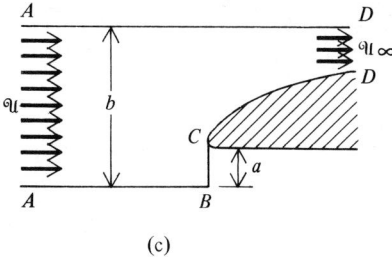

(c)

HINT: Summary of procedure: (1) Construct F plane (choose $\psi = 0$ along one boundary, $\phi = 0$ at some convenient point); (2) Construct $v = (\mathcal{U}/q)\,e^{i\theta}$ (reciprocal hodograph) plane by examining velocities along boundaries; (3) Construct $t = \log v$ plane; (4) Map t plane and F plane conveniently into upper half ζ plane; (5) Calculate $\mathcal{U}\,dz = e^t(dF/d\zeta)\,d\zeta$ or $\mathcal{U}\,dz = e^t(dF/d\zeta)(d\zeta/dt)\,dt$. Appropriate integration yields parametric equation of free streamline and value of σ.

Appendix A

Some Results from Real Calculus

In this brief appendix we shall outline the proofs of some results from the real calculus which are important for similar results in complex analysis. The proofs depend heavily on epsilon–delta arguments and the completeness of the real numbers; i.e., we shall assume without proof that a Cauchy sequence of real numbers has a limit.

Theorem 1 Let $\{x_n\}$ be a bounded sequence of real numbers, i.e., assume that there exist two real numbers m and M such that $m \leq x_n \leq M$ for all n. Then $\{x_n\}$ has a limit point, i.e., there is a number x, such that for each positive number ε the statement $|x_n - x| < \varepsilon$ is true for an infinite number of values of n.

Proof If $\{x_n\}$ is a convergent sequence, then the conclusion is already true. If not, then at least one of the two intervals $[m, (m + M)/2]$ and $[(m + M)/2, M]$ contains an infinite number of x_n's. Call this interval I_1 and let x_{n_1} be one of the x_n's in it. We then bisect I_1. At least one of its halves contains an infinite number of x_n's with $n > n_1$. We call this interval I_2 and let x_{n_2} be one of the x_n's in it with $n_2 > n_1$. In this manner we construct a sequence of intervals $\{I_1, I_2, \ldots, I_k, \ldots\}$ where the length of I_k is $(m + M)/2^k$ and a subsequence of $\{x_n\}$ called $\{x_{n_k}\}$. Then for each pair of integers k and p,

287

x_{n_k+p} and x_{n_k} are both in I_k; hence

$$|x_{n_k+p} - x_{n_k}| \leq \frac{m+M}{2^k},$$

which can be made as small as desired by making k large. Hence $\{x_{n_k}\}$ is Cauchy, and, therefore, it has a limit x. This is the point whose existence was asserted in the theorem. QED.

Theorem 2 A function f which is bounded from above, i.e., for which there is a finite number M such that $f(x) \leq M$ for all x in the domain of f, has a least upper bound. That is, there exists a number L such that (a) $f(x) \leq L$ for all x in the domain of f; (b) if $f(x) \leq M$ for all x in the domain of f, then $L \leq M$. Finally, if f is continuous on a closed bounded interval $[a, b]$, then it actually takes on its least upper bound at some point in $[a, b]$.

Proof Let M be an upper bound of f, and let x_0 be an arbitrary point in the domain of f, and let $m = f(x_0)$. If m is an upper bound of f, then we are through, since it would have to be the least upper bound. If m is not an upper bound of f, then if $[m, (m+M)/2]$ contains an upper bound of f, call this number y_1; if not, then $[(m+M)/2, M]$ contains an upper bound of f; call this number y_1. Call the interval in which y_1 occurs I_1; bisect I_1. If an upper bound for f occurs in the left half of I_1, call it y_2; if not, then there is an upper bound in the right half, and call this one y_2. Let I_2 be the interval containing y_2 and continue the process. In this way, as in Theorem 1, we construct a sequence $\{y_n\}$ which is Cauchy and, therefore, has a limit y. We claim y is the least upper bound of f. For (a) $f(x) \leq y$ for all x; if not, then there is a point x_1 for which

$$f(x_1) - y = \delta > 0.$$

But given δ, there is an N so large that

$$|y_n - y| < \frac{\delta}{2}, \quad \text{i.e.,} \quad -\frac{\delta}{2} < y_n - y < \frac{\delta}{2};$$

hence

$$f(x_1) - y_n = f(x_1) - y + y - y_n > \delta - \frac{\delta}{2} = \frac{\delta}{2},$$

which contradicts the fact that y_n is an upper bound of f. (b) If there is a $y' \leq y$ which is an upper bound of f, then first, $y' \in I_1$. For, if I_1 is $[m, (m+M)/2]$, then placing y' in $[(m+M)/2, M]$ would make $y' \geq y$, since $y \in I_1$; and if I_1 is $[(m+M)/2, M]$, then placing y' in $[m, (m+M)/2]$ would contradict the way I_1 was chosen. By the same token, y' is in I_k for each k; but then

this implies that
$$|y' - y_k| < \frac{m + M}{2^k},$$
and hence $y' = \lim y_k$. Therefore, $y' = y$.

Finally, if f is continuous on a closed bounded interval, we need to show that f assumes its least upper bound there. Let y be the least upper bound of f and let $x_1 \in [a, b]$. If $f(x_1) = y$, we are through. If not, then let x_2 be a point, where
$$f(x_2) = \tfrac{1}{2}[f(x_1) + y],$$
and for each n define x_{n+1} to be that value for which
$$f(x) = \tfrac{1}{2}[f(x_n) + y].$$
The sequence $\{x_n\}$ thus defined has a convergent subsequence $\{x_{n_k}\}$ by Theorem 1, which converges to a point $x \in [a, b]$. $f(x)$ is defined and, in fact, because f is continuous,
$$f(x) = \lim_{k \to \infty} f(x_{n_k}) = y;$$
which completes the proof. Note that this argument also establishes that the least upper bound is finite.

Theorem 3 Let $f(x)$ be a continuous real-valued function on a closed finite interval $[a, b]$. Then
$$\int_a^b f(x)\,dx$$
exists.

Proof We need, of course, to define, first of all, exactly what we mean by the integral of f from a to b. If we divide $[a, b]$ up into n smaller intervals,
$$a = x_0 < x_1 < x_2 < \cdots < x_n = b,$$
then for each of them let us choose an $\tilde{x}_k$ such that $x_k \geq \tilde{x}_k \geq x_{k-1}$ as follows: In each interval $[x_{k-1}, x_k]$, f, being continuous, assumes its maximum at some point M_k and its minimum at some point m_k. We then consider the two sums
$$l = \sum_{k=1}^{n} f(m_k)\,\Delta x_k, \quad L = \sum_{k=1}^{n} f(M_k)\,\Delta x_k.$$
All sums of the form L are bounded from below by $m(b - a)$; hence, as in the previous proof, there is a greatest lower bound, which is defined to be

the *upper integral* of f from a to b:

$$\mathrm{glb}(L) = \overline{\int_a^b} f(x)\,dx$$

Likewise the collection of all sums of the form l is bounded from above by $M(b-a)$; hence there is a least upper bound, which is defined to be the *lower integral* of f from a to b:

$$\mathrm{lub}(l) = \underline{\int_a^b} f(x)\,dx.$$

It is clearly true that for any bounded function f the upper and lower integrals exist and that

$$\underline{\int_a^b} f(x)\,dx \leq \overline{\int_a^b} f(x)\,dx,$$

since no l is ever larger than an L (this is not easy to show; the reader is nevertheless encouraged to attempt to prove it). We then say that in case the upper and lower integrals happen to be equal, then f is integrable and the common value is denoted by

$$\int_a^b f(x)\,dx.$$

If f is continuous on $[a, b]$, then it is uniformly continuous there.* This means that for each $\varepsilon > 0$, there is a $\delta > 0$ such that $|x - y| < \delta$ implies that $|f(x) - f(y)| < \varepsilon/(b-a)$. [We shall see why $\varepsilon/(b-a)$ in a moment]. Choose a partition of $[a, b]$, such that $x_k - x_{k-1} < \delta$ for all k. Then for this subdivision,

$$0 \leq \overline{\int_a^b} f(x)\,dx - \underline{\int_a^b} f(x)\,dx \leq \sum_{k=1}^n f(M_k)\,\Delta x_k - \sum_{k=1}^n f(m_k)\,\Delta x_k$$

$$= \sum_{k=1}^n [f(M_k) - f(m_k)]\,\Delta x_k.$$

Now each M_k and m_k is in the same interval $[x_{k-1}, x_k]$ and, hence, $|M_k - m_k| < \delta$. Therefore $[f(M_k) - f(m_k)] < \varepsilon/(b-a)$ and

$$0 \leq \overline{\int_a^b} f(x)\,dx - \underline{\int_a^b} f(x)\,dx \leq \sum_{k=1}^n \frac{\varepsilon}{b-a}\,\Delta x_k = \frac{\varepsilon}{b-a} \sum_{k=1}^n \Delta x_k$$

$$= \frac{\varepsilon}{b-a}(b-a) = \varepsilon.$$

---

* This is by itself a theorem worthy of being proved, which we shall omit here both for brevity and to avoid impeding the flow.

Hence for any ε, the upper and lower integrals differ by less than ε. Therefore they must be equal. QED.

Theorem 4 *(Green's Theorem)* Suppose $P(x, y)$ and $Q(x, y)$ are two real-valued functions defined inside and on a simple closed contour C and possessing continuous first partial derivatives there. Then

$$\oint_C (P\,dx + Q\,dy) = \int_I \int_{(C)} \left(\frac{\partial Q}{\partial x} - \frac{\partial P}{\partial y}\right) dx\,dy.$$

Proof We shall first assume that C is a smooth curve and that it is convex (Figure A1), so that a vertical line will intersect it in either two points or none, except for the two values $x = a$ and $x = b$, where a vertical line is tangent to C (Figure A2). Then the upper portion of C can be parametrized by a single-valued function of $y = y_1(x)$ and the lower portion by $y = y_2(x)$.

Hence we have

$$\int_I \int_{(C)} \frac{\partial P}{\partial y}\,dy\,dx = \int_a^b \int_{y_2(x)}^{y_1(x)} \frac{\partial P}{\partial y}(x, y)\,dy\,dx$$

$$= \int_a^b [P(x, y_1(x)) - P(x, y_2(x))]\,dx.$$

The latter integral, by the definition of an integral along a curve, represents the integral of P along the upper portion from left to right and along the lower portion from right to left: Hence, it is simply the integral of P around C in the negative direction. Therefore

$$\int_I \int_{(C)} \frac{\partial P}{\partial y}\,dy\,dx = \oint_C P(x, y)\,dx.$$

The proof of the other half of Green's theorem is similar.

This proof can be reasonably and easily generalized to contours which

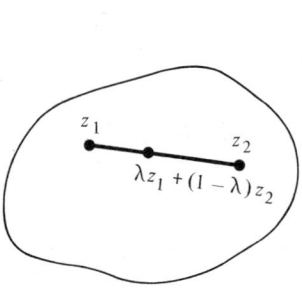

Figure A.1

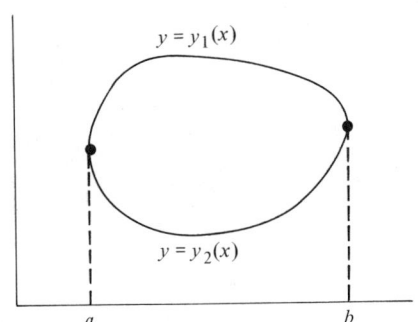

Figure A.2

292 SOME RESULTS FROM REAL CALCULUS

Figure A.3

are convex, but not strictly so (contours which can contain straight line segments), to piecewise smooth contours, and to nonconvex contours for which the interiors can readily be divided into convex domains by means of straight line segments (Figure A3). We shall not carry out these details, nor shall we trouble ourselves with more pathological cases.

Appendix B

Odd Numbered Answers

SECTION 1.1

A.1.
- **a** (4, 0), (5, 0)
- **b** (1, 5), (2, −29)
- **c** (−5, 14), (−30, −70)
- **d** (−8, 0), (25, 0)
- **e** (4, 4), (0, 4)
- **f** (−1/6, 17/12), (−2/3, −5/18)
- **g** (−8, 11/4), (53/4, −41/4)
- **h** (1, 0), (10/4, 0)
- **i** (5, 5), (0, 12)
- **j** (−4, 6), (−13, −12)
- **k** $(2\sqrt{2}/2, 0)$, (2, 0)
- **l** (8/3, −24/7), (−194/441, −96/21)

A.3.
- **a** $-i$
- **b** $\dfrac{1}{2} - \dfrac{1}{2}i$
- **c** $\dfrac{2}{13} - \dfrac{3}{13}i$
- **d** $\dfrac{1}{14}$
- **e** $\dfrac{14}{197} + \dfrac{1}{197}i$
- **f** $-\dfrac{3}{20} - \dfrac{1}{20}i$
- **g** $\dfrac{1}{\sqrt{2}} - \dfrac{1}{\sqrt{2}}i$
- **h** $\dfrac{1}{2} + \dfrac{\sqrt{3}}{2}i$
- **i** $-\dfrac{4}{5} - \dfrac{3}{5}i$
- **j** $\dfrac{5}{13} - \dfrac{12}{13}i$
- **k** $-1 + i$
- **l** $\dfrac{63}{441} + \dfrac{6}{441}i$

293

A.5. **a** $-\dfrac{109}{13} + \dfrac{90}{13}i$

    b $\cos\theta - i\sin\theta$

    c $7/25$

    d $\dfrac{4}{13}i$

    e $\cos^2\theta - \sin^2\theta + 2i\sin\theta\cos\theta$

    f 62

    g $321 - 33i$

    h $33 - 46i$

A.7. **a** $2 - i$

    b $2 + 2i$

    c $-1, -1$

    d $-2 + i, -2 - i$

SECTION 1.2

A.1. **a** $|z| = \sqrt{2}, \arg z = \pi/4$

       $|z| = \sqrt{2}, \arg z = -\pi/4$

       $|z| = \sqrt{2}, \arg z = 3\pi/4$

       $|z| = \sqrt{2}, \arg z = -3\pi/4$

    b $|z| = \sqrt{10}, \arg z = \tan^{-1}(1/3)$

       $|z| = 4, \arg z = -\pi/3$

       $|z| = 10, \arg z = 2\pi/3$

       $|z| = 4, \arg z = 5\pi/6$

    c $|z| = 17, \arg z = 0$

       $|z| = 23, \arg z = \pi$

       $|z| = 8, \arg z = \pi/2$

       $|z| = 12, \arg z = -\pi/2$

B.1. $\dfrac{z - \bar{z}}{2i} = a\left(\dfrac{z + \bar{z}}{2}\right) + b \qquad a = \text{Slope}, b = y - \text{Intercept}$

B.3. $|z - a| - |z - b| = k$

    a, b focus, $k = $ constant

B.5. **a** $|z - 2| + |z - 3i| > \sqrt{26.5}$

    b $\operatorname{Re}(z) > 0, 5 < \operatorname{Im}(z) < 27$

    c $\operatorname{Im}(z) \geqq -\operatorname{Re}(z)$

    d $\operatorname{Re}(z)\tfrac{2}{4} - \operatorname{Im}(z)\tfrac{2}{9} > 1$

SECTION 1.3

A.1. **a** $5, -5$

b $\dfrac{3\sqrt{2}}{2}(1+i), -\dfrac{3\sqrt{2}}{2}(1+i)$

c $2^{3/4}(\cos \pi/8 + i \sin \pi/8), 2^{3/4}(\cos 9\pi/8 + i \sin 9\pi/8)$
d $2^{3/4}[\cos(-\pi/8) + i \sin(-\pi/8)], 2^{3/4}(\cos 7\pi/8 + i \sin 7\pi/8)$
e $\sqrt{5}(\cos 1.288 + i \sin 1.288), \sqrt{5}[\cos(1.288 + \pi) + i \sin(1.288 + \pi)]$
f $\sqrt{13}(\cos 2.46 + i \sin 2.46), \sqrt{13}[\cos(2.46 + \pi) + i \sin(2.46 + \pi)]$
g $\sqrt{3}/2 - i\,1/2, -\sqrt{3}/2 + i\,1/2$
h $65^{1/4}(\cos .525 + i \sin .525),$
$\qquad 65^{1/4}[\cos(.525 + \pi) + i \sin(.525 + \pi)]$
i $40^{1/4}(\cos 2.18 + i \sin 2.18), 40^{1/4}[\cos(2.18 + \pi) + i \sin(2.18 + \pi)]$
j $\sqrt{13}(\cos 2.94 + i \sin 2.94), \sqrt{13}[\cos(2.94 + \pi) + i \sin(2.94 + \pi)]$
k $\sqrt{39}(-\sqrt{2}/2 - i\sqrt{2}/2), \sqrt{39}(\sqrt{2}/2 + i\sqrt{2}/2)$
l $20^{1/4}(\cos 1.012 + i \sin 1.012), 20^{1/4}[\cos(1.012 + \pi) + i \sin(1.012 + \pi)]$

B.1. $\dfrac{-(\sqrt{5}+1) + i(-\sqrt{5}+1)}{2}, \dfrac{(\sqrt{5}-1) + i(\sqrt{5}+1)}{2}$

3. $\tfrac{1}{2}[-i + \sqrt[4]{17}(\cos 2.13 + i \sin 2.13)]$
$\tfrac{1}{2}[-i - \sqrt[4]{17}(\cos 2.13 + i \sin 2.13)]$

5. $[1 + 4i + \sqrt[4]{319}(\cos 1.81 + i \sin 1.81)]$
$[1 + 4i - \sqrt[4]{319}(\cos 1.81 + i \sin 1.81)]$

7. $\tfrac{1}{4}[5 + 6i + \sqrt[4]{1348}(\cos 2.88 + i \sin 2.88)]$
$\tfrac{1}{4}[5 + 6i - \sqrt[4]{1348}(\cos 2.88 + i \sin 2.88)]$

SECTION 1.4

A.1. **a** Single-valued
Domain: Complex plane except $z = \tfrac{2}{5}$
Range: Complex plane except $w = -\tfrac{4}{5}$
b Single-valued
Domain: Complex plane except $z = 0$
Range: Complex plane
c Multiple-valued: Branch cut along imaginary axis between $z = \pm i/\sqrt{5}$
Domain: Complex plane except $z = 0$
Range: Complex plane
d Multiple-valued: $-\infty < \operatorname{Re}(z) \leq -1, \operatorname{Im}(z) = 0$
Domain: Complex plane
Range: Complex plane
e Single-valued

Domain: Complex plane
Range: Positive real axis and zero
- **f** Single-valued
  Domain: Complex plane except $z = i$
  Range: Complex plane except zero
- **g** Multiple-valued: $-\infty < \text{Im}(z) \leq -1$, $\text{Re}(z) = 0$
  Domain: Complex plane
  Range: Complex plane
- **h** Multiple-valued: $\text{Im}(z) = 0$, $-\infty < \text{Re}(z) \leq -1$, $1 \leq \text{Re}(z) < \infty$
  Domain: Complex plane
  Range: Complex plane
- **i** Multiple-valued: $0 \leq \text{Re}(z) < \infty$, $\text{Im}(z) = 0$
  Domain: Complex plane
  Range: $0 \leq w < 2\pi$
- **j** Single-valued
  Domain: Complex plane except $z = 0$
  Range: Complex plane

B.3. $\text{Re}(p(z)) = \sum\limits_{k=0}^{n} p_k r^k \cos(k\theta + \theta_k)$ $\text{Im}(p(z)) = \sum\limits_{k=0}^{n} p_k r^k \sin(k\theta + \theta_k)$

SECTION 1.5

A.1.
- **a** $\ln 1 + i(\pi/2 + 2n\pi)$
- **b** $\ln \sqrt{2} + i(\pi/4 + 2n\pi)$
  NOTE: $e^{f(z)}$ can be written $\exp(f(x))$, as an example $e^{-x^2} = \exp(-x^2)$. This notation will be used from now on throughout this answer key.
- **c** $\exp[-(\pi/4 + 2n\pi)]$
- **d** One answer is: $\exp[-\tan^{-1}(\tfrac{1}{2})][\cos(\ln\sqrt{5}) + i\sin(\ln\sqrt{5})]$
- **e** One answer is: $\exp(\sqrt{2}\ln 3)$
- **f** One answer is: $\exp[-\tfrac{1}{2}\tan^{-1}(\tfrac{4}{3})][\cos(\ln\sqrt{5}) + i\sin(\ln\sqrt{5})]$
- **g** One answer is: $\exp[\ln\sqrt{10} - \tan^{-1}(-3/1)]\{\cos[\tan^{-1}(-3/1) + \ln\sqrt{10}] + i\sin[\tan^{-1}(-3/1) + \ln\sqrt{10}]\}$
- **h** One answer is: $\exp[\sqrt{2}\ln\sqrt{5} + \tan^{-1}(1/-2)]\{\cos[\sqrt{2}\tan^{-1}(1/-2) - \ln\sqrt{5}] + i\sin[\sqrt{2}\tan^{-1}(1/-2) - \ln\sqrt{5}]\}$
- **i** $\exp\{(2-i)\exp[(2+i)\log(1+i)]\} = (1+i)^4 = -4$
- **j** $\exp[\log(3+i)\log(2-i)]$
  One answer is: $e^\alpha(\cos\beta + i\sin\beta)$
  where $\alpha = \ln\sqrt{10}\ln\sqrt{5} - \tan^{-1}(1/3)\tan^{-1}(-1/2)$
  $\beta = \ln\sqrt{5}\tan^{-1}(1/3) + \ln\sqrt{10}\tan^{-1}(-1/2)$
- **k** One answer is: $6 - 7i$
- **l** $\exp((i-1)\exp\{i[\log(3-2i)]\}) = (1+2i)^{-1-i}$
  One answer is: $\exp[-\ln\sqrt{5} + \tan^{-1}(2)][\cos(-\tan^{-1}2 - \ln\sqrt{5}) + i\sin(-\tan^{-1}2 - \ln\sqrt{5})]$

m $\exp\{-i\log[\log(3-2i)]\}$
One answer is: $e^{\alpha}(\cos\beta + i\sin\beta)$,
where $\alpha = \tan^{-1}\{[\tan^{-1}(-2/3)]/\ln\sqrt{13}\}$
$\beta = -\sqrt{\ln 13 + [\tan^{-1}(-2/3)]^2}$

n $e^{\log i \log i}$
One answer is: $\exp(-\pi^2/4)$

o $\log(e^{(1-3i)\log e})$
One answer is: $1 - 3i$

p $\exp(2i\log e)\exp(i\log 2)$
One answer is: $e^{(2+\ln 2)i}$

A.3. a $\text{Re}[\sinh(z)] = \frac{1}{2}(e^x \cos y - e^{-x} \cos y)$
$\text{Im}[\sinh(z)] = \frac{1}{2}(e^x \sin y + e^{-x} \sin y)$

b $\text{Re}[\cosh(z)] = \frac{1}{2}(e^x \cos y + e^{-x} \cos y)$
$\text{Im}[\cosh(z)] = \frac{1}{2}(e^x \sin y - e^{-x} \sin y)$

c $\text{Im}[\tan(z)] = -\dfrac{(e^{-2y} - e^{2y})(\cos^2 x + \sin^2 y)}{\cos^2 x(e^{-y} + e^y)^2 + \sin^2 y(e^{-y} - e^y)^2}$

$\text{Re}[\tan(z)] = \dfrac{4\cos x \sin y}{\cos^2 x(e^{-y} + e^y)^2 + \sin^2 y(e^{-y} - e^y)^2}$

d $\text{Re}[\cot(z)] = \dfrac{4\cos x \sin y}{\cos^2 x(e^{-y} - e^y)^2 + \sin^2 y(e^{-y} + e^y)^2}$

$\text{Im}[\cot(z)] = \dfrac{(\cos^2 x + \sin^2 y)(e^{-2y} - e^{2y})}{\cos^2 x(e^{-y} - e^y)^2 + \sin^2 y(e^{-y} + e^y)^2}$

e $\text{Re}[\sec(z)] = \dfrac{2\cos x(e^{-y} + e^y)}{\cos^2 x(e^{-y} + e^y)^2 + \sin^2 x(e^{-y} - e^y)^2}$

$\text{Im}[\sec(z)] = \dfrac{-2\sin x(e^{-y} - e^y)}{\cos^2 x(e^{-y} + e^y)^2 + \sin^2 x(e^{-y} - e^y)^2}$

B.1. $w = -i\log(iz + \sqrt{1-z^2})$
$w = -i\log(z + \sqrt{z^2-1})$

SECTION 2.1

A.1. a Continuous, discontinuous, discontinuous
b Continuous, discontinuous, discontinuous
c Discontinuous, discontinuous, continuous

A.3. a Complex plane, 2.1.8
b Complex plane except $z = 0$
c Complex plane except $z = -1$, 2.1.8

298 ODD NUMBERED ANSWERS

- **d** Complex plane except $z = -1$, 2.1.8
- **e** Complex plane except $z = (-1 + \sqrt{3}i)/2$ and $z = (-1 - \sqrt{3}i)/2$, 2.1.8
- **f** Complex plane, 2.1.9
- **g** Complex plane except $z = 1$, 2.1.9
- **h** Complex plane except $z = 0$, 2.1.8
- **i** Complex plane except $z = \pm 2i$, 2.1.4
- **j** Complex plane, 2.1.8 and 2.1.9
- **k** Complex plane, 2.1.8
- **l** Complex plane except $\text{Im}(z) = \pm \sqrt{\text{Re}^2(z) + \frac{1}{2}}$, 2.1.8

SECTION 2.2

A.1.
- **a** $2(z - 1/z^3)$
- **b** $(-z^4 - 8z^2 + 10z + 5)/(z^4 - 9z^3 + 13z^2 - 12z + 4)$
- **c** $(z^3 - 27)^2 (11z^4 + 9z^2 - 54z)$
- **d** $(z^3 + 10z - 1)^{-1/3} (2z^2 + 20/3)$
- **e** $\frac{7}{2}(z - 3)^{-1/2} \exp(7\sqrt{z - 3})$
- **f** $2 \dfrac{e^{2z+z}}{e^{2z+z^2}}$
- **g** $\log z + 1$
- **h** $(-3/2z^2 - 2z + 1)/[(z^2 + 2)^2 \sqrt{z + 1}]$
- **i** $2(z^2 - 3z - 1)/(z^2 - 1)^2$
- **j** $(ad - bc)/(cz + d)^2$

A.3.
- **a** 1
- **b** 0
- **c** 1
- **d** $-i\pi$
- **e** 0
- **f** -1

SECTION 2.3

A.1.
- **a** $1351/684 - i(\ln 6 + 5/2)$
- **b** $-55/18 - i\,110/18$
- **c** $-i\pi$
- **d** $14/3 - i$
- **e** 0
- **f** $10\pi i$
- **g** $100 + 16/3\,i$
- **h** $-80 - 14i$

SECTION 2.4

A.1.
- **a** $5\pi i/2$
- **b** $-\pi$
- **c** $-48\pi i$
- **d** 0
- **e** $\pi/8(-3 - 27i)$
- **f** $-(\pi/2)(1 + 9i)$

A.3.
- **a** $8\pi i$
- **b** $-(2\alpha + i)e^{i\alpha}/8$
- **c** 0
- **d** π^2

e	$26\pi i$			f	$-6\pi i$
g	$8\pi i(e^2 - e)$			h	$-14\pi i$

SECTION 2.5

A.1.
- a Not harmonic
- b Harmonic, $3x^2y - y^3 + k$
- c Harmonic, $2xy + 2y + k$
- d Not harmonic
- e Harmonic, $ay - bx + k$
- f Harmonic, $2e^{2x} \sin y \cos y + k$
- g Harmonic, $\cosh x \sin y + k$
- h Not harmonic
- i Not harminic
- j Harmonic, $e^{3x^2 - 3y^2 - x} \sin(6xy - y) + k$
- k Harmonic, $e^{x^2 - y^2} \sin 2xy + k$
- l Not harmonic

SECTION 3.1

A.1.
- a 0
- b $1 - 2i$
- c No limit
- d 0
- e 0
- f i
- g $i/2$
- h 0

A.3.
- a $-(3 + i)/(7 + 11i)$
- b $(7 - 9i)/13$
- c 1
- d $2/(-1 + 6i)$
- e $5/(3 + i)$
- f $3/4$

B.1.
- a $f(z) = \begin{cases} 1 & z = 0 \\ 0 & z \neq 0 \end{cases}$ uniformly convergent $|z| \geq \rho > 0$
- b $f(z) = \frac{2}{3}$ uniformly convergent $|z| \leq R < \infty$ or $|\operatorname{Im} z| \geq c > 0$.
- c $f(z) = 0$ as in Part b.
- d $f(z) = \begin{cases} 0 & z = 0 \\ \dfrac{2-z}{1-z} & 0 < |z| < 1 \\ \infty & |z| > 1 \end{cases}$ uniform for $0 < \alpha \leq |z| \leq \beta < 1$.
- e $f(z) = 0$ uniformly convergent $|z| \leq R < \infty$
- f $f(z) = z$ uniformly convergent $|z| \leq R < \infty$
- g 0 uniformly convergent $|\operatorname{Re}(z)| \leq M < \infty$
- h $f(z) = 0$ uniformly convergent $|z| \leq R < \infty$

B.5.
- a $-i/2$
- b $1/24$

SECTION 3.2

A.1. a $1/2$ b 1 c 1 d 1
      e 1 f ∞ g 4 h 1
      i ∞ j e k ∞ l 1

A.3. a ∞ b 1 c 1
      d 1 e 3 f $1/2$

B. a $\dfrac{e^2 - 1}{2e} + \dfrac{e^2 + 1}{2e} z + \dfrac{-e^2 + 3}{4e} z^2 + \cdots$

  b $\dfrac{e^2 - 1}{2e(1 + i)} + \dfrac{e^2 - 3 + 2i}{2e(1 + i)^2}(z - i) + \dfrac{9 - i}{8e}(z - i)^2 + \cdots$

  c $-5/2 - \tfrac{1}{4}(z - 2) - \tfrac{9}{8}(z - 2)^2 + \cdots$

  d $0 - \dfrac{1}{\sin 1}(z + 1) - \dfrac{\cos 1}{\sin^2 1}(z + 1)^2 + \cdots$

  e $e^{-i/3}\left(1 + \dfrac{1}{9}(z - 3i) + \dfrac{(1 + 6i)}{162}(z - 3i)^2 + \cdots\right)$

  f $\dfrac{5}{4 \sin (1/2)} + \dfrac{\sin (1/2) - (5/4) \cos (1/2)}{\sin^2 (1/2)}(z - 1/2)$

  $+ \dfrac{(13/4) \sin 2 (1/2) - 2 \cos (1/2)\,[\sin (1/2) - (5/4) \cos (1/2)](z - 1/2)^2}{2 \sin^3 (1/2)} + \cdots$

SECTION 3.3

A.1. a $-\dfrac{1}{2z} - \displaystyle\sum_{n=0}^{\infty} \dfrac{(-z)^n}{2^{n+2}}, \quad |z| < 2$

  b $\dfrac{1}{z^3} + \dfrac{1}{z} + \displaystyle\sum_{n=0}^{\infty} \dfrac{z^{2n+1}}{(n + 2)!}, \quad 0 < |z|$

  c $\displaystyle\sum_{n=1}^{\infty} (-1)^n \dfrac{\pi^{2n-1}}{(2n - 1)!}(z - 1)^{-(2n-1)}, \quad 0 < |z|$

  d $\dfrac{1}{2}\left(\dfrac{1}{(z - i)} + \displaystyle\sum_{n=0}^{\infty} \dfrac{(-1)^n}{(2i)^{n+i}}(z - i)^n\right), \quad 0 < |z - i| < 2$

  e $-\dfrac{1}{z} + \displaystyle\sum_{n=0}^{\infty} \dfrac{(-1)^n z^{2n+1}}{(2n + 3)!}, \quad 0 < |z|$

  f $-\dfrac{1}{3(z + 1)^2} + \dfrac{1}{9(z + 1)} - \displaystyle\sum_{n=0}^{\infty} \dfrac{(z + 1)^n}{3^{n+3}}, \quad 0 < |z + 1| < 3$

g $-\dfrac{2}{z^2} - \dfrac{2}{z} - \sum\limits_{n=0}^{\infty} 3z^n, \quad 0 < |z| < 1$

h $\dfrac{1}{z} + \sum\limits_{n=0}^{\infty} (2+n) z^n, \quad 0 < |z| < 1$

i $\dfrac{1}{4z} + \sum\limits_{n=0}^{\infty} \dfrac{z^n}{4^{n+2}}, \quad 0 < |z| < 4$

j $-\dfrac{1}{4(z-4)} + \sum\limits_{n=0}^{\infty} \dfrac{(z-4)^n}{4^{n+2}}, \quad 0 < |z| < 4$

B.1. $e\left(1 + \sum\limits_{n=0}^{\infty} \dfrac{(z-1)^{-n}}{n!}\right)$

B.3.   **a** $\dfrac{1}{z} + \dfrac{z}{6} + \dfrac{19z^4}{120} + \cdots,\quad$ POLE

     b $-\dfrac{1}{y} + \dfrac{y}{3} - \dfrac{y^3}{45} + \cdots,\quad$ POLE

     c $\dfrac{1}{z} + \dfrac{2z}{3} + \dfrac{14z^3}{45} + \cdots,\quad$ POLE

     d   The entire series is $\sum\limits_{l=-\infty}^{\infty} \sum\limits_{n=0}^{\infty} \dfrac{(-1)^n}{(2n+1)!} \binom{2n+1}{n+l} z^{4l-2}$

     e $\dfrac{1}{z} - \dfrac{1}{2} + \dfrac{12}{z} + \cdots,\quad$ POLE

     f $\dfrac{1}{\pi(z-1)^2} + \dfrac{\pi}{6} + \dfrac{7\pi^3}{360}(z-1)^2 + \cdots,\quad$ POLE

     g $\dfrac{1}{z} + 1 + \dfrac{5z}{6} + \cdots,\quad$ POLE

     h $\sum\limits_{0}^{\infty} \dfrac{(-1)^{n+1}}{2n!} \pi^{2n} (z-1)^{2n-1},\quad$ POLE

     i $\dfrac{1}{z^3} - \dfrac{1}{3z} + \dfrac{z}{5} - \dfrac{z^3}{7} + \cdots,\quad$ POLE

SECTION 3.4

A.1. 4th order pole at $z = 0$, Res $= -1/6$

A.3. 2nd order pole at $z = 1$, Res $= 5/4$
       1st order pole at $z = -1$, Res $= -1/4$

302 ODD NUMBERED ANSWERS

A.5. 3rd order pole at $z = -1$, Res $= -6$

A.7. 1st order pole at $z = 0$, Res $= -2/3$

A.9. 1st order pole at $z = i$, Res $= -(i/2)(\cos 1 + i \sin 1)$
1st order pole at $z = -i$, Res $= (i/2)(\cos 1 - i \sin 1)$

B.1. $-\pi i$

B.3. $2\pi i/e$

B.5. a $\pi i/8$
b 0
c $-15\pi i/32$

SECTION 3.5

A.1. a $\pi\sqrt{3}/3$ b $\pi/ab(b+a)$ c $\pi\sqrt{2}/2$
d $-2\pi\sqrt{7}$ e -12π f $6\pi/32$

B.1. a $\dfrac{\pi}{4 \sinh^2 \pi} - \dfrac{1}{2}$

b $\pi^4/90$
c $\pi^6/945$
d $-\pi^2/12$

e $\dfrac{\pi \cosh a\pi}{a \sinh a\pi} - \dfrac{1}{2a^2}$

f $\dfrac{\pi \cosh a\pi}{a \sinh a\pi} \sinh^2 a\theta$, assuming $a > 0$.

SECTION 4.1

B.1. The line $x = \tfrac{1}{2}$; the half plane where $x > \tfrac{1}{2}$; everywhere except $z = 1$.

SECTION 4.2

A.1. $w = a\dfrac{e^{-i\gamma}z - 1}{e^{-i\alpha}z - 1}$ where $\alpha \neq \gamma$.

A.3. $w = a\dfrac{cz - b}{z - b}$ where a, c, b are real and $ab(1 - c) > 0$.

A.5. If the strip is $\{z: a < y < b\}$ then $w = \exp \pi(z - ia)/(b - a)$.

B.1. Lines $x = $ constant go into ellipses with foci on the v axis; lines $y = $ constant go into hyperbolas with foci on the u axis.

SECTION 4.3

A.1. a $w = B\left(\dfrac{1}{3}, \dfrac{1}{3}\right)^{-1} \displaystyle\int_0^z t^{-2/3}(t-1)^{-2/3}\,dt,$

where in the w plane $A = -1,\ B = 1,\ C \doteq \sqrt{3}$, and in the z plane $A = 0,\ B = 1,\ C = \infty$, the real branches being chosen for $0 < z < 1$.

b $w = c\displaystyle\int_0^z t^{\alpha/\pi - 1}(1-t)^{\beta/\pi - 1}\,dt,$

where in the w plane A (angle α) $= a$, B (angle β) $= b > a$, $C = i$, and in the z plane $A = 0,\ B = 1,\ C = \infty$, the real branches being chosen for $0 < z < 1$.
Here $c = a/B(\alpha/\pi, \beta/\pi)$.

SECTION 5.1

A.1. In $0 < \arg z < \pi/3$, the streamlines are $3x^2 y - y^3 = c$, the velocity vector is $3(x^2 - y^2) - 6iy$.
3. Dipole flow described in Section 5.2.
5. Same as Example 1, origin translated to $z = -1$.
7. The streamlines and equipotential curves are interchanged.

SECTION 5.5

A. a $z = ia - \dfrac{\sigma a}{\pi} \displaystyle\int_{\pi/2}^{\phi} \dfrac{e^{-i\phi} 2 \sin 2\phi\, d\phi}{1 - \cos 2\phi}$

is the parametric equation of the free streamline; $\sigma = \pi/(\pi + 2)$.

b $z = \displaystyle\int_1^\zeta \dfrac{\sigma a}{\zeta\pi}\left(\sqrt{\dfrac{\zeta - 1}{\zeta - k}} + i\sqrt{\dfrac{1 - k}{\zeta - k}}\right) d\zeta + i(b - a)$

is the parametric equation of the free streamline; σ and k are given by

$$\sigma = \dfrac{b}{a}\left(\dfrac{1 - \sqrt{1-k}}{1 + \sqrt{1-k}}\right)^{1/2}; \quad \dfrac{1-\sigma}{\sigma} = \dfrac{\sqrt{1-k}}{\pi}\left(\dfrac{\pi}{\sqrt{k}} - \dfrac{2}{\sqrt{k}}\arctan\sqrt{\dfrac{1-k}{k}}\right).$$

c $z = -\dfrac{\sigma a}{\pi} \displaystyle\int_{-1}^{\zeta} \left(\dfrac{\zeta + k + 2}{k - \zeta} + \dfrac{2i\sqrt{-1-\zeta}\sqrt{k+1}}{k - \zeta} \right) d\zeta,$

$k = 2 + 2\sqrt{2}; \quad \sigma = \dfrac{\sqrt{3 + 2\sqrt{2}} - 1}{\sqrt{3 + 2\sqrt{2}} + 1}.$

Index

Absolute convergence, 120, 127
Addition of complex numbers, 5
Airfoil, 263, 265
Algebraic functions, 24
Analytic functions, 64
  uniform convergence of, 116
Antiderivative, 76
Aperture, flow through, 239
Arc, 70
  length of, 81
Arc-wise connected, 46
Argument, 11

Bent channel flow, 249, 265
Bernoulli's law, 225
Beta function, 216
Bilinear function, 30, 196
Binary operations, 3
Binomial theorem, 135
Blasius's theorem, 261
Borda mouthpiece, 274
Bounded set, 48
Branch, 27

Branch cut, 27
Branch point, 27

Cauchy criterion, 119
Cauchy integral formula, 94
Cauchy principle value, 244
Cauchy-Riemann equations, 61
Cauchy sequence, 119
Cauchy's theorem, 89
Chain rule, 60
Channel, flow through, 232
Circle, complex equation of, 16
"Circles," preservation of, 197
Circular cylinder, flow past, 253, 256
Circulation, 221, 254
Comparison test, 123
Complex number, 5
  addition, 5
  argument of, 11
  conjugate of, 15
  modulus of, 11
  multiplication, 6, 13
  multiplicative inverse of, 7, 14

306 INDEX

polar form of, 11
powers of, 19, 37
roots of, 20
sequences of, 111
vector interpretation of, 11–12
Complex potential, 228
Conformal mapping, 188
Conjugate, 15
Conjugate harmonic functions, 104
Connected set, 45
Conservation of momentum, 225
Continuity, 26, 46, 50
    of a uniform limit, 115
    uniform, 46
Contour, 70
Contour integration, 165
Contraction coefficient, 274
Convergence
    absolute, 120, 127
    of a sequence, 112
    of a series, 118
    radius of, 127
    uniform, 114, 129
Convergence tests, 123

d'Alembert's principle, 110
Derivatives, 57, 95
    chain rule for, 60
    of product, sum, etc., 59
Differentiable, 57
Disjoint sets, 45
Dipole flow, 237
Domain, 45

Ellipse, complex equation of, 16
Ellipse, flow around, 240, 260
Entire functions, 64
Essential singularity, 151
Euler, gamma function of, 83
Euler's formula, 36
Exponential function, 34–36
    derivative of, 63
    uniquely defined, 65

Field, 3, 5
Flux, 227
Free streamline, 268
Functions of a complex variable, 25
    algebraic, 24
    analytic, 64
    bilinear, 30
    entire, 64

    integral of, 73
    meromorphic, 154
    Möbius, 30
    multiple-valued, 25
    polynomial, 24
    rational, 30
    sequences of, 113
    single-valued, 25
    transcendental, 34
    univalent, 185
Fundamental theorem of algebra, 103
Fundamental theorem of calculus, 76

Gamma function, 83, 216
Geometric series, 118
Goursat's theorem, 108
Green's theorem, 88, 291

Harmonic functions, 104
Hodograph plane, 269
Hyperbolic functions, 39–40

Identity theorem for power series, 132
Imaginary axis, 11
Imaginary part, 5, 16
    of a function, 31
Infinite series, 118
    See also Series
Integral, 73
Inverse functions, 40
Inversion mapping, 196
Irrotational, 220
Isolated singular point, 138
Isolated zero, 184

Jordan curve theorem, 74
Joukowski transformation, 259

Kutta-Joukowski theorem, 263

Laplace's equation, 103
Laurent's theorem, 142
Length of an arc, 81
L'Hôpital's rule, 66
Lift, 263
Limit of a function, 50
Liouville's theorem, 102
Logarithm, 36
    derivative of, 63
    of a function, 79

Maclaurin's theorem, 131

Maximum modulus theorem, 100
Maximum principle, 105
Mean value principle, 203
Meromorphic functions, 154
Möbius function, 30, 196
Modulus, 11
Morera's theorem, 98
Multiplication of complex numbers, 6, 13
Multiplicative inverse, 7, 14

Nowhere differentiable function, 57

Obstacle, flow past, 256
Open disk, 45
Open set, 45
Ordered pairs, 2–3

P-series, 123
Partial fractions, 151
Partial sums of a series, 118
Piecewise smooth arc, 71
Poisson formula, 203
Polar form of a complex number, 11
Pole, 144
Polynomial, 24
 derivative of, 60
 integral of, 77–78
Potential, 228
Power series, 126
 identity theorem for, 132
Powers of a complex number, 19, 37
Primitive roots of unity, 23
Principal part, 150
Principal value, 244

Quaternions, 10

Radius of convergence, 127
Ratio test, 123
Rational function, 30
Rational numbers, 3
Real axis, 11
Real numbers, 3
Real part, 5, 16
 of a function, 31
Rearranging terms of a series, 121
Reciprocal hodograph plane, 269
Reflection principle, 254
Removable singularity, 54, 144
Residue, 151
Residue theorem, 156
Riemann mapping theorem, 204

Root test, 123
Roots of a complex number, 20
Roots of unity, 21
 primitive, 23
Rotation, 195

"Sandwich" principle, 51
Scale change, 195
Schlicht functions, 185
Schwarz-Christoffel transformation, 204, 207
Sequences, 111
 Cauchy, 119
 of functions, 113
Series, 118
 absolute convergence, 119, 127
 power, 126
 rearranging terms, 121
 summation of, 176
Simply connected, 79
Simple closed contour, 73
Single-valued logarithm of a function, 79
Singular point, 138
 removable, 144
 essential, 151
Sink, 233, 237
Smooth arc, 71
Source, 233, 237
Source-free, 220
Stagnation point, 229
Stream function, 226
Streamline, 226
 free, 268
Step, flow over, 266
Sum of a series, 118
Summation of series, 176

Taylor's theorem, 130
Terms of a series, 118
Transcendental functions, 34
Translation, 195
Triangle inequality, 17
Trigonometric functions, 38–39

Uniform continuity, 46
Uniform convergence, 114, 129
Univalent functions, 185

Vector interpretation of complex numbers, 11–12
Velocity potential, 228
Velocity vector of a flow, 223